PENGUIN BOOKS

DOES GOD PLAY DICE?

Ian Stewart was born in Folkestone in 1945. He graduated in mathematics from Cambridge and obtained a Ph.D. from the University of Warwick, where he is now Professor of Mathematics. He has also held visiting positions in Germany, New Zealand, Connecticut and Texas. He is an active research mathematician with over a hundred published papers, and he takes a particular interest in problems that lie in the gaps between pure and applied mathematics.

Ian Stewart has written or co-authored over sixty books, including *Nature's Numbers*, shortlisted for the 1996 Rhone-Poulenc Prize for Science Books; *The Collapse of Chaos*; *Fearful Symmetry*; *From Here to Infinity*; *Game, Set and Math*; *Another Fine Math You've Got Me Into*; *The Problems of Mathematics*; the bestselling *Does God Play Dice?*; *Nature's Numbers*; *From Here to Infinity*; *Figments of Reality*; and *Life's Other Secret* (several of which are published in Penguin); and three mathematical comic books published in French. He has written for a wide range of newspapers and magazines in the UK, Europe and the USA, including *Nature*, *Focus*, *Discover* and *The Sciences*. He is mathematics consultant for *New Scientist* and writes the 'Mathematical Recreations' columns in *Scientific American*. He also writes science fiction stories and has made numerous radio and television appearances. In 1995 the Royal Society awarded him the Michael Faraday Medal for the year's most significant contribution to the public understanding of science, and he has been selected for the 1997 Communicator Award of the Joint Policy Board for Mathematics in the USA.

Does God Play Dice?

THE NEW MATHEMATICS OF CHAOS

Second edition

Ian Stewart

PENGUIN BOOKS

PENGUIN BOOKS

Published by the Penguin Group
Penguin Books Ltd, 80 Strand, London WC2R 0RL, England
Penguin Putnam Inc., 375 Hudson Street, New York, New York 10014, USA
Penguin Books Australia Ltd, 250 Camberwell Road, Camberwell, Victoria 3124, Australia
Penguin Books Canada Ltd, 10 Alcorn Avenue, Toronto, Ontario, Canada M4V 3B2
Penguin Books India (P) Ltd, 11 Community Centre, Panchsheel Park, New Delhi – 110 017, India
Penguin Books (NZ) Ltd, Cnr Rosedale and Airborne Roads, Albany, Auckland, New Zealand
Penguin Books (South Africa) (Pty) Ltd, 24 Sturdee Avenue, Rosebank 2196, South Africa

Penguin Books Ltd, Registered Offices: 80 Strand, London WC2R 0RL, England

www.penguin.com

First published by Basil Blackwell 1989

Published in Penguin Books 1990
Second edition published 1997

7

Set in 9/11 pt Monotype Baskerville
Typeset by Rowland Phototypesetting Ltd,
Bury St Edmunds, Suffolk
Printed in England by Clays Ltd, St Ives plc

Contents

Preface to the Second Edition

The first edition of *Does God Play Dice?*, published in 1989, didn't have a preface. I was going through a period when I didn't write prefaces because I thought that nobody ever reads them, so it started with a prologue instead. The prologue remains, but now there's a preface to go with it – after all, two prologues would be overkill. If you already own the first edition and are wondering whether this one is different enough to be worth buying, you should either read this bit or thumb through Chapters 14–17 where most of the new stuff lives. If not, just buy it now, OK? You can decide whether to read the preface when you get the book home.

'Chaos' is not just a trendy word for random. In the sense now prevalent in science, it is an entirely new and different concept. Chaos occurs when a deterministic (that is, non-random) system behaves in an apparently random manner. It may sound paradoxical, but 'apparently' hides many sins. The big discovery of the last decade is that chaos is just as common as traditional types of regular behaviour, such as steady states and periodic cycles. In retrospect, there's nothing very surprising about chaos. From today's viewpoint, and with the benefit of 20/20 hindsight, it's very easy to understand how chaos arises and why it often occurs. Despite this lots of people, many of them scientists, still talk about chaos as if it is something weird and exotic. Sorry, but it isn't. Chaos is just as down-to-earth as periodic cycles. But over several centuries we've got used to periodic cycles, whereas we've only just stumbled upon the existence of chaos and we haven't got used to it yet. That's not a surprise either: chaos is much more subtle and intricate.

Chaos has come a long way since 1989. It has transmuted, especially in the popular press, into something called 'chaos theory'. I think it's a mistake to view chaos as a theory in its own right, but I appreciate that journalists need a catchy phrase to sum up the area, and what else would you call it? So I sometimes use that phrase myself, though on the whole I use it to refer to the popular image of chaos, as distinct from that of practising scientists. However, chaos *isn't* a theory. It's a concept, and one that cannot sensibly be separated from the rest of dynamics. It's an idea that cuts across all of the traditional subject boundaries of science. It's a missing piece from a vast jigsaw puzzle. It's a far-reaching unification of order and disorder. But whatever it is, chaos no more deserves to be isolated as a theory in its own right than 'skeleton theory' deserves to be isolated from zoology.

There *is* a new theory, which goes by such names as nonlinear systems theory, dynamical systems theory, or nonlinear dynamics: new not in the sense that nothing like it ever existed before, but in the sense that it has really 'taken off' and deserves to be considered as a theory in its own right. Often this is what people really mean when they say 'chaos theory'. In fact there are at least two meanings to the word 'theory'. There is the sense used in the phrase 'quantum theory' or 'relativity theory' – a statement about how nature behaves. The utility of such a theory depends upon it matching nature sufficiently well. Nonlinear systems theory is a theory in the other sense, a coherent body of mathematical knowledge with a clear and consistent identity. As such, there is no serious question about its correctness: when mathematics is wrong the mistakes are usually glaring. The big question is: does the concept of chaos afford new scientific insights? (If not, then we'll have to scrub nonlinear dynamics: you can't divorce one from the other.) Can chaos theory in the mathematical sense become the basis for new theories in the scientific sense?

This is where chaos can become controversial, because it is no longer just a matter of checking the mathematics and making sure there aren't any errors. Analogously, calculus is a valid mathematical theory, but that doesn't imply that every application of calculus to science must be right. If you set up the wrong differential equations when you model the motion of the Moon, then no matter how correctly you apply calculus, you'll get nonsense. The same is true if your theoretical model generates chaos: the link from the model to chaos may be impeccable, but what about the link from reality to the model?

Does God Play Dice? has two distinct themes. One is to explain the mathematical concept of chaos, and why it is both natural and inevitable. The other is to ask: does chaos occur in the real world? In order for that question to make sense, it has to be reformulated. No mathematics *occurs* in the real world. What it does is model the real world in a useful manner. The geometry of the circle helps us understand why wheels roll smoothly, but you won't find a genuine mathematical circle on a car. You can find two sheep, two apples, or two bookends in the real world, but you'll never encounter the number two as such. So the question should be: 'does the mathematical concept of chaos model the real world in a useful manner, and does it help us understand some of the things we see?'

If you look at what's appearing in the scientific journals, it is absolutely clear that the answer is 'yes'. In 1995 I went to a conference in Utah on Applications of Dynamical Systems, run by the Society for Industrial and Applied Mathematics. SIAM is the premier professional body for the applied mathematicians of the most technologically advanced country in the world, not some Mickey Mouse conglomeration on the lunatic fringe. Some five hundred mathematicians attended over a period of four days, and there were over two hundred research talks (in parallel sessions). About half of them were either about chaos, or about issues that arise from it, such as new methods of data analysis. So if anyone tries to tell you that chaos is nothing but media hype, they're wrong. It's been around too long for that, and it now runs far too deeply in the scientific consciousness. Of course this level of activity does not imply that every proposed application of chaos is correct. The assumption that once chaos theory is 'proved' in one area then you are automatically forced to swallow it in all areas – which I think is one reason why some critics are so relentlessly negative – stems from the confusion about the two meanings of 'theory' that I've just mentioned. Each application must prove its worth in its own right and within its own area of science.

This new edition of *Does God Play Dice?* differs from its predecessor mainly by including new material on applications. I've left the original edition virtually untouched: nothing has happened since it was published to require major surgery. There are three completely new chapters inserted near the end. The first is on prediction in chaotic systems, which is perfectly possible, depending on what you want to predict; it also discusses various related issues. I've included several new applications, ranging from the pulsations of variable stars to quality control in the spring-making industry. The second new chapter is about the control of chaotic systems, a potential source of practical applications and a case study in what advantages accrue when you learn how to use chaos instead of trying to pretend that it doesn't exist. Among the applications here are ways to steer artificial satellites more economically, and work heading in the direction of intelligent heart pacemakers.

The third new chapter is much more speculative. It is an attempt to explain how the concept of chaos might lead to a new answer to Einstein's famous question, the title of this book. Einstein was worrying about quantum mechanics, which is generally held to be irreducibly probabilistic. Is it possible that the apparent randomness of the quantum world is actually deterministic chaos? Would the course of physics have been different if chaos had been discovered before quantum mechanics? In 1989 there wasn't a great deal to say about these questions, but today there is. There is one quite specific proposal in the scientific literature: speculative, but based on solid discoveries, some of them very new. It's a fascinating story, and all of the ingredients are good science: it is only the overall mix that is speculative. And if you don't speculate, you don't accumulate.

I have also brought the earlier chapters up to date. That at least some instances of turbulence in fluids are due to chaos is now absolutely certain. There are new results on the dynamics of the solar system, which it seems will not survive in its present form for much more than another billion years or so. The universe is clumpier on even larger scales than we thought. Chaos in at least some ecosystems is close to being an established fact. Fractal geometry has acquired serious commercial uses. Mathematical technique has advanced to the point that we can now prove, in all rigour, that the model set up by the meteorologist Edward Lorenz definitely does lead to chaos. This is bad news for those stalwarts of orthodoxy who maintained that the appearance of chaos was due to computer error, but good news for the logical underpinnings of nonlinear dynamics.

Finally, a companion to chaos theory has now appeared, known as complexity theory. Chaos theory tells us that simple systems can exhibit complex behaviour; complexity theory tells us that complex systems can exhibit simple 'emergent' behaviour. No discussion of chaos nowadays would be complete without some mention of complexity theory, so I've put it in the final chapter. Complexity theory definitely *is* controversial, but it brings a welcome breath of fresh air to a sackful of stuffy overblown old-fashioned linear theories. I'm absolutely convinced that over the next few decades the kind of thinking towards which complexity theorists are currently groping will turn out to be of fundamental importance in nearly every area of scientific activity. I don't think complexity theory holds the answers – yet – but I do think it offers a much more interesting angle on the questions, which in turn suggests new ways to look for the answers.

I'm not trying to *sell* you chaos. I'm not a prophet of a new religion seeking converts. I don't want your *faith* – perish the thought. What I'm trying to do is to set before you, in as comprehensible a form as I can manage, the information that you need to make your own judgements about chaos's present achievements and future potential. I've tried to make it clear when I'm being speculative. The rest of the time I'm presenting ideas or results that have been published in the serious scientific and mathematical literature. That doesn't mean that they are necessarily right, but it does mean that they are respectable . . .

I'm now becoming horribly aware of the reason why people don't read prefaces: they do go on, don't they? And I haven't told you about all the new applications of chaos that I had to leave out for lack of room – chaos in the Earth's molten interior, in the aurora borealis, in the deep structure of spacetime, in ant colonies, in coding theory and communication, in the voices of opera singers . . .

I think I'll stop.

Ian Stewart
Coventry
January 1996.

Prologue

Clockwork or Chaos?

You believe in the God who plays dice, and I in complete law and order.

Albert Einstein, Letter to Max Born

There is a theory that history moves in cycles. But, like a spiral staircase, when the course of human events comes full circle it does so on a new level. The 'pendulum swing' of cultural changes does not simply repeat the same events over and over again. Whether or not the theory is true, it serves as a metaphor to focus our attention. The topic of this book represents one such spiral cycle: chaos gives way to order, which in turn gives rise to new forms of chaos. But on this swing of the pendulum, we seek not to destroy chaos, but to tame it.

In the distant past of our race, nature was considered a capricious creature, and the absence of pattern in the natural world was ascribed to the whims of the powerful and incomprehensible deities who ruled it. Chaos reigned and law was unimaginable.

Over a period of several thousand years, humankind slowly came to realize that nature has many regularities, which can be recorded, analysed, predicted, and exploited. By the 18th century science had been so successful in laying bare the laws of nature that many thought there was little left to discover. Immutable laws prescribed the motion of every particle in the universe, exactly and forever: the task of the scientist was to elucidate the implications of those laws for any particular phenomenon of interest. Chaos gave way to a clockwork world.

But the world moved on, and our vision of the universe moved with it. Today even our *clocks* are not made of clockwork – so why should our world be? With the advent of quantum mechanics, the clockwork world has become a cosmic lottery. Fundamental events, such as the decay of a radioactive atom, are held to be determined by chance, not law. Despite the spectacular success of quantum mechanics, its probabilistic features have not appealed to everyone. Albert Einstein's famous objection, in a letter to Max Born, is quoted at the head of this chapter. Einstein was talking of quantum mechanics, but his philosophy also captures the attitude of an entire age to classical mechanics, where quantum indeterminacy is inoperative. The metaphor of dice for chance applies across the board. Does determinacy leave room for chance?

Whether Einstein was right about quantum mechanics remains to be seen. But we do know that the world of classical mechanics is more mysterious than even Einstein imagined. The very distinction he was trying to emphasize, between the randomness of chance and the determinism of law, is called into question. Perhaps God can play dice, and create a universe of complete law and order, in the same breath.

The cycle has come full turn – but at a higher level. For we are beginning to discover that systems obeying immutable and precise laws do not always act in predictable and regular ways. Simple laws may not produce simple behaviour. Deterministic laws can produce behaviour that appears random. Order can breed its own kind of chaos. The question is not so much *whether* God plays dice, but *how* God plays dice.

This is a dramatic discovery, whose implications have yet to make their full impact on our scientific thinking. The notions of prediction, or of a repeatable experiment, take on new aspects when seen through the eyes of chaos. What we thought was simple becomes complicated, and disturbing new questions are raised regarding measurement, predictability, and verification or falsification of theories.

In compensation, what was thought to be complicated may become simple. Phenomena that appear structureless and random may in fact be obeying simple laws. Deterministic chaos has its own laws, and inspires new experimental techniques. There is no shortage of irregularities in nature, and some of them may prove to be physical manifestations of the mathematics of chaos. Turbulent flow of fluids, reversals of the Earth's magnetic field, irregularities of the heartbeat, the convection patterns of liquid helium, the tumbling of celestial bodies, gaps in the asteroid belt, the growth of insect populations, the dripping of a tap, the progress of a chemical reaction, the metabolism of cells, changes in the weather, the propagation of nerve impulses, oscillations of electronic circuits, the motion of a ship moored to a buoy, the bouncing of a billiard ball, the collisions of atoms in a gas, the underlying uncertainty of quantum mechanics – these are a few of the problems to which the mathematics of chaos has been applied.

It is an entire new world, a new kind of mathematics, a fundamental breakthrough in the understanding of irregularities in nature. We are witnessing its birth.

Its future has yet to unfold.

1

Chaos from Order

Lo! thy dread empire, Chaos! is restor'd;
Light dies before thy uncreating word;
Thy hand, great Anarch! lets the curtain fall,
And universal darkness buries all.

Alexander Pope, *The Dunciad*

The eternal battle between order and disorder, harmony and chaos, must represent a deeply felt human perception of the universe, for it is common to so many creation myths and so many cultures. In the cosmology of ancient Greece, chaos was both the primaeval emptiness of the universe, and the underworld where dwelt the dead. In Old Testament theology 'the Earth was without form, and void, and darkness was upon the face of the deep'. In an early Babylonian epic the universe arises from the chaos that ensues when an unruly family of gods of the deep is destroyed by its own father. Chaos is the original formless mass from which the creator moulded the ordered universe (Figure 1). Order is equated with good and disorder with evil. Order and chaos are seen as two opposites, poles upon which we pivot our interpretations of the world.

Some innate impulse makes humankind strive to understand the regularities in nature, to seek the laws behind the wayward complexities of the universe, to bring order out of chaos. Even the earliest civilizations have sophisticated calendars to predict the seasons, and astronomical rules to predict eclipses. They see figures in, and weave legends around, the stars in the sky. They invent pantheons of deities to explain the vagaries of an otherwise random and senseless world. Cycles, shapes, numbers. Mathematics.

Figure 1 History of the Earth (clockwise from top right): *the chaotic liquid, the pristine Earth, Earth during the Flood, modern Earth, Earth during the conflagration to come, Earth during the Millennium, and Earth's ultimate fate as a star. (From Thomas Burnet:* Tellurus theoria sacra, *1681.)*

Unreasonable Reasoning

The physicist Eugene Wigner wrote of the 'unreasonable effectiveness of mathematics' in describing the structure of the physical world. Mathematics arises from questions about the physical world, and earns its keep by supplying some of the answers. But the process is seldom a direct one. Often a mathematical idea must take on a life of its own, existing as it were in limbo, developed and discussed for its own sake as a purely mathematical object, until its inner secrets are dissected out and its physical significance is perceived. Perhaps mathematics is effective because it represents the underlying language of the human brain. Perhaps the only patterns we can perceive are mathematical because mathematics is the instrument of our perception. Perhaps mathematics is effective in organizing physical existence because it is inspired by physical existence. Perhaps its success is a cosmic delusion. Perhaps there are no real patterns, only those that we feeble-mindedly impose. These are questions for philosophers. The pragmatic reality is that mathematics is the most effective and trustworthy method that we know for understanding what we see around us.

It is now a little over three hundred years since the publication of a work without historical parallel – the *Mathematical Principles of Natural Philosophy* of Isaac Newton (Figure 2). The book still sells some 700 copies every year – mainly to students in liberal arts colleges who study the masters from primary sources. Its longevity is astonishing, but it is no longer a bestseller. Instead, its message has been absorbed into the very foundations of our culture.

That message is: *Nature has laws, and we can find them.*

Newton's law of gravity is a simple thing. Every two particles of matter in the universe attract each other, with a force that depends in a precise and simple manner on their masses and on the distance between them. (It is proportional to the product of the two masses, divided by the square of the distance separating them.) The law can be condensed into a brief formula in algebra. When coupled with another of Newton's laws, this time the law of motion (the acceleration of a body is proportional to the force acting on it), it explains a wealth of astronomical observations ranging from the paths of the planets through the Zodiac to the wobbles of the Moon on its axis, from the resonant locking of Jupiter's satellites to the light-curves of binary stars, from the gaps in Saturn's rings to the birth of galaxies.

Simple. Elegant. Elusive.

Order from chaos.

Newton was an ambitious man. He sought nothing more nor less than 'the system of the world'. The Theory of Everything.

In the terms of his era he succeeded beyond his wildest dreams. For more

Figure 2   Isaac Newton (engraving based on a painting by Godfrey Kneller).

than two centuries Newton's laws reigned supreme as the ultimate description of nature. Only in the microscopic domains of the atom, or the vast reaches of interstellar space, do minuscule discrepancies between nature according to Newton and nature according to Nature make themselves known. In those domains Newton has been displaced by quantum mechanics and relativity. Now physicists, once more questing for the holy grail of a Theory of Everything, talk of supergravity and superstrings, quarks and chromodynamics, broken symmetries and Grand Unified Theories. We are living in a world of twenty-six dimensions (or perhaps a mere ten), all but four of which are curled up tightly like a terrified armadillo and can be detected only by their shivering. A passing fad or a vision of our future? We cannot yet tell. But as theory supplants theory, paradigm overturns paradigm, one thing remains onstant: the relevance of mathematics. The laws of nature are mathematical. God is a geometer.

Clockwork World

The revolution in scientific thought that culminated in Newton led to a vision of the universe as some gigantic mechanism, functioning 'like clockwork', a phrase that we still use – however inappropriate it is in an age of digital watches – to represent the ultimate in reliability and mechanical perfection. In such a vision, a machine is above all predictable. Under identical conditions it will do identical things. An engineer who knows the specifications of the machine, and its state at any one moment, can in principle work out exactly what it will do for all time. Let us leave to one side, noted but as yet unelaborated, the question of what is possible in *practice* rather than high principle, and first understand why the scientists of the 17th and 18th centuries found themselves led to what at first sight appears such a barren and sterile view of this universe of wonder and surprise.

Newton cast his laws in the form of mathematical equations, which relate not just quantities, but also the rates at which those quantities change. When a body falls freely under constant gravity, it is not its position that remains constant – if that were so it would hover improbably, unsupported. Nor is it the velocity – the rate of change of position – that is constant. The longer the body continues to fall, the faster it does so: this is why it's more dangerous to fall off a high building than a low one. No, it is the acceleration – *the rate of change of the rate of change of position* – that is constant. Perhaps we can now see why it took so many centuries for this dynamical regularity to be noticed: the law is simple only for those who acquire a new conception of simplicity.

Equations that involve rates of change are referred to as *differential* equations. The rate of change of a quantity is determined by the difference between its values at two nearby times, and the word 'differential' consequently permeates the mathematics: differential calculus, differential coefficient, differential equation, and just plain differential. Solving algebraic equations, not involving rates of change, is not always easy, as most of us know to our cost: solving differential equations is an order of magnitude more difficult. Looking back from the end of the 20th century the big surprise is that so many important differential equations *can* be solved, given enough ingenuity. Entire branches of mathematics have sprouted from the need to understand a single, crucial, differential equation.

Despite the technical difficulties in solving particular equations, some general principles can be established. The key principle, for the present discussion, is that the solution of the equations describing the motion of some dynamical system is *unique* if the initial positions and velocities of all components of the system are known. A bicycle has some five or six essential moving parts: if we know *now* what each is doing, we can predict the motion of the bicycle from the moment it is pushed off down the road until it falls into the wayside ditch. More ambitiously, if at some fixed instant we know the positions and velocities of every particle of matter in the Solar System, then all subsequent motions of those particles are uniquely determined.

This statement assumes, for simplicity, that there are no outside influences on the motion. Trying to take those into account too leads to the interpretation that the positions and velocities of every particle of matter in the entire universe, taken at some fixed instant, completely determine its future evolution. The universe follows a unique, predetermined dynamical path. *It can do only one thing.* In the eloquent words of Pierre Simon de Laplace (Figure 3), one of the leading mathematicians of the 18th century, in his *Philosophical Essays on Probabilities*:

> An intellect which at any given moment knew all the forces that animate Nature and the mutual positions of the beings that comprise it, if this intellect were vast enough to submit its data to analysis, could condense into a single formula the movement of the greatest bodies of the universe and that of the lightest atom: for such an intellect nothing could be uncertain; and the future just like the past would be present before its eyes.

This is rather an awe-inspiring statement to get out of a straightforward uniqueness theorem in mathematics. Later I'll try to bring into the open some of the intellectual sleight-of-hand involved in the transition, because it's really quite outrageous; but for the moment let's allow the interpretation to stand. What we must realize, when considering statements such as Laplace's, is the atmosphere of excitement that prevailed in the science of the time, as phenomenon after phenomenon – mechanics, heat, waves, sound, light,

Figure 3   Pierre Simon de Laplace reading his Celestial Mechanics *(19th-century lithograph).*

magnetism, electricity – was brought under control by the selfsame technique. It must have looked like the big breakthrough to ultimate truth. *It worked.* The paradigm of classical determinism was born: if the equations prescribe the evolution of the system uniquely, without any random external input, then its behaviour is uniquely specified for all time.

Voyage to Hyperion

We time-shift back to 5 September 1977. A gigantic Titan III-E/Centaur rocket waits in readiness on the pad at Launch Complex 41, Air Force Eastern Test Range, Kennedy Space Center, Cape Canaveral, Florida. In its topmost stage, dwarfed by the giant but the reason for its existence, is a tiny triumph of engineering, the *Voyager 1* spacecraft (Figure 4).

The countdown reaches its final seconds. Twin solid-fuel boosters, filled with aluminium powder and ammonium perchlorate, ignite with a roar that can be heard fifteen kilometres away. The rocket, tall as a fifteen-storey building and weighing 700 tonnes, drags itself skyward from the bottom of Earth's deep gravity well. At first its motion is painfully slow, and it burns a substantial proportion of its fuel in the first hundred metres. Yet within ten hours *Voyager 1* is further away than the Moon, *en route* for the distant planets: Mars, Jupiter, Saturn (Figure 5).

Sixteen days earlier a sister craft, *Voyager 2*, has already made its departure: the launch of *Voyager 1* has been delayed by technical faults. In compensation, *Voyager 1* follows a faster trajectory, so that by the time it nears Jupiter it is four months ahead of its sister craft. *Voyager 1*'s mission terminates after its close encounter with Saturn; but *Voyager 2* continues to Uranus and Neptune. Only Pluto evades scrutiny, for Pluto is in the wrong part of its orbit and the 'Grand Tour' cannot reach it.

The journey of the *Voyager*s is a miracle of engineering. It is also a miracle of mathematics, here playing its role as the servant of technology. Mathematics governs the design of the probe and of its launch-vehicle. Mathematics computes the loads and stresses on its metal frame, the combustion patterns of its fuel, the dynamics of the air that streams past the vehicle's skin during its brief traverse of the Earth's atmosphere. Mathematics governs the electronic impulses that course through the computers as they anxiously watch every tiny step in the spacecraft's progress. Mathematics even decides the coding of the radio messages by which the earthbound controllers communicate their instructions to the probe, which in the fullness of time will transmit back to Earth breathtaking images of our Solar System.

But, above all, mathematics governs the stately celestial dance of the planets, their moons, and the paths of the *Voyager*s as they make their heavenly

WIDE ANGLE TV
NARROW ANGLE TV
TV ELECTRONICS
ULTRAVIOLET SPECTROMETER
INFRARED INTERFEROMETER SPECTROMETER AND RADIOMETER

PLASMA

COSMIC RAY

PHOTOPOLARIMETER

LOW ENERGY CHARGED PARTICLES

THRUSTERS (16)

ELECTRONIC COMPARTMENTS

SCIENCE INSTRUMENT CALIBRATION PANEL AND SHUNT RADIATOR

PROPULSION FUEL TANK

PLANETARY RADIO ASTRONOMY AND PLASMA WAVE ANTENNA

HIGH–GAIN DIRECTIONAL ANTENNA

RADIOISOTOPE THERMOELECTRIC GENERATORS (3)

MAGNETOMETER (1 OF 4)

EXTENDABLE BOOM

PLANETARY RADIO ASTRONOMY AND PLASMA WAVE ANTENNA

Figure 4   The Voyager spacecraft.

Figure 5 Saturn and some of its satellites (photo-montage from Voyagers 1 and 2).

rendezvous. A single, simple law – Newton's law of gravitation. No need for Einstein's improvements – at the comparatively slow speeds that prevail in the Solar System, Newton suffices.

Were the Solar System inhabited by Sun and Earth alone, Newton's law would predict that they move in ellipses about their mutual centre of gravity – a point buried deep within the Sun, because the star is so much more massive than the planet. Effectively the Earth should move in an ellipse with the Sun stationary at one focus. But the Earth is not alone in the Solar System – why else dispatch the *Voyager* craft? Each planet travels along its own ellipse – or would, were it not for the others. These perturb it away from its ideal orbit, speeding it up or slowing it down. The cosmic dance is intricate and elaborate: saraband to a score by Newton, *Largo con gravità*.

The law prescribes each step of the dance, completely, exactly. The calculations are not easy, but they can be performed with persistence and a fast computer, to an accuracy enough for *Voyager*'s purpose. Using Newton's mathematical laws, astronomers have predicted the motion of the Solar System over 200 million years into the future: a few years is child's play in comparison.

Past Jupiter, a banded, swirling enigma. On to Saturn, a planet dominated by rings. But Saturn has other features of interest, notably its moons. From earthbound observations, the planet was known to have at least ten satellites: *Voyager* raised the total to fifteen.

One moon, Hyperion, is unusual. It is irregular in shape, a celestial potato. Its orbit is precise and regular; but its attitude in orbit is not. Hyperion is tumbling. Not just end over end, but in a complex and irregular pattern. Nothing in this pattern defies Newton's laws: the tumbling of Hyperion obeys the laws of gravitation and dynamics.

It is time for a hypothetical exercise. Suppose that *Voyager 1* had been able to measure the tumbling of Hyperion to an accuracy of ten decimal places. It didn't, but let's be generous. Suppose, on this basis, that earthbound scientists were to make the best possible prediction of Hyperion's future motion, predetermined according to Newton's law. Then only a few months later, when *Voyager 2* passed by Hyperion, they could compare their predictions with actuality. And they would expect to find . . .

. . . that the prediction was totally wrong.

A failure of prediction?

Not exactly.

A failure of Newton's law?

No. The prediction is expected to be wrong *because* of Newton's law.

Indeterminacy? Random outside effects, such as gas clouds, magnetic fields, the solar wind?

No.

Something much more remarkable. An inherent feature of mathematical equations in dynamics. The ability of even simple equations to generate motion so complex, so sensitive to measurement, that it appears random. Appropriately, it's called *chaos*.

Chaos

Like all buzzwords, this one doesn't have the same connotations that it would in everyday use. Compare the dictionary:

> **chaos** ('keios) *n.* 1 (Usu. cap.) The disordered formless matter supposed to have existed before the ordered universe.
> 2. Complete disorder, utter confusion.

To these, the makers of new dictionaries will have to append the buzzword definition. The one below was proposed, after some initial discomfort, at a prestigious international conference on chaos held by the Royal Society in London in 1986. Although everybody present knew what they thought 'chaos' meant – it was their research field, so they really ought to have known – few were willing to offer a precise definition. This isn't unusual in a 'hot' research area – it's hard to define something when you feel you still don't fully understand it. At any rate, here it is:

> 3. (Math.) Stochastic behaviour occurring in a deterministic system.

That's two more buzzwords – 'stochastic' and 'deterministic'. Laplacian determinism is already familiar to us. 'Stochastic' means 'random'. To understand the phenomenon of chaos we shall need to discuss their meanings further, because in its present form the definition is a paradox. Deterministic behaviour is ruled by exact and unbreakable law. Stochastic behaviour is the opposite: lawless and irregular, governed by chance. So chaos is 'lawless behaviour governed entirely by law'.

Like Hyperion.

Calculator Chaos

Why *does* Hyperion behave that way? The answer must wait until Chapter 12, but at this point I can show you a more accessible example of chaos which you can experiment with for yourself. All you need is a pocket calculator. If you've got a home computer, you can easily program it to do the same job and save yourself a lot of work.

The equation that governs the motion of Hyperion is a differential equation. Effectively what it tells you is this. Suppose that, at a given instant, you know the position and velocity of Hyperion. Then there is a fixed rule, which you apply to these numbers, to get the position and velocity at the next instant. Then you just apply it again, and keep going until you reach whatever time you want.

You may object that time is infinitely divisible, so there's no such thing as an instant, let alone a next one. You may be right, though Zeno of Elea and several modern physicists would disagree; certainly you're stating the conventional position. But, in a sense that can be made precise in several different ways, the above description is morally correct. In particular, the way a computer solves a differential equation is precisely like that, where by 'instant' we now mean 'the time-step used in the calculation'. The method works because very small time-steps give a good approximation to a continuously flowing time.

The equations for Hyperion involve many variables – position, velocity, angular rotation. You *could* put them on your calculator, but life is short. Instead, we'll choose a much simpler equation. Let me emphasize that it has nothing whatsoever to do with the motion of Hyperion; but it does illustrate the phenomenon of chaos.

My calculator has an x^2 button, and I'll assume yours has too. If not, $\times$ followed by $=$ has the same effect. Pick a number between 0 and 1, such as 0.54321, and hit the x^2 button. Do it again, over and over, and watch the numbers. What happens?

They shrink. By the ninth time I hit the button on my calculator, I get zero, and since $0^2 = 0$ it's no surprise that after that nothing very interesting happens.

This procedure is known as *iteration*: doing the same thing over and over again. Try iterating some other buttons on your calculator. Below, I've always started with 0.54321, but you can use other starting values if you want to. Avoid 0, though. On my calculator, in 'radian' mode, after pressing the **cos** button about forty times I get the mysterious number 0.739085133 which just sits there. Can you guess what special property this number has? At any rate, once again the iteration just settles down to a single value: it *converges* to a steady state.

The **tan** button looks like it does the same kind of thing. Appearances are deceptive. I've iterated it 300,000 times by computer and it never converges, nor does it go periodic. It does however, get 'stuck' in places where it increases very, very slowly – say by 0.0000001 per iteration. This effect is called *intermittency*, and it explains why at first sight the numbers may appear to be converging.

There are also infinitely many starting values for which the **tan** sequence

Figure 6   Iteration of x² − 1 *leads to regular oscillations. The value of* x *is plotted vertically, and the number of iterations runs horizontally.*

just repeats the same number over and over again, but 0 is the only one you're likely to run into by accident. The 'typical' behaviour is intermittency.

The **e**^{x} button blows up to 268 point something and then gives an error message because it's got too big: it's heading happily off into infinity. The $\sqrt{}$ button converges to 1.

The **1 / x** button does something more interesting: the number switches alternately from 0.54321 to 1.840908673 and back again. The iteration is *periodic* of period 2; that is, if you hit the button twice you get back where you started. You can probably work out why this is.

Push all the buttons you've got: you'll find that the above seem to exhaust the possible types of behaviour.

But that may be because the buttons on a calculator are designed to do nice things. To get round that, you can invent new buttons. What about an **$x^2 - 1$** button? To simulate it, hit the **x^2** button and then **−1=**. Keep doing it. You soon find you're cycling between 0 and −1, over and over again (Figure 6). That makes sense:

$$0^2 - 1 = -1$$
$$(-1)^2 - 1 = 0.$$

But cycles are nothing new either.

One last try: a **$2x^2 - 1$** button. Start with a value somewhere between 0 and 1 equal to neither. Looks pretty harmless, can't see why anything special should happen. Hmmm . . . Jumps around a lot. Let's wait for it to settle down . . . Taking its time, isn't it? Can't see much of a pattern . . . Looks pretty chaotic to me (Figure 7).

Aha!

Figure 7   Iteration of 2x² − 1 *leads to chaos.*

A simple equation: just iterate $2x^2 - 1$. But the results don't look so simple: in fact *they look random*.

Now try the $2x^2 - 1$ button again, but start with 0.54322 instead of 0.54321. It still looks random – and after fifty or so iterations it also looks completely different.

What you're seeing is a sort of Hyperion-in-microcosm. Deterministic equation: patternless output. Slight change in the starting value: lose track completely of where it's going. What makes this all the more remarkable is that while $2x^2 - 1$ is so weird, the superficially similar button $x^2 - 1$ is perfectly well behaved.

I don't suggest you try the following on a calculator, unless you enjoy long calculations; but if you've got a home computer, here's a program to run.

```
INPUT k
x = 0.54321
FOR n = 1 TO 50
x = k*x*x - 1
NEXT n
FOR n = 1 TO 100
x = k*x*x - 1
PRINT x
NEXT n
STOP
```

You can pretty it up if you want. I won't include any more programs, but computer buffs will find it instructive to write their own programs to experiment with other aspects of chaos. This iterates a $kx^2 - 1$ button for any choice of k. The first FOR loop gives the sequence of iterations time to settle down to the 'long-term' behaviour, without the numbers being printed out. For example, if you put k = 1.4 you'll get a $1.4x^2 - 1$ button. That does a rather complicated cycle through *sixteen* different values! Chaos sets in around k = 1.5. After that, the bigger you make k, the more chaotic things get.

Or so it may seem. But it's not that easy.

At k = 1.74, you see well-developed chaos. At k = 1.75, it looks like that – to begin with. Except that after about fifty iterations it settles into a cycle of length *three*, with numbers around.

$$0.744 \qquad -0.030 \qquad -0.998.$$

Out of chaos emerges pattern. The two are inextricably related.

I hope you find this mysterious and stimulating.

If you do, I'd encourage you to explore the behaviour in the range $k = 1$ to 1.40155 and beyond. You may need to use a longer loop in lines 30 or 60 to see the full pattern – when there is one.

A word about computers and chaos. We tend to think of computer calculations as being the pinnacle of accuracy. Actually, they're not. The limitations of memory mean that numbers can be held in the computer to very limited accuracy, say eight or ten decimal places. Furthermore, the 'private' internal code that the computer uses to represent its numbers and the 'public' one that gets printed on the screen are different. This introduces two sources of error: rounding error in internal calculations and translation error from private code to public. Usually these errors don't matter much but one of the characteristic features of chaos is that tiny errors propagate and grow.

Life would at least be straightforward if all computers used the same codes. But of course they don't. This means that *the identical program run on two different makes of computer can produce different results*. The same goes for the same machine running different versions of the 'same' software. Occasionally I will tell you some numerical results that I get on my computer. Be warned that yours may not give exactly the same numbers! But, if you explore numbers close to those that I'm using, you ought to be able to find the same kind of behaviour that I do.

What have we discovered?

A miracle. Order and chaos, intimately intertwined, emerging from a formula as simple as $kx^2 - 1$. Some values of k lead to iterations that are ordered, others – not noticeably different – to chaos. Which? Ah, now you're talking research mathematics.

We started out not understanding Hyperion; now we don't even understand $2x^2 - 1$. In mathematical terms that constitutes stunning progress.

It's progress because we're beginning to learn *where the problem lies*. Before messing about on the calculator, we could be forgiven for assuming that there's just something pretty complicated about Hyperion. Now we know that isn't so. Complication has very little to do with it. Something very subtle, very fundamental, and utterly fascinating is going on.

All this makes me feel very unhappy about cosmologists who tell us that they've got the origins of the universe pretty well wrapped up, except for the first millisecond or so of the Big Bang. And with politicians who assure us that not only is a solid dose of monetarism going to be good for us, but they're so certain about it that a few million unemployed must be just a minor hiccup. The mathematical ecologist Robert May voiced similar sentiments in 1976. 'Not only in research, but in the everyday world of politics and economics, we would all be better off if more people realized that simple systems do not necessarily possess simple dynamical properties.'

Hinduism and the Art of Mechanical Maintenance

We shall shortly observe how Western civilization came to view the universe as a regular clockwork machine, and deluded itself into thinking that deterministic equations always lead to regular behaviour. The oriental mind tends to have a different philosophical outlook. The Hindus, for example, ascribe to chaos a more subtle role than mere formless confusion, and recognize the underlying unity of order and disorder. In classical Hindu mythology the cosmos passes through three major phases: creation, maintenance, and destruction – mirroring birth, life, and death. Brahma is the god of creation, Vishnu the god of maintenance (order), and Shiva the god of destruction (disorder). But Shiva's personality is multifaceted. Shiva is he who walks on the wild side, the lone hunter, the dancer, the yogin who withdraws from human society, the ascetic covered in ash. The untamed. The distinction between the order of Vishnu and the disorder of Shiva is not that between good and evil. It represents instead two different ways in which divinity makes itself manifest: benevolence and wrath; harmony and discord.

In the same way, mathematicians are beginning to view order and chaos as two distinct manifestations of an underlying determinism. And neither exists in isolation. The typical system can exist in a variety of states, some ordered, some chaotic. Instead of two opposed polarities, there is a continuous spectrum. As harmony and discord combine in musical beauty, so order and chaos combine in mathematical beauty.

2

Equations for Everything

So, I for one, think it is gratuitous for anyone to enquire into the causes of the
motion towards the centre when once the fact that the Earth occupies the middle
place in the universe, and that all weights move towards it, is made so patent by
the observed phenomena themselves.

Ptolemy, *Almagest*

The metaphor of a clockwork world goes back a very long way, and it's
important that we appreciate just how deep-seated it is. Before grappling
with chaos, we must first study law.

A good place to start from is ancient Greece, with Thales of Miletus. He
was born around 624 BC, died in about 546 BC, and is famous for having
predicted an eclipse of the Sun. He probably appropriated the method from
the Egyptians or the Chaldeans, and it was accurate only to within a year or
so. Be that as it may, the eclipse occurred at a propitious moment, halting a
battle between the Lydians and the Medians, and the Sun was almost totally
obscured. These chance circumstances no doubt enhanced Thales' reputation
as an astronomer. One of the frustrations of being a historian is the way in
which, almost by accident, some events can be dated accurately while others
remain conjectural. Our knowledge of Thales' date of birth is based on
writings of Apollodorus; that of his death to Diogenes Laërtius: both dates
are unreliable. It is almost certain that the eclipse was that of 28 May 585
BC. So reliably does the cosmic clock tick that, two and a half millennia later,
we can calculate not just the times of ancient eclipses, but the positions on
the Earth's surface from which they could have been seen. Solar eclipses are
rare, and this particular one is the only one that Thales might reasonably
have witnessed. Astronomical happenings still provide historians with one of
their best methods for dating events.

Thales, it is said, was walking one evening, and became so absorbed in his
study of the night sky that he fell into a ditch. A female companion remarked,
'How can you tell what's going on in the heavens, when you can't see what

lies at your own feet?' In many ways the tale sums up the attitudes that gave rise to classical mechanics. The philosophers of ancient Greece could calculate the motions of the planets with breathtaking accuracy, but they still believed that heavy objects fall faster than light ones.

Dynamics only began to make progress when mathematicians dragged their eyes from the cosmos and looked more closely – and more critically – at what was happening at their own feet. Ptolemy imagined the Earth to be stationary at the centre of things because he took the evidence of his own senses too literally and failed to question its meaning. But cosmology provided the spur, and we may doubt whether more down-to-earth questions would have provided sufficient inspiration.

Cosmic Revolution

Early cosmology is strong on mythological imagination but short on factual content. We pass by visions of a flat Earth supported by an elephant, the sun-god riding his chariot across the sky, and stars that – in anticipation of electric lighting – hang on cords and are switched off during the daytime. The Pythagorean view was no less mystical, but it was strong on the mystic significance of numbers, and inadvertently let mathematics on the scene. Plato suggested that the Earth lies at the centre of the universe, with everything else revolving about it on a series of hollow spheres. He also thought that the Earth was round, and his Pythagorean-inspired belief that everything, even the motion of the heavens, was a manifestation of mathematical regularity, was to prove highly influential.

Eudoxus, a powerful mathematician who also invented the first rigorous theory of irrational numbers, realized that the observed motion of the planets against the stars didn't fit the Platonic ideal. The paths followed by the planets are tilted, and every so often they appear to move backwards. Eudoxus conceived a mathematical description in which the planets were considered to be mounted on a series of twenty-seven concentric spheres, each revolving about an axis borne by its neighbour. His successors improved the fit with observation by adding additional spheres. By 230 BC Apollonius had supplanted this system with a theory of epicycles, in which planets moved in small circles whose centres in turn moved in large circles. Claudius Ptolemaeus, otherwise Ptolemy, who lived at Alexandria in AD 100–160, refined the system of epicycles until they agreed so well with observations that nothing supplanted them for 1,500 years. It was a triumph of empirical mathematics.

Gears from the Greeks

The metaphor that the heavens move 'like clockwork' may have a more literal basis. Our ideas on ancient Greek culture have largely been derived from its intellectual side – philosophy, geometry, logic. Technology has received less attention. In part this is because few examples of Greek technology have survived. We are told that the Greeks valued logic – intellectual mathematics – above logistics – practical mathematics. But our sources for this view are not unbiased, and similar statements might well be heard today in the corridors of departments of mathematical logic. The full story of Greek technology may never be known, but the hints that we have are intriguing.

In 1900 some fishermen were searching for sponges off the coast of the tiny Greek island of Antikythera (opposite the larger island of Kythera, between the Greek mainland and Crete). They found the wreck of a ship that had been sunk in a storm in 70 BC while travelling from Rhodes to Rome. Their haul included statues, pottery, wine-jars, and coins, together with a rather dull lump of corroded metal. When the lump dried it split into pieces, revealing traces of gear wheels. In 1972 Derek de Solla Price had the lump X-rayed; he was able to reconstruct a complicated arrangement of thirty-two gear wheels (Figure 8). But what was it for? Analysing its structure, he decided that it must have been used to compute the positions of the Sun and Moon against the background of the stars.

The Antikythera mechanism has many interesting features, among which is the earliest-known example of a differential gear. Such gears are now used in the rear axles of cars to allow the wheels to move at different speeds, for example when cornering. In the Antikythera mechanism a differential gear was needed to compute the phases of the Moon by substracting the Sun's motion from the Moon's. The device is intricate and made with considerable precision, arguing the existence of a long tradition of gear-cutting and geared machines in ancient Greece. No other examples have survived – probably because old and broken machines were melted down to recycle their metal.

In his article 'Gears from the Greeks' (*Proceedings of the Royal Institution*, 58 (1986)) the British mathematician Christopher Zeeman has speculated about the influence of such devices on Greek science:

> First came the astronomers observing the motions of the heavenly bodies and collecting data. Secondly came the mathematicians inventing mathematical notation to describe the motions and fit the data. Thirdly came the technicians making mechanical models to simulate those mathematical constructions. Fourthly came generations of students who learned their astronomy from these machines. Fifthly

Figure 8   Gearing in the Antikythera mechanism, an ancient Greek planetary calculator.

came scientists whose imagination had been so blinkered by generations of such learning that they actually believed that this was how the heavens worked. Sixthly came the authorities who insisted upon the received dogma. And so the human race was fooled into accepting the Ptolemaic System for a thousand years.

The Central Sun

In 1473 Nicolaus Copernicus noticed that the Ptolemaic theory involves a great number of *identical* epicycles, and found that he could eliminate them if the Earth is deemed to go round the Sun. The identical epicycles are traces of the motion of the Earth, superimposed upon the motions of the remaining planets. At a stroke, this *heliocentric* theory reduced the number of epicycles to thirty-one.

Johannes Kepler was equally dissatisfied with Copernicus's revision of Ptolemy. He had inherited a series of new and highly accurate astronomical observations made by Tycho Brahe, and he was looking for the mathematical patterns behind them. He kept an open mind – so open that some of his ideas, such as the relation between the spacing of planetary orbits and the regular polyhedra (Figure 9), now appear rather ridiculous. Kepler later abandoned this theory when it conflicted with observations; we still have no theory of planetary formation that prescribes correctly the sizes and distances of the planets.

Eventually he was forced, almost against his will, to his *First Law*: planets move in elliptical orbits about the Sun. Buried in his work are two other laws which later acquired enormous significance. The *Second Law* states that the orbit of a planet sweeps out equal areas in equal times. The *Third Law* holds that the cube of the planet's distance from the Sun is proportional to the square of its orbital period.

Kepler's theory is aesthetically far more appealing than a jumble of epicycles but, like its predecessors, it is purely descriptive. It says *what* the planets do, but gives no unifying rationale. Before cosmology could go beyond Kepler, it had to bring itself down to earth.

Swing of the Pendulum

For a student at the University of Pisa in the 1580s, life must have been exciting, for it was a period of dramatic advances in human knowledge. But excitement cannot be sustained all the time. During a church service one student must have become bored, for his attention wandered and he began

Figure 9   Kepler's model of the spacing of planetary orbits, based on the five regular polyhedra (published in 1596).

to watch a large lamp, swinging in the breeze. It swung erratically, but he noticed that when it made a wide swing its speed increased, so that the time taken remained constant. Accurate clocks or watches had not then been invented, so he timed the lamp using his pulse.

The student was Galileo Galilei (Figure 10), who entered the University at the age of seventeen to study medicine, taking private lessons in mathematics. Galileo was born in Florence in 1564 and died in 1642. As well as being a scientist of the first order he was also a major literary figure, and his writings are elegant and skilful. He had a mechanical bent and made his own telescopes: he discovered that Jupiter has four moons, the first celestial bodies *known* not to revolve around the Earth. He had a talent for clear thinking, preferring simple logical explanations to flowery arguments designed to complicate and

Figure 10   Galileo Galilei, founder of theoretical and experimental mechanics. (Reproduced by permission of John Wiley & Sons Ltd., © 1968.)

obscure. He lived in an age that accepted explanations of events in terms of religious purpose. For example, rain falls *because* its purpose is to water crops; a stone thrown in the air falls to the ground *because* that is its proper resting place.

Galileo realized that enquiries into the purposes of things give humankind no control over natural phenomena. Instead of asking *why* the stone falls, he asked for an accurate description of *how* it falls. Instead of the motion of the Moon, which he could not influence or regulate, he studied balls rolling on inclined planes. And, in a stroke of genius, he confined his attention to a few key quantities – time, distance, velocity, acceleration, momentum, mass, inertia. In an age that concerned itself with qualities and essences his choice showed a remarkable grasp of essentials, especially since many of his chosen variables did not immediately lend themselves to quantitative measurement.

Time, in particular, caused Galileo much headache. You can't time a falling stone by watching the change in length of a burning candle. He used water clocks and his own pulse rate, and according to Stillman Drake he probably hummed tunes to himself, dividing the beat in the way that a musician would. To slow down dynamic phenomena and improve the accuracy of his timing he studied a ball rolling on a shallow slope, rather than one falling freely. And by a mixture of thought experiments and real ones he came to an elegant description of how bodies fall under gravity.

In character with the spirit of Greek geometry – in which all objects are idealized, so that a line has no breadth, a plane has no thickness – Galileo idealized his mechanics, choosing to neglect such effects as air resistance when seeking the underlying simplicities. In order to disentangle the web of interrelated influences that control the natural world, it is best to begin by studying a single strand at a time.

In mediaeval times it was thought that the path of a projectile came in three parts: an inital straight line motion, a portion of a circle, and a final vertical drop (Figure 11). Galileo discovered that the speed of a falling body increases at a constant rate, that is, its *acceleration* is constant. From this he deduced the correct path, a parabola. He also showed that a cannon-ball will travel the greatest distance if it is projected at an angle of 45°. He found laws for the composition of forces. He realized that, in the absence of air resistance, a heavy mass and a light one will fall with equal speed. These may seem simple things today, scarcely worth mentioning; but they were the first solid evidence that the rule of natural law might be read by humankind. Galileo had a dry sense of humour, as when he espoused the heliocentric theory in his *Dialogue on the Two Chief World Systems*:

> I should think that anyone who considered it more reasonable for the whole universe to move in order to let the Earth remain fixed would be more irrational than one who should climb to the top of a cupola just to get a view of the city and its environs, and then demand that the whole countryside should revolve around him so that he would not take the trouble to turn his head.

One system of natural law for matters celestial; another for those mundane. Kepler with his eyes on heaven and Galileo with his ear to the ground. That there should be a connection between the two realms was almost unthinkable.

Figure 11   Mediaeval theory of the motion of a projectile as a mixture of straight line and circular movements: the trajectory diagram is due to Tartaglia; here it has been superposed on a landscape in Walter H. Ruff's Der Geometrischen Buchsenmeisterey.

Heaven was pure, unsullied, the home of God and his angels; Earth was the home of sinful Man.

A single stroke of insight changed that perception forever.

Gravity and Geometry

Some great scientists are child prodigies, but the young Isaac Newton was a relatively ordinary child, save for a knack for making gadgets. The family cat, which is said to have disappeared in a hot air balloon, learned this to its cost. Newton was born in 1642 in the village of Woolsthorpe, a sickly and premature baby. As an undergraduate at Trinity College, Cambridge, he made no particular impact. But when the great plague struck, he returned to his home village, away from academic life, and almost single-handedly created optics, mechanics and calculus. In later life he became master of the Royal Mint and president of the Royal Society. He died in 1727.

Galileo had discovered that a body moving under the Earth's gravity undergoes a constant acceleration. Newton was after bigger game: a code of laws that would govern the motion of a body under all combinations of forces.

In a sense, the problem was geometrical rather than dynamic. If a body moves at a uniform speed then the distance that it travels is the product of its velocity and the time that has elapsed. If it moves at a non-uniform speed there is no such simple formula. Mathematicians before Newton had made important progress, showing that various basic dynamical questions could be posed in a geometric form. However, the geometric problems were seldom easy to solve.

A graph showing how the body's speed varies with time takes the form of a curve. By geometric arguments it can be shown that the total distance travelled is equal to the *area* under the curve. Similarly the velocity is the slope of the *tangent* to another graph, this time plotting distance against time. But how do we find these areas and tangents? Newton, and independently Gottfried Leibniz, solved these problems by dividing time into tinier and tinier intervals. The area under a curve then becomes the sum of the areas of a large number of narrow vertical strips. They showed that the error made by such an approximation becomes very tiny as the time interval becomes smaller and smaller, and argued that 'in the limit' the error can be made to vanish altogether. In the same way, the slope of a tangent can be calculated by considering two nearby time values and letting the difference between them become arbitrarily small. Neither mathematician could supply a logically rigorous justification for his method, but both were convinced that it was correct. Leibniz talked of 'infinitesimal' changes in time; Newton had a more physical picture of quantities that flowed continuously, which he called *fluents* and *fluxions*.

These methods of the calculus, now known as *integration* and *differentiation*, solved the practical problems of determining distances from velocities or velocities from distances. They brought an enormous wealth of natural phenomena within the range of mathematical analysis.

The System of the World

The *Mathematical Principles of Natural Philosophy* (Figure 12), which contains the laws of motion, was published in three volumes. It owed much to Galileo, as Newton duly acknowledged, and was based on a similar scientific philosophy. In it he reduces all motion to three simple laws laid down in the first volume:

- If no forces are acting on a body then it remains at rest, or moves uniformly in a straight line.
- Its acceleration is proportional to the force that is acting.
- To every action there is always an equal and opposite reaction.

Newton also shows that Kepler's laws of planetary motion follow from these three laws, together with the inverse square law of gravity. But the true significance of Newton's conception of gravity is not so much that it can be described numerically. Newton's law is *universal*. Every particle of matter in the universe attracts every other particle according to the same law. The gyrations of Jupiter and the path of a cannonball are two manifestations of the *same* law. Man is in his Heaven and the universe is whole again.

The discovery was taken up and elaborated in the third book. 'I now,' said Newton, 'demonstrate the system of the World.' And he did. He applied his theory of gravitation to the motion of planets round the sun and satellites round their planets. He found the masses of the planets, and of the Sun, relative to that of the Earth. He estimated the Earth's mass to within 10 per cent of its true value. He showed that the Earth is flattened at the poles and obtained a fairly accurate estimate of the amount of that flattening. He discussed the variation of gravity over the Earth's surface. He calculated irregularities in the Moon's motion due to the pull of the Sun, and the orbits of comets – showing that these supposedly lawless harbingers of cosmic disapproval were governed by the same laws as the planets.

Aldous Huxley once said that 'Perhaps the men of genius are the only true men. In all the history of the race there have been only a few thousand real men. And the rest of us – what are we? Teachable animals. Without the help of the real man, we should have found out almost nothing at all.' It's not necessary to agree with Huxley to accept that some people have a

PHILOSOPHIÆ

NATURALIS

PRINCIPIA

MATHEMATICA.

Autore *JS. NEWTON*, *Trin. Coll. Cantab. Soc.* Matheseos
Professore *Lucasiano*, & Societatis Regalis Sodali.

IMPRIMATUR.
S. PEPYS, *Reg. Soc.* PRÆSES.

Julii 5. 1686.

LONDINI,

Jussu *Societatis Regiæ* ac Typis *Josephi Streater*. Prostat apud
plures Bibliopolas. *Anno* MDCLXXXVII.

Figure 12   Title-page of Newton's Mathematical Principles of Natural Philosophy.

disproportionate impact on history. Newton was a 'true man'. In the same way, calculus is 'true mathematics', and has had an equally disproportionate impact. But the importance of the calculus for Newton's dynamics was not immediately obvious to most of his contemporaries. The reason is simple: nowhere in the *Mathematical Principles of Natural Philosophy* was any explicit use made of it. Instead, Newton cast his proofs in the language of classical Greek geometry. But the calculus did eventually see the light of day in 1736, thanks to the urgings of Newton's scientific friends. By the end of the 17th century mathematicians throughout Europe were in full possession of the methods of the calculus, and had received a strong hint from Newton that Nature's pages were open for any with the wit to read them. They needed no further encouragement.

Bells and Whistles

The label 'analysis' is used today to describe calculus in its more rigorous form: the theory behind it rather than the computational technique. It acquired that connotation during the 18th century, when the theoretical side of calculus was being substantially extended. The chief architect of this development was Leonhard Euler, the most prolific mathematician of all time. Euler was also responsible for large parts of the application of calculus to mathematical physics. Born in Switzerland in 1707, he was at first trained in religion, but he soon turned to mathematics and was publishing by the age of eighteen. At nineteen he won a major mathematical prize awarded by the French Academy of Sciences, on a problem about the masting of ships. In 1733 he was appointed to the academy of St Petersburg in Russia. In 1741 he moved to Berlin, but returned to Russia in 1766 at the request of Catherine the Great. In consequence Switzerland remembers him as a great Swiss mathematician, Russia as a great Russian mathematician, and Germany as a great German mathematician. His eyesight began to fail, and by 1766 he was totally blind. This had no noticeable effect on his prodigious output of original mathematics.

The first extensive flowering of the Newtonian seed was the subject of analytical mechanics: mechanics based fully and explicitly on the calculus, in which the objective was first to find the differential equations that governed the motion of the system concerned, and then to solve them. But entirely new areas of mathematical physics were soon to be opened up. The ancient Pythagoreans had sought harmony in number – or, more accurately, number in harmony, for the numerology of music was their greatest discovery. Many have professed to detect an affinity between mathematics and music. Be that as it may, an amazing amount of important mathematics has been derived

Figure 13   Vibrations of a violin string: sinusoidal fundamental (1) and its second and third harmonics (2, 3) superpose to create a more complex waveform (heavy line).

from the problem of a vibrating violin string. It can for instance be argued that without it, we wouldn't have radio and television.

By solving an appropriate differential equation, Brook Taylor discovered in 1713 that the fundamental form of a vibrating string is a sine curve (Figure 13(1)). In 1746 Jean Le Rond d'Alembert noticed that other shapes were also possible. D'Alembert was the illegitimate child of Madame de Tencin, a famous socialite, and her lover the Chevalier Destouches. The fruit of their liaison was abandoned on the steps of the church of S.Jean-le-Rond in Paris, whence his unusual christian names.

Let it never be said that all mathematicians lead dull and ordinary lives.

D'Alembert carried out a general analysis of the vibrating string. Assuming that the amplitude (size) of the vibration is small (to eliminate undesirable terms from the equations, a practice to which we shall return later), he wrote down a differential equation that must be satisfied by the string. But this was a new type of equation, a *partial differential equation*. Such equations involve the rates of change of some quantity with respect to *several* different variables. For the violin string, these variables are the position of a point on the string, and time. D'Alembert proceeded to show that the equation is satisfied by the superposition of two waves, of *arbitrary* shape, one travelling to the left and the other to the right.

Euler was quick to follow up this discovery. It occurred to him that Taylor's single sinusoidal waveform can be combined with its higher harmonics – waves with the same shape, but vibrating at twice, three times, four times . . . the fundamental frequency (Figure 13 (2, 3)). In *A New Theory of Music* he analysed the vibrations of bells and drums. Daniel Bernoulli extended the results to organ pipes.

Out of music came physics. In 1759 Joseph-Louis Lagrange, a young man just beginning to make a name for himself, applied the ideas to sound waves, and within ten years a comprehensive and successful theory of acoustics was well on the way.

Wind and Waves

The 18th century was an age of sea power, demanding knowledge of the way water and other fluids flow. In 1752 Euler turned his attention to the dynamics of fluids, and by 1755 he had set up a system of partial differential equations to describe the motion of a fluid without viscosity ('stickiness'). He considered both incompressible fluids (water) and compressible (air). He modelled the fluid as a continuous, infinitely divisible medium, and described its flow in terms of continuous variables that depend on the position of fluid particles: velocity, density, pressure.

One by one, the various branches of physics came under the sway of mathematical law. Joseph Fourier developed an equation to describe heat flow, and came up with a new and powerful method to solve it, now known as *Fourier analysis*. The main idea is to represent any waveform as a superposition of sine curves, like Figure 13 but more complicated.

The deformation of materials under stress, fundamental to engineering, led to the equations of elasticity. Deeper analysis of gravitation led to equations now named in honour of Pierre Simon de Laplace and Simeon-Denis Poisson. The same equations appeared again in hydrodynamics and electrostatics, and a common generalization evolved, known as potential theory. Potential theory let mathematicians attack problems such as the gravitational attraction due to an ellipsoid. This is important in astronomy, because most planets aren't spheres – they're flattened at their poles. The 18th century (and the early 19th) was the period in which most of the great theories of classical mathematical physics were forged, the main exceptions being the Navier–Stokes equations for the flow of a viscous fluid, and James Clerk Maxwell's equations for electromagnetism, which came a little later. From Maxwell's equations came the discovery of radio waves.

One overwhelming paradigm emerged. The way to model nature is through differential equations.

Abandoned by Analysis

But there's a price to pay. The mathematicians of the 18th century ran headlong into a problem that has plagued theoretical mechanics to this day: to *set up* the equations is one thing, to *solve* them quite another. Euler himself said: 'If it is not permitted to us to penetrate to a complete knowledge concerning the motions of fluids, it is not to mechanics, or to the insufficiency of the known principles of motion, that we must attribute the cause. It is analysis itself which abandons us here.' The 18th century's main achievements were in setting up equations to model physical phenomena. It had much less success solving those equations.

Despite this, there was boundless optimism and a general feeling that the problems of Nature had been cracked wide open. The successes of the differential equation paradigm were impressive and extensive. Many problems, including basic and important ones, led to equations that *could* be solved. A process of self-selection set in, whereby equations that could not be solved were automatically of less interest than those that could. The textbooks from which new generations learned the techniques, of course, contained only the soluble problems. Zeeman's remarks on the Antikythera mechanism spring to mind. Clockwork models, belief in a clockwork world. Deterministic mathematical models, belief in a deterministic world.

Mathematics in Pawn

The process was not universal. Some unanswered questions, such as the motion of three bodies under gravity, became notorious for their impenetrability. But somehow such equations became seen as exceptions when a more honest appraisal would have exhibited them as the rule.

And in fact, even the *mathematical* determinism of the equations of motion had loopholes. One of the common idealizations of Newtonian mechanics is to consider hard elastic particles. If two such particles collide, they bounce off at well-determined angles and speeds. But Newton's laws are not enough to fix the outcome of the simultaneous collision of *three* such particles (Figure 14). The claims were magnificent but the delivery was faulty, even in the heyday of Laplacian determinism. As Tim Poston and I wrote in *Analog* (November 1981):

> So the 'inexorable laws of physics' on which – for instance – Marx tried to model his laws of history, were never really there. If Newton could not predict the behaviour of three balls, could Marx predict that of three people? Any regularity in the behaviour of large assemblies of particles or people must be *statistical*, and that has quite a different philosophical taste . . . In retrospect we can see that the determinism of pre-quantum physics kept itself from ideological bankruptcy only by keeping the three balls of the pawnbroker apart.

At any rate, mathematics thought it had struck the mother-lode, and was

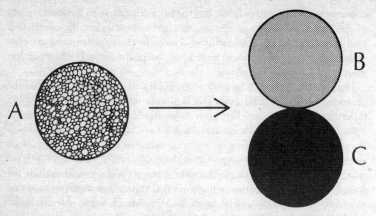

Figure 14   Where do they go? According to Newton's laws of motion, and assuming the spheres are perfectly elastic, the result depends on whether A hits B or C first. If it hits both exactly together, Newton's laws do not specify what happens.

busy gathering all the gold it could. To point out from the lofty heights of the 20th century that some of it was fools' gold is a nasty case of 20/20 hindsight.

The Reformulation Period

In 1750 Lagrange took up Euler's ideas and produced from them an elegant and far-reaching reformulation of dynamics. Two important ideas crystallized out of his work. Both had been around, as half-baked ideas, for decades, but Lagrange baked tham golden brown, took them out of the oven, and placed them on the bakery counter for all to admire, buy, and consume.

The first was the Principle of Conservation of Energy. Classical mechanics recognized two forms of energy. *Potential energy* is the energy that a body has by virtue of its position. For instance, in a gravitational field, potential energy is proportional to height. A body on top of a hill has more potential energy than one in a valley, which is why a hill climb is more tiring than a walk along a canal bank. *Kinetic energy* is the energy that a body has by virtue of its speed: you have to work a lot harder to slow down a runaway horse than you do when you trot it round the meadow.

During motion, and in the absence of friction, these two forms of energy can be converted into each other. When Galileo dropped his celebrated cannon-ball off the leaning tower of Pisa, it started out with a lot of potential energy but no kinetic, and traded potential for kinetic as it fell. That is, it *got lower* and *speeded up*. Mother Nature is a scrupulous accountant: the balance in her ledger – the total energy, potential plus kinetic – doesn't change. When a cannon-ball falls off a parapet it loses potential energy, and therefore must gain kinetic energy. That is, it speeds up. Newton's second law of motion effectively expresses this qualitative argument in precise quantitative form.

Lagrange's second idea was to introduce 'generalized coordinates'. Coordinates are a trick to convert geometry into algebra, by associating with each point a set of numbers. Mathematicians had found it convenient to work with various systems of coordinates, depending on the problem being tackled. Lagrange must have decided that it was inconvenient to cart this sort of computational baggage around in a mathematical theory. He began by assuming *any system of coordinates whatsoever*. Then, with stunning simplicity, he derived the equations of motion in a form that *does not depend upon the coordinate system* chosen. Lagrange's formulation has numerous advantages over Newton's. Many of them are technical – it is easier to apply when there are constraints on the motion, it avoids messy coordinate transformations. But above all, it is more general, more abstract, more elegant, and *simpler*.

The ideas were taken up by William Rowan Hamilton (1805–1865), the great Irish mathematician. He reformulated dynamics yet again, with still greater generality. In Hamilton's version of the theory, the state of a dynamical system is specified by a general set of position coordinates (like Lagrange's) together with a related set of momentum coordinates (the corresponding velocities, multiplied by the mass). A single quantity, now called the *Hamiltonian* of the system, defines the total energy in terms of these positions and momentums. The rates of change of the position and momentum coordinates with time are then expressed in terms of the Hamiltonian, in a simple, elegant, unified system of equations. Today's advanced dynamics texts often *start* with Hamilton's equations.

Trouble in the Marketplace

In the marketplace of mathematical physics, the wares of the deterministic stall are now set out. Nature obeys a relatively small set of fundamental laws. The laws are differential equations, and *we know what they are*. Given the state of any natural system at a given time, and knowing the laws, in principle all future motion is uniquely determined. In practice, the equations can be solved in many cases. Wind and waves, bells and whistles, the motion of the moon.

If the stall-owner could see into the future, he would be astonished by the technological marvels that will flow from his wares. Radio, television, electronics. Automobiles. Telephones. Radar. Wide-bodied jets. Digital watches. Computers. Vacuum-cleaners. Washing-machines. Personal stereos.

Suspension bridges. Synthesizers. Hang-gliders. Communications satellites. Compact discs. And, to be even-handed: machine guns, tanks, anti-personnel mines, cruise missiles, MIRVed nuclear warheads, and pollution. Let us not underestimate the effect of the classical deterministic paradigm of mathematical physics on our society.

But let us not be misled. Technology is our own creation. In technology we don't so much understand the universe as build tiny universes of our own, which are so simple that we can make them do what we want. The whole object of technology is to produce a controlled effect in given circumstances. We *make* our machines so that they will behave deterministically. Technology creates systems to which the classical paradigm applies. No matter that we can't solve the equations for the motion of the Solar System – we don't build any machines whose operation relies on knowing those answers.

The stall-keeper polishes his shiny new equations, oblivious to such matters, and dreams of a glittering future. The customers flock around him, clamouring, hunting for bargains.

But what's this? Another stall? There's no need for another stall. The local council must be mad to allow such a scruffy-looking bunch into the market! And what are they selling?

Dice?

Look, if you're going to allow gambling in the market, the whole place will go to the . . .

Oh. They're not for gambling. What else have you got on that stall?

Life insurance? The efficacy of prayer? The heights of human beings? The sizes of crabs? The petals of buttercups? The frequency of paupers per poor-law union? *The divorce rate?*

It'll be crystal-gazing next. The marketplace has gone to the dogs already. This is supposed to be a *scientific* market. Can this twaddle possibly be science?

Oh yes.

3

The Laws of Error

> The huger the mob, and the greater the apparent anarchy, the more perfect is its sway. It is the supreme law of Unreason. Whenever a large sample of chaotic elements are taken in hand and marshalled in the order of their magnitude, an unsuspected and most beautiful form of regularity proves to have been latent all along. The tops of the marshalled row form a flowing curve of invariable proportions; and each element, as it is sorted into place, finds, as it were, a preordained niche, accurately adapted to fit it.
>
> Francis Galton, *Natural Inheritance*

For all the impressive gains made by classical mathematical physics, entire areas of the natural world remained untouched. Mathematics could calculate the motion of a satellite of Jupiter, but not that of a snowflake in a blizzard. It could describe the growth of a soap bubble but not that of a tree. If a man were to leap from the Eiffel Tower, mathematics could predict how long it would take him to hit the ground, but not why he chose to jump in the first place. And for all the proofs that 'in principle' a small number of laws predict the entire future of the universe, in practice such concepts as the pressure of a gas or the temperature of a lump of burning coal were immeasurably beyond the frontiers of what could rigorously be deduced from the laws that were actually known.

Mathematicians had finally managed to pin down at least some of the order in the universe, and the reasons for it, but still they lived in a disordered world. They believed, with some justification, that much of the disorder obeyed the same fundamental laws; their inability to apply those laws to any effect was just a matter of complexity. The motion of two point masses under mutual forces could be calculated precisely. That of three was already too difficult for a complete solution, but in specific cases approximate methods could be brought to bear. The long-term motion of the fifty or so major bodies of the Solar System was impossible to grasp in its entirety, but any specific feature could be understood reasonably well by making a big enough computational effort. But a milligram of gas contains roughly a hundred trillion particles. Even to *write down* the equations of motion would take a piece of paper comparable in size to the area enclosed by the Moon's orbit. To think seriously about solving them is ridiculous.

A method which in theory solves everything, but in practice is as much use as a spider's web against an avalanche, isn't likely to win many devotees, no matter how impeccable its philosophical credentials may be. Science wasn't going to throw up its hands in despair at the problem of a gas, just because it was impossible to describe the individual motions of every single particle. The detailed complexity of large numbers of particles may be unimaginable; but progress might still be made by setting more realistic goals. Experiment suggests that, complexity notwithstanding, gases behave in a pretty regular way. If the detailed behaviour of large systems is unknowable, can we find regularities in the coarse, average behaviour? The answer is 'yes', and the mathematics needed is the theory of probability and its applied cousin, statistics.

Gambling Gain

Probability theory originated in a supremely practical topic – gambling. Every gambler has an instinctive feeling for 'the odds'. Gamblers *know* that there are regular patterns to chance – although not all of their cherished beliefs survive mathematical analysis. Girolamo Cardano (Figure 15), the gambling scholar, an intellectual genius and an incorrigible rogue, was the first to write about probability. In 1654 the Chevalier de Meré asked Blaise Pascal how best to divide the stakes in a game of chance that is interrupted. The same names that crop up in the development of deterministic mathematics also appear in that of the mathematics of chance: Pascal wrote to Fermat and between them they found an answer. It saw print in 1657 in the first book to be devoted entirely to probability theory, *On Reasoning in Games of Chance* by Christian Huygens.

Probability as a subject in its own right stems from the publication of the *Analytic Theory of Probabilities* of Laplace in 1812. According to Laplace, the probability of an event is the number of ways in which it can occur, divided by the total number of things that can happen – on the assumption that all of the latter are equally likely. For example, the probability of a family of seven consisting entirely of girls is 1/128, because of the 128 possible boy/girl sequences, exactly one goes GGGGGGG. (This assumes that a boy or girl is equally likely; in fact boys are slightly more likely than girls. It isn't hard to take this into account.)

Figure 15   Girolamo Cardano, the gambling scholar. (Reproduced by permission of John Wiley & Sons Ltd., © 1968.)

Figure 16   Binomial approximation to a normal distribution (Quetelet, 1846). Horizontal scale is height; vertical scale, the number of people having that height.

The Average Man

The practical arm of probability theory is statistics. The most striking feature in the development of statistics is that both 'hard' and 'soft' sciences played decisive roles, and that important ideas and methods were repeatedly transferred between them. Over the next few pages we will pursue a typical instance. Much of statistics centres around the so-called *normal distribution* (Figure 16). This is a bell-shaped curve that closely models the proportions of a population that have some particular characteristic. For example, if 1,000 men are drawn at random from the population of Outer Mongolia, and a graph is drawn showing how many of them have a given height in centimetres, it will closely resemble the bell-shaped curve of the normal distribution. The same goes if you plot the wing-span of a population of ducks, the burrowing ability of a population of moles, the sizes of sharks' teeth, or the number of spots on a leopard.

The normal distribution was originally called the *error law*, and it arose from the work of 18th-century astronomers and mathematicians who, in trying to calculate the orbits of celestial bodies, were forced to take account of the effect of observational error. The error law describes how observed values cluster around their average, and provides estimates for the chances of an error of a given size. It was imported into social science by Adolphe

Quetelet (Figure 17), who applied the method to everything he could think of: measurements of the human body, crime, marriage, suicide. His *Social Mechanics* was so titled as a deliberate parallel to Laplace's *Celestial Mechanics*. Quetelet was quick to draw general conclusions from the supposed constancy of average values of social variables, and came up with the tantalizing notion of the 'average man'. Not only does Quetelet think of the human condition as a kind of social dynamic: he wants to deal with it in the manner of a control systems engineer. Tune it, stabilize it, damp out oscillations. To Quetelet, the 'average man' wasn't just a mathematical abstraction, but a moral ideal.

Hereditary Genius

The social sciences differ from the physical sciences in many ways, an important one being that controlled experiments are seldom feasible in the social sciences. If a physicist wishes to examine the effect of heat on a metal bar, he can heat it to various temperatures and compare the results. If an economist wishes to examine the effect of fiscal policy on a country's economy, he can either try it or not; but he doesn't have the luxury of trying several different taxation regimes on the *same* economy under the same conditions. Around 1880 the social sciences began to evolve a substitute for controlled experiment, derived from the early work of Quetelet. The most important work was done by three men: Francis Galton, Ysidro Edgeworth, and Karl Pearson. Each was prominent in a traditional field: Galton in anthropology, Edgeworth in economics, Pearson in philosophy. Between them they converted statistics from a controversial ideology into a more or less exact science. We follow only Galton's career in any detail.

Francis Galton (1822–1911) was trained in medicine, but abandoned it when he received an inheritance, and set out to see the world. In 1860 he turned his attention to meteorology, and by graphical methods extracted the existence of anticyclones from a mass of irregular data. He dabbled in psychology, education, sociology, and fingerprinting, but by 1865 his main interest had emerged – heredity. Galton wanted to understand how inherited characteristics are passed on to succeeding generations. In 1863 he came across the writings of Quetelet, and became an instant convert to the ubiquity of the normal distribution. However, the way in which he used it was quite different from what Quetelet was advocating. Galton saw the normal distribution not as a moral imperative, but as a method for classifying data into groups of different origin. For example, consider a mixed population of pygmies and giants. The heights of the pygmies conform to a normal distribution, and so do the heights of the giants. However, these two curves are quite different; in particular their peaks will be in different places. The heights

Figure 17   Adolphe Quetelet (portrait by J. Odevaere, 1822).

Figure 18   Superposing two normal distributions can produce a two-peaked curve.

of the *combined* population cannot possibly form a normal distribution – for the mathematical reason that superposing two independent normal distributions does not, in general, produce another. Instead, it produces a two-peaked curve (Figure 18). Galton reasoned that the normal distribution applies only to 'pure' populations; that in a mixed population it would fail; and that the mixed population might be separated into its pure constituents by analysing the manner of its failure. One peak for giants, another for pygmies.

But this very picture caused Galton considerable headaches when thinking about heredity. Suppose that the first generation of a pure population has its heights normally distributed. Each individual produces offspring, whose heights are presumably also normally distributed. However, the peak height of the offspring depends on that of the parent – otherwise how could the characteristic 'height' be inherited? Thus the heights of the succeeding genera-tion are described by the superposition of many different normal distributions. But superposing normal distributions, as we have just seen, does not in general lead to a normal distribution. Conclusion: *when a pure population produces the next generation, the resulting population is no longer pure.* But this is absurd: after all, the original 'pure' population is itself a succeeding generation from the previous one!

It was not until 1877 that Galton resolved the paradox. By then he had extensive data on sweet peas showing that successive generations *did* in fact conform to the normal distribution; and he also had a curious experimental device known as a quincunx which simulated the mathematics by allowing lead shot to fall through an array of metal pins, bouncing at random to left or right. His resolution of the paradox runs as follows. Because the parents come from a pure population, the separate normal distributions for their descendants are *not independent*. Their behaviour under superposition is thus special. In fact there is a miniature mathematical miracle: they are related in just such a manner that upon superposing them all, a normal distribution again results.

Figure 19   Francis Galton's diagram of the relation of children's heights to those of their parents, showing a pattern of concentric ellipses.

Galton was struck by the crispness of this result, and it led him to the idea of *regression*. The children of tall parents are, on average, shorter; the children of short parents are, on average, taller. This does not prevent the children of the tall parents being taller than those of the short ones, but the height of the offspring is just displaced slightly towards the average.

In 1855 Galton plotted a diagram showing the heights of 928 adult children against those of their parents (Figure 19). In the diagram, the numbers in a given row and column show how many children in the sample have parents whose mean height is given at the left-hand end of that row, and whose own height differs from that of the parents by the amount at the head of the column. Galton noticed that numbers in a given range, say 3–5 or 6–8, arrange themselves along approximate ellipses, centred on the average height of the entire population. This picture fitted Galton's theory of regression perfectly, and from it emerged the method of *regression analysis*, which can deduce underlying trends from random data,

Galton did not couch his ideas in precise mathematical terms, preferring to rely on graphical descriptions and demonstrations with his quincunx. Mathematical solidity was supplied by Edgeworth, broadening the ideas and making them far more widely applicable. Pearson, a competent mathematician but mathematically less talented than Edgeworth, was a popularizer with the drive and ambition needed to sell the methods to the world. Visionary, technician, salesman: it took all three for statistics to make its impact.

Technology Transfer

Statistics, as already observed, is remarkable for way its ideas ebb and flow between the physical and social sciences. Starting with error analysis in astronomy, the social scientists developed mathematical tools for spotting patterns in random data. But now the hard sciences were to borrow back those tools, with a very different aim in view: the mathematical treatments of physical systems so complex that they appeared random.

In 1873 the great physicist James Clerk Maxwell proposed using statistical methods to a meeting of the British Association for the Advancement of Science:

> The smallest portion of matter which we can subject to experiment consists of millions of molecules, none of which ever becomes individually sensible to us. We cannot, therefore, ascertain the actual motion of any of these molecules; so we are obliged to abandon the strict historical method, and to adopt the statistical method of dealing with large groups of molecules. The data of the statistical method as applied to molecular science are the sums of large numbers of molecular quantities. In studying the relations between quantities of this kind, we meet with a new kind of regularity, the regularity of averages, which we can depend upon quite sufficiently for all practical purposes, but which can make no claim to that character of absolute precision which belongs to the laws of abstract dynamics.

Physicists repeatedly cited the success of statistical methods in the social sciences as justification for their probabilistic procedures. In their hands, the statistical method blossomed, and the kinetic theory of gases grew into a major – and fundamental – area of scientific activity. Nor was it just the fruit of a loose analogy between molecules and individuals in a population: there were close mathematical correspondences. In particular Maxwell tackled a basic question: what is the statistical distribution of the randomly varying velocity of a molecule? He began with two plausible physical assumptions:

- The component of the velocity in any given direction is independent of the component in any perpendicular direction.
- The distribution is spherically symmetric, that is, treats all directions equally.

From these abstract principles alone, without any appeal to the laws of dynamics, he advanced a solely mathematical argument to prove that the distribution must be the three-dimensional analogue of Quetelet's error law.

Dutch Chaos

The word 'gas' was invented by the Dutch chemist J. B. Van Helmont in his *Ortus Medicinae* of 1632, with a deliberate similarity to the Greek word 'chaos'. It was a very perceptive choice.

In the physics of gases, randomness and determinacy first come face to face. But in principle a gas is a purely deterministic aggregate of moving molecules obeying precise dynamical laws. Where does the randomness come from?

The answer – and it was one that any scientist worth his salt would automatically give until the 1970s, and most would still have given at the start of the 1980s – is *complexity*. The detailed motion of a gas is just too complex for us to grasp.

Suppose that you possessed a device capable of tracking a reasonable number of individual gas molecules as they move. No such device yet exists, and, even if it did, you'd need to use a computer to slow down the motion by many orders of magnitude to see what was happening; but suppose. What would you see? Concentrate your attention on a small group of molecules. They follow straight-line paths for a short time, then begin to bounce off each other in ways that you can predict from the previous geometry of the paths. But just as you're beginning to see the pattern of the motion, a new molecule comes whizzing in from outside and crashes into your nicely organized group, breaking the pattern. And before you can work out the new pattern, along comes another molecule, and another, and another . . .

If all you see is a tiny part of an enormously complicated motion, it will *appear* random, *appear* structureless.

In a sense, this is the same mechanism that makes social science so difficult. You can't study a living economy, or a nation, or a mind, by isolating a small part. Your experimental subsystem will be constantly perturbed by unexpected and uncontrollable outside influences. Even in the physical sciences, most of the day-to-day effort of experimental method goes into eliminating outside influences. The flashing neon signs of Broadway are effective in attracting the night-life and the low-life to the strip clubs and bars, but they play havoc with an astronomer's telescope. A sensitive seismometer will record not just earthquakes, but also the footsteps of the tea-lady pushing her trolley down the corridor. Physicists go to vast extremes to eliminate such unwanted effects. They perch telescopes on mountain-tops instead of the roof-tops of Manhattan, they bury neutrino-counters miles underground instead of putting them in the office. But the social scientist, denied even this luxury, must use statistical methods to model, or filter out, these outside effects. Statistics is a method for panning precious order from the sand of complexity.

The scientists of a hundred years ago were well aware that a deterministic system can behave in an apparently random way. But they knew that it wasn't *really* random; it just *looked* that way because of imperfect information. And they also knew that this phoney randomness only occurred in very large, complicated systems – systems with enormously many degrees of freedom, enormously many distinct variables, enormously many component parts. Systems whose detailed behaviour would forever be beyond the capacity of the human mind.

Spare a Paradigm?

By the end of the 19th century science has acquired two very different paradigms for mathematical modelling. The first, and older, was high-precision analysis by way of differential equations: in principle determining the entire evolution of the universe but in practice applicable only to relatively simple and well-structured problems. The second, a brash young upstart, was statistical analysis of averaged quantities, representing coarse features of the motion of highly complex systems.

There was virtually no contact, at a mathematical level, between the two. The statistical laws were not calculated as mathematical consequences of the laws of dynamics: they were an extra layer of structure imposed upon the mathematical models employed in physics, and they were based on physical intuition. The rigorous deduction of the behaviour of bulk matter from the laws of dynamics remains a challenging problem for mathematical physicists, even today: only recently has anyone come close to a proof that (in a suitably defined model) gases exist. Crystals, liquids, and amorphous solids remain firmly out of reach.

As the 20th century unrolled, statistical methodology took its place alongside deterministic modelling as an equal partner. A new word was coined to reflect the realization that even chance has its laws: *stochastic*. (The Greek word *stochastikos* means 'skilful in aiming' and thus conveys the idea of *using* the laws of chance for personal benefit.) The mathematics of stochastic processes – sequences of events determined by the influence of chance – flourished alongside the mathematics of deterministic processes.

No longer was order synonymous with law, and disorder with lawlessness. Both order and disorder had laws. But the laws were two distinct codes of behaviour. One law for the ordered, another for the disordered. Two paradigms, two techniques. Two ways to view the world. Two mathematical ideologies, each applying only within its own sphere of influence. Determinism for simple systems with few degrees of freedom, statistics for complicated systems with many degrees of freedom. Either a system was random, or it wasn't. If it was, scientists reached for something stochastic; if not, they polished up their deterministic equations.

The two paradigms were equal partners, equally accepted in the scientific world, equally useful, equally important, equally mathematical. Equal. But different. Totally, irreconcilably different. Scientists *knew* they were different, and they knew why: simple systems behave in simple ways, complicated systems behave in complicated ways. Between simplicity and complexity there can be no common ground.

But what one generation of scientist *knows*, beyond any shadow of doubt, with a knowledge that is built into the very fabric of their world, is precisely what the succeeding generation will challenge and overturn. If you *know* something that strongly, you don't question it. If you don't question it, you're living by faith, not by science.

But this one is a very difficult question. Can a simple deterministic system behave like a random one? It ran counter to almost everybody's intuition even to ask it. The whole progress of science was based on the belief that the way to seek simplicity in nature is by finding simple equations to describe it. What a silly question!

At the point in history at which we have now arrived, only one dissident voice could be discerned, and then only faintly, uncertainly, just a tremulous hint of future trouble, a voice raised only once, then silent; a voice that – if it was heard at all – was ignored. It was the voice of a man who was arguably the greatest mathematician of his age, another revolutionary of the turbulent science of dynamics, who created an entire new field of mathematics as a by-product. The voice of a man who touched chaos . . .

And was horrified by it.

4

The Last Universalist

Unknowing, mathematics writhed in the grip of Catch-22.

If you can solve an equation by a formula, then its solutions will *ipso facto* behave in a regular and analysable way. That's what formulas tell you. And if you think the name of the game in dynamics is finding formulas for the solution of differential equations, your mathematics will only be able to study regular behaviour. You will actively seek out problems to which your methods apply, and ignore the rest. Not even sweep them under the carpet: to do that, you must at least acknowledge their existence. You're living in a fool's paradise, or at least, you would be if you weren't too clever by half to be a fool.

It takes a very special combination of circumstances to get out of such a bind. The time, the place, the people, the culture – all must be right.

There was nothing wrong with the place: France was among the very top rank of mathematical nations. It still is.

The person had the mild, confused look of an absent-minded professor, but he was an intellectual giant. With one foot in the 19th century and one in the 20th, he straddled one of the pivotal points in mathematical history, when mathematics began its love-affair with generality and abstraction – an affair that many, enamoured of the concrete, neither understood nor approved, and still do not. His name was Henri Poincaré (Figure 20), perhaps the last mathematician able to roam at will throughout every nook and cranny

Figure 20   Portrait of Henri Poincaré, illustrating his discovery of 'Poincaré recurrence'. If a transformation is applied repeatedly to a mathematical system, and the system cannot leave a bounded region, it must return infinitely often to states near its original state.

of his subject. After Poincaré came the specialists – and the explosion in mathematical information that made them necessary owed its existence, in no small measure, to his breadth and depth of mathematical insight. Among his innumerable discoveries and inventions, Poincaré founded the modern qualitative theory of dynamical systems.

The place, the person. But the time was not quite right, and the culture even less so. When scientists first began to probe the ocean depths, their nets brought up the remains of strange monsters, dull in colour, and ugly as sin. Only when bathyscaphes equipped with searchlights were able to explore the deep sea trenches did the often delicate beauty and colour of these remote regions manifest itself. It is hard to judge beauty from a corpse. It was the same for Poincaré. He gazed into the abyss of chaos, he discerned some of the forms that lurked within; but the abyss was still dark and he mistook for monstrosities some of the most beautiful things in mathematics. Poincaré had the depth, but he lacked the means of illumination. It took another age, armed with Poincaré's own qualitative theory of differential equations, together with computers and other technological assistance, to shine some light into the chaotic depths and reveal that beauty.

But they could never have done it if Poincaré hadn't pioneered the way to the abyss's edge.

Absent-minded Dreamer

Henri Poincaré was born on 29 April 1854 at Nancy in north-east France. His father was a physician: astonishingly little is known of his mother. He was an unusually intelligent but physically uncoordinated child – a condition not helped by a severe bout of diphtheria at the age of five – and throughout his life his coordination remained poor. He first showed serious leanings towards mathematics at the age of fifteen. In 1871 he passed the examinations for his first degree – almost failing in mathematics when he managed to confuse himself over a simple question about geometric series. He put matters right soon after in the examinations for the School of Forestry, when he gained first prize in mathematics without having taken any lecture notes. He moved on to the École Polytechnique, the hot-bed of French mathematics, acquiring a reputation as a mathematical whiz-kid. Several attempts to bring him down a peg by setting him hard mathematical problems misfired when Poincaré sailed through them effortlessly.

In 1875 he entered the School of Mines, planning to become an engineer. But in his spare time he made some discoveries in the field of differential equations, and three years later presented them as a doctoral thesis to the University of Paris. This resulted in his being appointed as Professor of

Mathematical Analysis at Caen in 1879. By 1881 he was firmly entrenched at the University of Paris, from whence he reigned as the undisputed leader of French – and arguably world – mathematics.

The traditional stereotype of the mathematician is the absent-minded dreamer – bearded, bespectacled, forever searching for those spectacles, unaware that they are perched on his nose. Few of the great (or ordinary) mathematicians actually fit this stereotype; but Poincaré was one who did. More than once he forgetfully took hotel linen with him on departure.

Poincaré was a unifier, a seeker of general principles, the last of the traditionalists and the first of the moderns. He ranged over virtually all of the mathematics of his age: differential equations, number theory, complex analysis, mechanics, astronomy, mathematical physics. His collected works include over 400 books and papers, often lengthy. His greatest creation was topology – the general study of continuity. He called it *analysis situs* – the analysis of position. And he applied it to one of the hardest problems at the frontiers of dynamics.

An Oscar for Mathematics

In 1887 King Oscar II of Sweden offered a prize of 2,500 crowns for an answer to a fundamental question in astronomy. *Is the Solar System stable?* We now see that it was a major turning-point in the development of mathematical physics.

A state of rest or motion is stable if it does not change much under the effect of small disturbances. A pin lying on its side is stable (Figure 21). In theory a pin can balance on its tip, but in practice it will topple over if a bug flaps its wings in a neighbouring room. In principle it will even topple over if a bug-eyed monster flaps its wings in a neighbouring galaxy; but it may take a little time for the effect to be noticed because the pin starts to topple infinitely slowly, and before it gets very far some disturbance much nearer home will mask the gravitational attraction of Worsel of Velantia's scaly reptilian wings.

Whether or not a particular state of rest or motion *exists* can be studied just by looking at that state. If a pin is balanced perfectly vertically, then the downward force of its weight passes exactly through the point of support, and is cancelled by the upward reaction at that point – which, by Newton's third law, must be equal and opposite. That's all you need to know. But whether the state is stable can be deduced only by analysing *nearby* states as well. Tilt the pin slightly: the centre of mass swings a little to one side, and now the reaction and the weight form a couple that, while equal in magnitude, are no longer exactly opposed. The couple causes the pin to continue to

Figure 21   A pin balanced on end is unstable and in practice will topple. One lying on its side is stable.

rotate in the direction of the tilt. The initial displacement is magnified; the position is unstable.

Stability is thus a more complicated question than existence. Stability is also extremely important. A jumbo jet must not only fly; its flight must be stable, or it will drop out of the sky. When a car rounds a corner it must not tip over on its side: it must remain stable on the road. Theoretically, stable and unstable states are solutions to the same basic dynamical equations: the mathematics finds one as easily as the other. But experimentally, an unstable state of rest will never be observed at all, because tiny outside influences will destroy it. An unstable state of motion *can* be observed, but only as a *transient* phenomenon – while the system is *en route* from its original unstable state to wherever it will finally end up. The motion of a bicycle between the time you give it a push and the moment it falls into the ditch in a final, tangled, stable rest state.

Actually there's another way to observe an unstable state: take special action to stabilize it, by sensing and correcting any motion away from it. That's how a tightrope walker defies gravity. But such considerations belong more to control theory than to dynamics.

The Solar System is a very complicated piece of dynamics. Its motion certainly *exists*, and by the deterministic nature of Newton's laws it is unique (unless there are collisions – the three balls of the pawnbroker – or other types of singular behaviour, which possibilities we ignore here). The Solar System does its own thing, but once set going, it can be doing only *one* thing. But is that thing stable? Will all the planets continue to move in roughly their current orbits, or could the Earth wander off into the cold and dark or Pluto crash into the Sun? Will the Solar System hold the road, or will it skid sideways and crash into the cosmic ditch?

You must admit it's an intriguing question. Just how important it is in practice is moot, however. In celestial mechanics instabilities often take a very long time to manifest themselves, as in the tale of the man who, when told the universe would end in a hundred billion years, replied: 'You had me worried for a moment there . . . I thought you said a hundred *million!*'

In any case, the Sun will probably blow up first.

In King Oscar's day, much of this additional layer of physical complexity went unsuspected, and the stability of the Solar System was a serious, practical problem. Today it's not terribly important in itself: but like all good physical problems, its mathematics lives on long after its physics has died. It encapsulates in concrete form a far-reaching general problem: find out how to deal with questions of stability in complicated dynamical systems.

Rubber Sheet Dynamics

Poincaré has been described as the 'last universalist', the last of the great mathematicians who was capable of working in every area of the subject. He was the last because the subject grew too big, rather than its practitioners too stupid or specialist. Today there are signs of a new unification in mathematics: the day of the universalist may yet return. Naturally, Poincaré had a go at King Oscar's problem. He didn't solve it: that came much later, and the solution was not of the kind originally anticipated. But he made such a dent in it that he was awarded the prize anyway; and to do it he invented a new brand of mathematics: *topology*.

Topology has been characterized as 'rubber sheet geometry'. More properly, it is the mathematics of continuity. Continuity is the study of smooth, gradual changes, the science of the unbroken. Discontinuities are sudden, dramatic: places where a tiny change in cause produces an enormous change in effect. A potter, moulding a lump of clay in his hands, is deforming it in a continuous fashion; but when he breaks a lump of clay off, the deformation becomes discontinuous. Continuity is one of the most fundamental mathematical properties of them all, so natural a concept that its basic role only became clear a hundred years or so ago, so powerful a concept that it is transforming mathematics and physics, so elusive a concept that even the simplest questions took decades to answer.

Topology is a kind of geometry, but a geometry in which lengths, angles, areas, shapes are infinitely mutable. A square can be continuously deformed

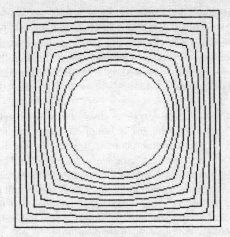

Figure 22   To a topologist, squares and circles are the same, since each can be continuously deformed into the other.

into a circle (Figure 22), a circle into a triangle, a triangle into a parallelogram. All of the geometrical shapes that we are taught so assiduously as children: to a topologist, they are one. Topology studies only those properties of shapes that are unchanged under reversible continuous transformations. By 'reversible' I mean that undoing the transformation must also be continuous. Adding more clay is a continuous transformation; but the reverse – pulling some off – is not. So, to a topologist, two lumps of clay are not the same as one: some things that we normally think of as different stay different.

What are the archetypal topological properties? To the untutored ear they sound nebulous, abstract, woolly. Connectedness, just alluded to, is an example. One lump or two? Knottedness is another: a knot is a loop that cannot be undone *no matter how it is deformed*. Put that way, it even *sounds* topological. Holes are topological objects: you can't get rid of a hole by a reversible continuous deformation. To a topologist, a doughnut is the same as a coffee-cup, because each has one hole. (You can guess that it was an American who coined the phrase: British doughnuts don't have holes, they have – provided you avoid certain supermarkets – jam.)

You can't develop topology, as a technical tool, in this sort of language. The hole 'in' a doughnut really surrounds it, and the doughnut surrounds the hole: hole and doughnut are linked. Considerable rethinking is in order. It requires new concepts, concepts not part of everyday experience, concepts for which no words exist. So mathematicians invent new words or borrow old ones, attach meanings to them with the necessary hairsplitting logic –

such as my insistence on reversibility – and build a new world. If you pick up a textbook on topology you may read about doughnuts or rubber sheets in the introduction, but when it gets down to the hard stuff, the terminology is less friendly. Continuous mapping. Compact space. Manifold. Triangulation. Homology group. Excision axiom. And the whole towering edifice, *the* major creation of 20th-century mathematics, is ultimately the brainchild of Henri Poincaré.

Topology, at first encounter, appears abstract in the extreme. Like a baby warthog, pretty to the few who love it, but of no interest to anyone else. But Poincaré could see the beautiful mentality beneath the warthog's skin. He had the breadth of mathematical experience, both pure and applied, to see the potential for a rigorous theory of the continuous. Sometimes it takes a universalist to see what is really important: nobody else has all the pieces. In every direction he turned, Poincaré ran into questions that only topology could answer. In his work on number theory. On complex analysis. On differential equations. And on King Oscar's problem.

Madly in All Directions

Poincaré devoted several years of his life to topology, creating most of its central themes. Others took it up: more definitions, more theorems, more jargon, more abstraction. Less contact with nature. By the 1950s topology, along with much of mainstream mathematics, had emulated Stephen Leacock's hero and ridden off madly in all directions: it seemed to many outsiders to have lost touch with reality. In his book *Chaos* James Gleick reports the words of Ralph Abraham, a mathematician at Santa Cruz, describing his own experience:

> The romance between mathematicians and physicists had ended in divorce in the 1930s. These people were no longer speaking. They simply despised each other. Mathematical physicists refused their graduate students permission to take math courses from mathematicians. *Take mathematics from us. We will teach you what you need to know. The mathematicians are on some kind of terrible ego trip and they will destroy your mind.* That was 1960. By 1968 this had completely turned around.

It turned around because Poincaré, and the mathematicians that followed him in droves, really were on to a fundamental idea. But it was such a difficult one to get working effectively, and it took so long, and the path led so far into the abstract wilderness, that even many of the mathematicians had forgotten that Poincaré had started with a problem in physics, and had become so enamoured of the new kind of mathematics that it was enough, for them, in splendid intellectual isolation.

It was like an expedition to cross an unscalable mountain range. At the outset, you can see the peak that must be conquered. But there's no way to climb it. And so the expedition heads off into the desert, trying to go round the mountain, and bypass the peak. Now, the techniques you need to survive in the desert are not those that help you climb mountains. So you end up with specialists on cacti and rattlesnakes and spiders, and the flow of sand-dunes in the wind, and the causes of flash-flooding, and nobody cares any more about snow, ropes, crampons, or peg-hammers. So, when a moun-taineer asks the sandunologist why he's studying sand-dunes, and is told 'to get past that mountain', he doesn't believe a word of it. And it gets worse when the answer is 'I don't give a hoot about mountains – sand-dunes are much more fun.'

But the mountain's still there, and the desert still goes round it. And if the desertologists do their stuff well enough – *even if they've forgotten about the mountain* – then one day the mountain will cease to be a barrier.

In the middle of the 1960s, under the guidance of a group of American mathematicians and another group of Russians, mathematics finally crossed the Desert of Topology. The main problems in topology fell into line and everything came together. Many individual mathematicians and physicists – though not all – had forgotten that topology came from physics. Mathematics and physics had not.

Eternal Triangle

Which brings us back to King Oscar. In human affairs, two's company and three's a divorce. In the same way, in celestial mechanics the interaction of two bodies is well behaved, but that of three is fraught with disaster (Figure 23). As for the dozen or more major bodies in the solar system – well, anyone who wanted King Oscar's crowns was going to have to work hard to get them.

Poincaré's prizewinning memoir is called (in French) *On the Problem of Three Bodies and the Equations of Dynamics*. It was published in 1890 and ran to 270 pages in the original. The first part establishes general properties of dynamical equations; the second applies the results to the problem of arbitrarily many bodies moving under Newtonian gravitation.

The motion of two bodies – a universe consisting only of the Earth and the Sun, say – is periodic: it repeats over and over again. By hallowed tradition, the period – the time taken for the motion to repeat – is a year. This immediately proves that the Earth can't fall into the Sun or wander off

Figure 23   The complexities of three-body motion: here a dust particle orbits two fixed planets of equal mass.

into the outer reaches of infinity; if it did, it would have to fall into the Sun every year, or wander off to infinity every year. Those aren't things you can do more than once, and they didn't happen last year, so they never will. In other words, periodicity gives you a very useful handle on stability. In a real universe other bodies can shatter this cosy scenario, but periodicity – or related concepts – may still be applicable.

In Chapter 3 of his memoir, Poincaré comes to grips with the question of the existence of periodic solutions to differential equations. He begins in the classic mould, and shows how to obtain such solutions by expanding the variable concerned as an infinite series, each term being a periodic function of time. 'From that it results,' says he, 'that there exist series whose coefficients are periodic and which formally satisfy the equations.'

Poincaré uses the word 'formally' for a good reason. The procedure *appears* to make sense, but he's worried that appearances may be deceptive. An infinite series only has a meaningful sum if the sum of large numbers of terms settles down towards a unique value – behaviour known as *convergence*. Poincaré is well aware of this, saying: 'It remains to demonstrate the convergence of this series.' But here Analysis, fickle as ever, abandons him. He affirms his belief that it *could* be done directly, but declines to embark on such a calculation – either because he knows it will be an impenetrable mess, or because he doesn't actually know how to do it. 'Be that as it may,' Poincaré tells us, 'I'm not going to do it, for I shall, by looking at the question again from another point of view, rigorously demonstrate the existence of periodic solutions, which implies the convergence of the series.'

A Question for Topology

Here's Poincaré's idea. Suppose that at some particular time the system is in some particular state; and that at a certain time later it's again in the identical state. All positions and velocities are exactly the same as before, simultaneously. Then the uniqueness of solutions to differential equations means that it must repeat, over and over again, the motion that took it from that state to itself. That is, *the motion is periodic*.

Imagine that the state of the system is described by the coordinates of a point in some huge-dimensional *phase space*. As the system evolves in time, this point moves, tracing out a curve. To get from some state back to itself again, this curve must close up into a loop (Figure 24). 'When is a curve a closed loop?' The question asks nothing about the shape or size or position of the loop: it's a question for topology. The existence of periodic solutions depends on topological properties of the relation between the position of a point *now* and its position one period later.

Poincaré doesn't put it in quite this language, but this is the underlying geometric idea; and elsewhere he says as much. Now it's easier to pose a problem in a new way than to solve what it then becomes, but Poincaré even has an idea how you might go about finding such closed loops. Let me describe it in fanciful terms. You're a Russian space engineer and you've put yet another *Cosmos*-series spy-satellite into orbit round the Earth and you want to know whether its orbit is periodic. Rather than track the satellite all the way round, you point your telescope so that it scans a plane running north–south from horizon to horizon and pointing straight up from the centre of the Earth. Every so often the satellite crosses this plane. Note where it first does so, and how fast and in what direction it's moving. Keep observing, but only when the satellite crosses the plane. If its motion is periodic then it must eventually hit the plane at the same point, with the same speed, and travelling in the same direction, as the numbers you wrote on your notepad.

In other words, instead of looking at all initial states, you can look at just

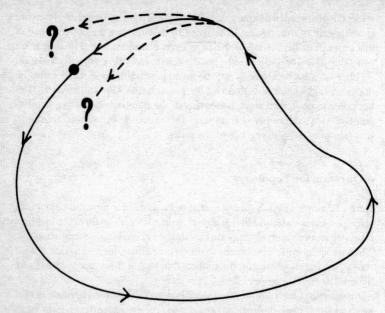

Figure 24   If a point in phase space traces out a closed loop, then it will repeat the same motion periodically for ever.

a few. Imagine a whole surface of initial states, and follow the evolution of each until (if it ever does) it comes back and hits the surface again (Figure 25). Can you find one state that returns exactly to where it started? If so, you've bagged a periodic solution.

Nowadays such a surface is called a *Poincaré section*. Its great virtue is that it throws away a lot of confusing junk, thereby simplifying the problem of observing the dynamics. And in this game, you need every simplification you can get. For example, the mere *existence* of a Poincaré section can sometimes force, for topological reasons, the occurrence of a periodic solution.

Celestial Chaos

So powerful an idea was this, that it opened Poincaré's eyes to a totally new kind of behaviour. Nobody had ever thought of anything like it before. In fact, you have to think topologically, or at least geometrically, to have a hope of spotting it: you'll never get it from a formula.

Poincaré was looking at an idealized three-body problem, called *Hill's*

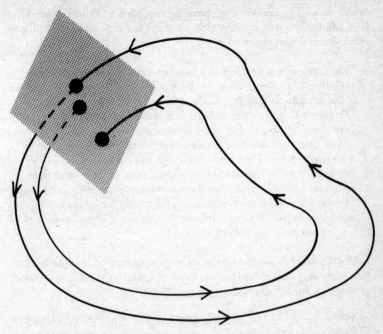

Figure 25   Detecting a periodic motion using a Poincaré section. For periodicity, the curve must return to the section at its exact starting point.

reduced model. This applies when one of the three bodies has so small a mass that it does not affect the other two – but, paradoxically, they do affect it. Imagine a universe containing only Neptune, Pluto, and a grain of interstellar dust. Neptune and Pluto are pretty ignorant about the particle of dust and you can imagine that it doesn't disturb their motion much; so they think they're in a two-body universe. 'Aha!' says Neptune, waving his trident, 'Newton worked that out – I move in an ellipse!' Pluto, wagging his tail, agrees, and the two revolve in stately fashion about their mutual centre of gravity.

The dust particle, on the other hand, is well aware of the gravitational pull of both Neptune and Pluto because they tug it all over the place. It moves within the rotating mutual gravitational field of the two planets. It thinks of itself not as a member of a three-body system, but as a tiny ball rolling around on a rotating but fixed landscape.

That's Hill's reduced model.

Poincaré decided to apply his surface-of-section method to Hill's reduced model, looking for periodic motions of the dust particle. What he found has been admirably summarized by Otto Rössler: I've edited his words a little to cut down on technicalities.

> When trajectories intersect in a two-dimensional dynamical system, they do so in singular points. These points had been classified by Poincaré, for example the 'saddle' and the 'node'. When the 'same' thing happens in a two-dimensional cross-section, where the trajectories correspond to sheets, then the intersection may indeed again be a saddle, node, etc. But there is a second possibility now: intersection in a *nonsingular* point. The trajectory through this point is, like any other nonsingular point, bound to hit the cross-section at some other point next time. Only – there are two sheets now. Both sheets therefore have to cross each other again and again. So a 'grid' of infinitely many intersection points is formed [Figure 26]. All this is a bit complicated and counterintuitive, as Poincaré noted.

Indeed Poincaré found the behaviour so complicated and counterintuitive that, as he says in the third volume of his *New Methods of Celestial Mechanics*, he made no attempt to draw it:

> When one tries to depict the figure formed by these two curves and their infinity of intersections, each of which corresponds to a doubly asymptotic solution, these intersections form a kind of net, web, or infinitely tight mesh; neither of the two curves can ever cross itself, but

Figure 26   Footprints of chaos in the sands of time . . . Homoclinic tangles in the three-body problem. Poincaré was horrified.

must fold back on itself in a very complex way in order to cross the links of the web infinitely many times. One is struck with the complexity of this figure that I am not even attempting to draw. Nothing can give us a better idea of the complexity of the three-body problem.

Poincaré's discovery – which, as June Barrow-Green recently discovered, appears only in the published version of his prizewinning memoir; he overlooked it in the version actually submitted – means that very complicated dynamics indeed can occur in something as simplified as Hill's reduced model. A system that starts at an intersection point of the web traces out a curve which, when it returns to the Poincaré section, hits the web at another intersection point, then another, then another. But the web is stretched and folded in such a complicated way that effectively the system passes through the Poincaré section at a random sequence of points. It's a bit like a bus which tours a city, repeatedly passing through the central square, but each time choosing at random from a million different bus-stops in the square itself. You can see the bus coming round again, you know it will stop in the square – but you've got no idea at all which stop to wait at.

In his grid of intersecting sheets, now known as *homoclinic tangles*, Poincaré was gazing at the footprints of chaos. Like Robinson Crusoe, staring at five toes neatly imprinted in the sand, he knew the importance of what he had seen. Like Robinson Crusoe, he was less than overjoyed at the prospect.

5

One-way Pendulum

PROS. COUN.	What was it that made you take it up in the first place?
MR G.	I was at a loose end at the time, sir.
	(*The Judge looks sharply up.*)
PROS. COUN.	You were at a loose end. Would you tell the court, Mr Groomkirby, as clearly as you can in your own words, exactly how loose this end was?
MR G.	It was worn right down, sir.
JUDGE	(*intervening*): Worn right down. That tells us very little. Was it swinging loose? Was it rattling about?
	(*Counsel, with a barely perceptible sigh and the briefest of glances towards Counsel for the Defence, sits down.*)
MR G.	It was practically hanging off, m'lord.

N. F. Simpson, *One Way Pendulum*

N. F. Simpson's farce *One Way Pendulum* was first performed at the Theatre Royal, Brighton, on 14 December 1959. If you've never seen it, do so. It's hilarious.

I mention it here because the pendulum, which itself hangs at a loose end, plays a pivotal role (pun intended) in the history of mechanics. We've already seen how it inspired Galileo. It's amazing how many good ideas have originated from such a humble mechanism. A light string, a heavy bob on the end, and a pin from which to hang the thing: simplicity itself. But then, the best mathematics always is simple, if you can only look at it the right way. In order to understand chaos, we must first take a closer look at how topologists view more regular dynamics. The pendulum is a good place to start.

N. F. Simpson's main character, Kirkby Groomkirby, won't eat his dinner until someone rings a cash-register bell, the house is full of weighing-machines, and Mr Gantry stands in the garden next to Parking-meters and 'once he's put his sixpence in there's no budging him till his time is up'. Sylvia, the daughter of the house, is upset because her arms won't reach her knees without bending: her mother suggests she acquires a set of monkey glands.

A *one-way* pendulum? Simpson must have thought that pretty bizarre, to use it as his title, given what's actually *in* the play. Perhaps he thought that if a two-way pendulum moves to-and-fro, then a one-way pendulum must move one-and-fro.

But pendulums *can* go one way, you know. Ever watched a small boy whirling a conker round and round on the end of a string? That's a one-way pendulum. And it's just as much a part of what makes a pendulum tick as Galileo's belief that church lamps swing with the same period no matter what arc they swing through. But, as Simpson's title drives home, it's an aspect of the pendulum that we tend to forget.

I want to contrast Poincaré's qualitative view of dynamics with the traditional bash-out-a-formula approach, and the pendulum – its one-way aspects as well as its two-way – is an ideal subject. In line with the wish for simplicity, I hasten to add that this will be a stripped-down, ideal mathematical pendulum, designed to capture the essence of pendulumnity as economically as possible. Our ideal pendulum will swing not in three-dimensional space, but in a vertical plane. There will be no friction at the pivot, and no air resistance. The string will be replaced by a perfectly rigid rod of zero mass. Gravity will act vertically downward and be constant. You won't find a pendulum like that in any laboratory; but science has often made progress by studying simple abstractions when more realistic models are too complicated and confusing. One step at a time; crawl before you hot-dog it on the ski slopes.

If You Can't Win, Cheat

The traditional treatment of the pendulum goes something like this. The state of the pendulum is adequately described if we know the angle at which it hangs at any given time. Write down Newton's law of motion for the pendulum system. This is a differential equation involving the second derivative – the rate of change of the rate of change – of that angle, together with some other variables such as the length of the string and the acceleration due to gravity.

Next step: solve the equation. You may be surprised to learn that this is desperately hard, involving tricky items called elliptic functions. Few undergraduate courses in mechanics actually run through this material. You now see what Euler meant when he said 'analysis abandons us'. The time-honoured gambit at this point is to cheat.

The reason that the equations are hard to solve is that the force acting on the pendulum is almost, but *not quite*, proportional to the angle between it and the vertical. If it were *exactly* proportional, then all you'd need would be a little trigonometry, and you'd be home and dry. But it isn't (and thereby hangs a tale as well as a pendulum, one to be taken up when the time is ripe).

Now mathematics, reputedly, is an exact science. 'Not quite proportional' is *not* the same as 'proportional', no matter how small the discrepancy. Too bad: let's lower our standards of rigour in the interests of progress and pretend that the tiny discrepancy isn't there at all. ('*It's a fiddle!*' we all used to cry in the physics class, when confronted with this ploy. The teacher agreed that the method also applies to a vibrating violin string.) If we can't solve the equations for our already idealized pendulum, let's analyse equations for a fake pendulum which, for small angles, is acted on by forces very close to those in the ideal model. In this fake pendulum – dubbed a *simple harmonic oscillator* in the hope of making it sound more respectable – the force is exactly proportional to the angle.

Now we can solve the equation. Imagine that at time zero we pull the pendulum sideways to make an angle A with the vertical, and then let go. The result is that the angle at time t is

$$A \cos \left(\sqrt{\frac{g}{l}} t \right)$$

where:

t = time
g = acceleration due to gravity
l = length of the pendulum
A = initial displacement.

The mass of the pendulum doesn't come into it – for much the same reason that Galileo observed: light and heavy bodies fall at the same speed.

We know what the cosine curve does: it wiggles between 1 and -1 and repeats itself every 2π radians (360°). So the angle of the pendulum wiggles in like fashion, between A and $-A$. Negative angles mean 'to the left of vertical' and positive angles mean 'to the right', so the pendulum swings periodically from left to right, between angles A and $-A$, repeating the same motion over and over again. How long does it take to repeat? From the formula we can extract the period: it's

$$2\pi \sqrt{\frac{l}{g}}$$

You can learn a lot from this formula. Longer pendulums take longer to swing: four times the length doubles the period of swing, nine times the length trebles it, and so on. You can use it to do experiments to find the force of gravity: just measure the length and the period, and use the formula to solve for g. If you were on Jupiter you could measure Jupiter's gravity and use it to deduce things about the planet's chemical composition by working out its mean density.

This analysis of the pendulum, then, is good physics. But it isn't, in its current form, good mathematics. Beautiful romances can be founded on a lie, but they tend to come unstuck when confronted by the dreadful truth. In the same way, apparently beautiful mathematics can be founded on a lie; and it too is liable to come unstuck when confronted by harsh reality.

There are several ways to make the pendulum analysis into good mathematics. The easy way was mentioned above: introduce an idealized form of motion, 'simple harmonic motion', in which the driving force is proportional to the displacement. Then you have some tricky footwork to explain what that might have to do with pendulums. A more honest approach is to state, and prove, a theorem that explains in just what sense this exact solution to an approximate problem can be viewed as an approximate solution to the exact problem. (No, Virginia, they are *not* the same thing: in mathematics there is no Santa Claus.) This can be done: the necessary theorem was proved in 1895 by the great Russian dynamicist Aleksandr Mikhaylovitch Liapunov. A great deal of beautiful mathematics has developed out of his Centre Theorem – all of which would have been missed if mathematicians had been content to assume, rather than prove, that small oscillations of a pendulum approximate simple harmonic motion.

On the other hand, you don't sit around moaning that you can't measure the acceleration due to gravity, just because nobody's proved the theorem yet. Science is a complex creature with mixed motives, and creative dishonesty works well on the right occasion.

But let's suppose you're not so much interested in using the pendulum to measure gravity, but in understanding what pendulums really do (Figure 27). 'Small oscillations? Twaddle! I want to know about large oscillations! To and fro? Look, I can make the thing whiz round and round like an aeroplane propeller! And it goes faster and faster the more energy I give it. *What was that you said about the period always being the same?*'

There's a classical answer to that, too; and, as I've said, it involves elliptic functions and a lot of complicated and advanced mathematics.

But there's also a very pretty geometric answer, which gets the main phenomena right with amazingly little effort and has the advantage that it provides some real insight into the dynamics. To this we turn.

Geometry on the Energy Surface

To know what a pendulum is doing you must have at your fingertips two quantities: its position and its velocity. Call these x and v. You'd like to know how they vary with time. To picture this, take a piece of graph paper, and draw x horizontally and v vertically. Now imagine the pendulum set going at time zero. Every hundredth of a second you measure x and v, and draw a

Figure 27   Waveforms of a nonlinear pendulum. Only oscillations of very small size are sinusoidal. (Reproduced by permission of John Wiley & Sons Ltd., ©1986.)

dot on the graph paper at that position. What will you see? Well, you'll get a lot of closely spaced dots; and they'll trace out a *curve* in the (x, v) plane. This is the *trajectory* corresponding to the chosen initial position and speed. Another name is the *orbit*, by analogy with the motion of planets.

Start with different initial conditions and you'll get a different trajectory. The trajectories form a family of curves, which cover the entire plane. For the 'fake' pendulum, the simple harmonic oscillator, these curves are concentric circles (Figure 28).

For a 'genuine' pendulum the picture has more structure: it looks a bit like an eye with eyebrows below as well as above (Figure 29). Wrinkles brought on by too much oscillation, maybe. You could confirm this picture with experiments – a laser to measure position and speed, a microcomputer to process the data, and a plotter to draw the graph – £10,000 or so would be more than enough. With about 5p worth of paper, a £12 scientific calculator, and half an hour's thought, you can get the same picture out of the dynamical equations for the pendulum, *without ever solving them completely.*

Let me show you how. One mathematical consequence of Newton's laws

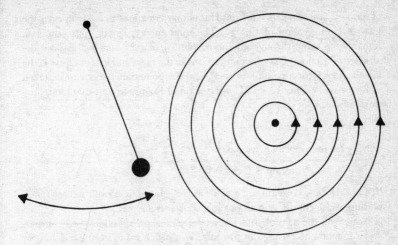

Figure 28   Phase portrait (right) *of an idealized linear pendulum* (left). *The horizontal coordinate is its position, the vertical its velocity. As time flows, the state of the pendulum describes a circle. Which circle depends on initial conditions.*

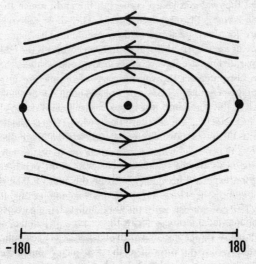

Figure 29   Phase portrait of a 'genuine' nonlinear pendulum.

of motion (you can prove it from Hamilton's equations in a few lines) is the Law of Conservation of Energy. The total energy, kinetic plus potential, remains the same throughout the motion. (It is here that we assume the absence of friction.) Choosing units to make the mass and length equal 1, the kinetic energy for a pendulum is $\frac{1}{2}v^2$, and the potential energy can be taken to be $-\cos x$. So the Law of Conservation of Energy tells us that along any trajectory

$$\frac{1}{2}v^2 - \cos x = \text{constant}$$

Solving for the velocity v, we have

$$v = \pm \sqrt{(\text{constant} + 2\cos x)}$$

(It's not the *same* constant – it's twice as big – but that doesn't matter much because we're considering all possible constants anyway.)

Now, with the help of your pocket calculator, or trig tables, you can plot v as a function of x using this formula. You pick a value for the constant in the above formula, say 1.5, and work out $\sqrt{(1.5 + 2\cos x)}$ for values of x running from $-180°$ to $+180°$. If the term under the square root sign goes negative, ignore it; otherwise plot two points on the vertical line through x: one at $\sqrt{(1.5 + 2\cos x)}$ and one at $-\sqrt{(1.5 + 2\cos x)}$.

In this particular case you get an oval shape. You'll find that if the constant is smaller than -2 there are no points at all; if it is -2 you get just a single point; at $+2$ the oval gets sharp corners at the ends; and if the constant is greater than 2 you get two separate curves. The whole system is exactly the 'eye' picture for the trajectories of the pendulum. The single point is the pupil, the ovals are the iris, the oval with sharp corners is the edge of the eye, and the separate lines are the eyebrows (above) and wrinkles (below).

You can also interpret the various parts of the picture in terms of the dynamics of the pendulum. The single isolated point, for example, represents the state when the pendulum just hangs vertically and doesn't move. Both position x and velocity v are constant: that's why you get a single point. The energy, -1, is the lowest possible energy of the system. (Potential energy can be negative: it depends on which level you choose to be the zero level.)

The closed ovals are the standard oscillations of the pendulum, the ones N. F. Simpson expected his audience to be thinking of. The ones that go *tick-tock* in a grandfather clock. To check this, imagine starting at the bottom of an oval. The position x is zero: the pendulum is hanging vertically downwards in the middle of a swing. The velocity is negative: it is swinging to the left (*tick!*). Further round the oval x is negative, so it has swung to the left, but v is now zero. At the furthest point of its swing, where the pendulum turns back to go the other way, its instantaneous velocity is zero. (The same is true of a ball thrown in the air: its velocity is zero at the top of its trajectory.)

Now *v* becomes positive, and the pendulum moves to the right (*tock!*) until *x* passes zero and the velocity reaches its maximum. The pendulum swings back to the right. Now the position reaches its furthest distance to the right and the velocity drops to zero: the pendulum has reached the right hand edge of its swing. It returns to its original position and the *whole cycle repeats over and over again*. The closed loop corresponds to a periodic state.

Now consider one of the eyebrows. Here *v* is always positive, while *x* runs from −180° (that is, 180° clockwise) to +180°, a complete revolution. This is the propeller-like trajectory, round and round forever in the same direction. The lower eyebrows are similar motions but clockwise instead of anticlockwise.

What about the edge of the eye, the oval with corners? This is the trajectory where the pendulum changes from side-to-side swings and turns into a propeller. How can that happen? Imagine the swings becoming slowly bigger and bigger. At first the pendulum stays near the bottom, but slowly the oscillations get larger – like a child on a playground swing, getting more and more energetic. Soon, to the alarm of all adults present, the motion becomes very violent; at her highest point the child is way up in the air above the pole from which the swing hangs. *If she swings much harder she'll* . . . What? *Go over the top*. From pendulum to propeller.

The edge of the eye is the path the pendulum would follow if it were held vertically and then released. Well, that's not quite right. If you did that, it would stay, exactly balanced, at a single point (the corner of the eye). But, like a pin balanced on end or student ballerina *sur les pointes*, this is an unstable state. The tiniest disturbance will cause the pendulum to topple. At first it topples infinitely slowly, but then it picks up speed, whizzes past the bottom, climbs up the far side, and creeps closer and closer to the top again. In theory the total motion takes infinitely long; in practice it takes a very long time indeed.

Do you see how well the picture fits our intuition about the way a real pendulum moves?

But we've paid a price. If you look at how we plotted the curves, you'll find that we did use a formula – but *we didn't solve the equations*. To solve the equations means to specify what *x* and *v* are for each time *t*. But *t* never appears!

If you want to keep things simple, there's usually a price to pay. Here the price is throwing away the precise time-dependence. The picture gives us no information at all about the sizes of the periods. In return for this omission, it does give a coherent and convincing qualitative description of *all possible motions* of a genuine – though idealized – pendulum.

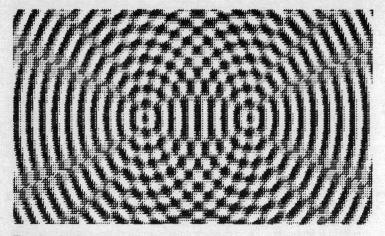

Figure 30   Interference fringes formed by superposing two waves.

Nonpachydermology

A lot of fuss about a pendulum, you're thinking. But there's a greater message.

Some years ago a colleague got married in North Wales, and my family spent the weekend driving round the area. In one forested section we found a lake, about a hundred metres across, which was almost perfectly flat and still. The boys, true to their breed, threw a stone in, and we watched as the ripples spread in perfect circles, almost the entire way across the lake. At that point more stones flew, and several more circular patterns superposed themselves on the first.

This non-laboratory experiment demonstrates the physical principle of *interference* (Figure 30). Where peak overlaps peak, or trough overlaps trough, the ripples are reinforced. Where peak meets trough, they cancel out.

It also demonstrates a mathematical property of differential equations known as *linearity*. An equation is linear if the sum of two solutions is again a solution. The motion of shallow waves on a liquid surface is very closely described by the wave equation, which – like most classical equations – is linear. The solution for a two-stone disturbance is just the sum of solutions for a one-stone disturbance, centred at appropriate points.

As that statement suggests, linear equations are usually much easier to solve than nonlinear ones. Find one or two solutions, and you've got lots more free. The equation for the simple harmonic oscillator is linear; the true equation for a pendulum is not. The classical procedure is to *linearize* the nonlinear by throwing away all awkward terms in the equation. For the pendulum this gives an approximate theory, which assumes the swings are very tiny.

It's tacitly assumed that since the neglected terms in the equations are small – which is true – the difference between the solution of the linearized equation and that of the true equation must also be small – which remains to be seen. For the pendulum, as I've said, there's a theorem that states the procedure works. On the other hand, we get a much more satisfactory picture by facing up to the full equations, even if we lose the luxury of a formula for the answer.

Formula? Who cares about formulas? Those are the surface of mathematics, not the essence!

In classical times, lacking techniques to face up to nonlinearities, the process of linearization was carried to such extremes that it often occurred *while the equations were being set up*. Heat flow is a good example: the classical heat equation is linear, even before you try to solve it. But real heat flow isn't, and according to at least one expert, Clifford Truesdell, whatever good the classical heat equation has done for mathematics, it did nothing but harm to the physics of heat.

Few asked themselves what the long-term future might be for a method which – to be brutal – *solves the wrong equations*. 'Give me an answer!' is the demand. So the linear theory obliges, hoping that nobody will notice when it's the *wrong* answer.

Today's science shows that nature is relentlessly *non*linear. So whatever it is that God deals in, it's not explicit formulas. God's got an analogue computer as versatile as the entire universe to play with – in fact it *is* the entire universe – and He finds little fascination in formulas designed for pencil and paper. Less blasphemously: it's no surprise that nature is nonlinear. If you draw a curve 'at random' you won't get a straight line. Similarly, if you reach into the lucky dip of differential equations, the odds against your emerging with a linear one are infinite.

Classical mathematics concentrated on linear equations for a sound pragmatic reason: it couldn't solve anything else. In comparison to the unruly hooligan antics of a typical differential equation, linear ones are a bunch of choirboys. (Is it coincidence that 'rule' means both 'law' and 'straightedge'?) So docile are linear equations that the classical mathematicians were willing to compromise their physics to get them. So the classical theory deals with *shallow* waves, *low*-amplitude vibrations, *small* temperature gradients.

So ingrained became the linear habit that by the 1940s and 1950s many scientists and engineers knew little else. 'God would not be so unkind,' said a prominent engineer, 'as to make the equations of nature nonlinear.' Once more the Deity was carrying the can for humanity's obtuseness. The engineer meant he didn't know how to solve nonlinear equations, but wasn't honest enough to admit it.

Linearity is a trap. The behaviour of linear equations – like that of choirboys – is far from typical. But if you decide that only linear equations are worth thinking about, self-censorship sets in. Your textbooks fill with triumphs of linear analysis, its failures buried so deep that the graves go unmarked and the existence of the graves goes unremarked. As the 18th century believed in a clockwork world, so did the mid-20th in a linear one.

And, to be fair, there are places where 'linear theory' gets you a long way. However, on most such occasions, the success has little to do with miraculous triumphs of physical intuition, or the remarkable relevance of the rules of thumbodynamics – it's because there are decent theorems that explain exactly why the linear theory works, and when.

But in some areas, it doesn't. It didn't in celestial mechanics, which is how Poincaré ran into chaos. It doesn't in other problems of mechanics, like the general motion of a free body in three dimensions. It doesn't in something as simple as a pendulum. Increasingly, physicists and engineers are finding that at research level it is the *non*linear phenomena that control the game. Ohm's Law provides a simple example. It states that the current flowing through a circuit is equal to the applied voltage divided by the resistance of the circuit. This is the linear relation: according to Ohm's Law, if you add two voltages, thus 'superposing' two circuits, then the currents that correspond also add together to give the current in the combined circuit. But transistors work because they *don't* obey Ohm's Law.

Really the whole language in which the discussion is conducted is topsy-turvy. To call a general differential equation 'nonlinear' is rather like calling zoology 'nonpachydermology'. But you see, we live in a world which for centuries acted as if the only animal in existence was the elephant, which assumed that holes in the skirting-board must be made by tiny elephants, which saw the soaring eagle as a wing-eared Dumbo, the tiger as an elephant with a rather short trunk and stripes, and whose taxonomists resorted to corrective surgery so that the museum's zoological collection consisted entirely of lumbering grey pachyderms.

So 'nonlinear' it is.

To Wrap It Up . . .

Back to the pendulum. We can play some mathematical games with the pendulum picture, to bring out other features. When discussing the propeller-like motion I said that motion from $-180°$ to $+180°$ completes a full circle, and so it does: these values represent the identical position of the pendulum.

As it stands, the picture doesn't show that very clearly: the right-hand edge at +180° looks a long way removed from the left-hand edge at −180°. How can we make −180° and +180° appear to be in the same place?

The problem here isn't with the pendulum; it's with our coordinate system. The pendulum knows that −180° = +180°, and it proves it by going round and round smoothly rather than leaping wildly across this imaginary chasm every time it gets back to the top. We are the victims of a peculiarity of the way we measure angles. We're trying to represent an *angle*, which lives on a circle, by a *number*, which lives on a straight line. We do it by (conceptually) wrapping the line around the circle, so that by the time we get to 360° we've got back to where we started at 0°. This means that adding 360°, and hence any multiple of it, to the numerical measure of an angle, represents the same *angle*. Since −180° + 360° = +180°, those two angles are the same.

Incidentally, you can't divide by 180 and deduce that −1° = +1°. You may care to ponder why.

How does a geometric circle 'know' that −180° = +180°? It knows because it wraps round and joins itself. This gives the circle a very different topology from the line, and explains why we're having problems: we're trying to use numbers, which live on a line, to represent an object that has the wrong topology. No wonder we have to wriggle on the hook a bit!

To get a more faithful picture of the pendulum motion – one whose geometry accurately reflects reality – we do the same thing. We *wrap the whole picture up* in the horizontal direction, to bring the left- and right-hand edges together, and physically force −180° and +180° to coincide. In other words, we roll the sheet of paper up into a cylinder (Figure 31).

I should add that there's no such problem with the velocity of the pendulum. An angular velocity of 180° per second is *not* the same as an angular velocity of −180° per second. The first represents a propeller moving anticlockwise; the other, one moving clockwise. If you muse long and hard on this curious difference between angular *position* and angular *velocity* many mysteries will reveal themselves to your gaze, including – if you have the perceptivity of an Euler or a Hamilton – the entire modern topological approach to Hamiltonian dynamics as 'symplectic structure on the cotangent bundle'. This is a topic seldom encountered below postgraduate level; but in a very real sense it's all there in the pendulum. In mathematics, big theories from little examples grow. Don't worry about that – but do remember that position and velocity have very different mathematical properties.

Fine. So now the dynamics of the pendulum lives on a cylinder, and periodic motions actually *look* periodic. What else can we do?

Some motions are more energetic than others. But at the moment, it's hard to see the 'energy levels'. The picture should make it clear that the pupil of the eye is the motion with lowest energy, and that as the energy increases,

Figure 31   Rolling the phase plane of a pendulum into a cylinder to represent position – which is an angle – more faithfully.

the pendulum passes through the iris, past the edge of the eye, and up on to the eyebrows and wrinkles. Or, dynamically, the oscillations grow until it goes over the top and starts whizzing.

The solution is to bend the cylinder into a U-tube (Figure 32). If you do this in just the right way, you get a picture that shows the motions of the pendulum and the corresponding energy levels, all at once. If you slice horizontally through the U-tube at a given energy level, the resulting curve depicts the corresponding motion.

You also see why, at high enough energy, there are *two* distinct types of periodic motion (clockwise and anticlockwise), whereas at low energies there is only *one* (to-and-fro). You can't make a distinction between 'to-and-fro clockwise' and 'to-and-fro anticlockwise'. A U-tube has *two* branches at the top but they come together at the bottom. If they didn't, it wouldn't be a U-tube. It would be a II-tube.

You may be asking what the point of all these machinations is. They

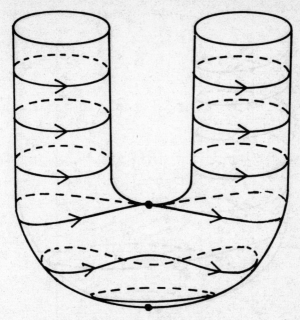

Figure 32   A geometric view of the conservation of energy. If the cylindrical phase space of a pendulum is bent into a U-tube, trajectories remain at a constant height.

illustrate that virtually all of the qualitative dynamical features of a pendulum – not just near its rest state, but globally, everywhere, at high or low energy – can be captured in a single geometric picture.

This picture can be formalized, put into proper mathematical language, and used to study not just pendulums, but (at least in principle) any dynamical system, however complex. Because geometry and topology are very powerful techniques, you can use such a picture to obtain information about dynamics that is totally inaccessible from the classical bash-out-a-formula viewpoint. There may not *be* a formula. But geometry, like poverty, is always with us.

Stranger than Friction

The power of this geometric viewpoint becomes apparent if we now ask 'What happens if there's a tiny amount of friction?' I suppose you *might* get an answer by calculating with elliptic functions. I've never seen it done – it would be a real *tour de force*, or perhaps a *tour de farce* because it's utterly pointless. But using the geometry, it's simplicity itself.

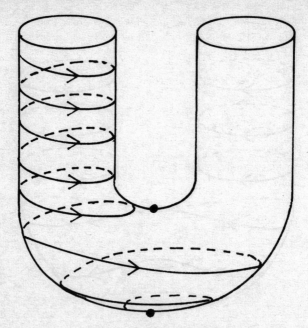

Figure 33   Damping dissipates energy: a damped pendulum spirals down the energy levels.

What is the effect of friction? It causes a loss of energy. In practice the energy turns into heat, which causes a little rejigging of the Law of Conservation of Energy. This is why you rub your hands to keep warm.

In our U-tube picture, loss of energy corresponds to descent to a lower level. Imagine starting with a propeller-like motion at high speed: the moving point on the cylinder that represents the motion of the pendulum goes round and round, very fast, some way up one branch of the U-tube. Add in a little friction to force a slow descent, and it begins to spiral down the tube (Figure 33). That represents a gradual slowing of the pendulum's revolutions, but it continues to rotate in the same direction because it's still on the same branch of the tube.

But eventually the spiral reaches the bend in the tube, and passes on to the lower region of to-and-fro motion, spiralling down that. Dynamically, the pendulum rotates slower and slower, until it just fails to make it to the top, hesitates, and falls back. Now rotating the other way, it gets near the top on the other side, but fails to get there by a slightly greater margin. Now it oscillates to and fro, the size of swing slowly decreasing, and ultimately it comes to rest at the bottom.

All this is physically intuitive, and emerges naturally from the U-tube picture. But as I've said, it's horribly difficult to extract this behaviour from the true dynamical equations. So here's a simple case where solving the equations by formula isn't a practical prospect, but we can get the answer from the geometry with hardly any effort.

Romance of Many Dimensions

In 1884 an English clergyman named Edwin A. Abbott published the second edition of his charming book *Flatland: a Romance of Many Dimensions*. The dedication runs:

TO
The Inhabitants of SPACE IN GENERAL
And H. C. IN PARTICULAR
This work is Dedicated
By a Humble Native of Flatland
In the Hope that
Even as he was Initiated in the Mysteries
Of THREE Dimensions
Having been previously conversant
With ONLY TWO
So the Citizens of that Celestial Region
May aspire yet higher and higher
To the Secrets of FOUR FIVE OR EVEN SIX Dimensions
Thereby contributing
To the Enlargement of THE IMAGINATION
And the possible Development
Of that most rare and excellent Gift of MODESTY
Among the Superior Races
Of SOLID HUMANITY.

The hero, 'A Square', inhibits a space of two dimensions. Enlightened by a visiting sphere from outer space as to the existence of a third dimension, he enrages his visitor by seeking still higher dimensions and ends up being gaoled for heresy by his compatriots.

Nowadays the notion of multidimensional space has become so widespread in the mathematical sciences that it is taken almost for granted. The heresy would be to deny its existence, not to assert it. Physicists are currently speculating that space-time may actually have ten dimensions: three of space, one of time, and six extras curled up so tightly that you can't see them.

The six extras do vibrate, though, whence all the complexities of particle physics.

The concept of a multidimensional space plays a crucial but behind-the-scenes role in the development of topological dynamics and the discovery of chaos. The idea is simple; the mental pictures involved perhaps less so.

It all hinges on a natural generalization of coordinate geometry. Start in one dimension: a line. Every point on a line can be described by one number x: how far it is from a given fixed point. Similarly every point in the plane can be described by its two coordinates x and y relative to a pair of fixed axes. And every point in three-dimensional space can be described by three coordinates x, y, and z.

But why stop there?

Well, it *is* the end of the alphabet, but somehow that doesn't seem to be the real obstacle. What about points described by *four* coordinates, say w, x, y, z? Presumably they correspond to some sort of four-dimensional space. Coordinates v, w, x, y, z yield a five-dimensional space, and so on.

In a sense, that's it. There's nothing more to be said. We have now defined what we mean by a five-dimensional space, *Finis*.

Of course, there's some small print that ought to be taken care of. Let us acknowledge that there is something less than spatial about these new 'spaces'. We don't – it appears – live in any of them: we live in good old three-dimensional space. (Four if you include time: see later.) *Why* our physical space limits itself in this way is a mystery. But it means that our minds have a certain amount of trouble *visualizing* spaces with four or more dimensions.

To some extent that's where the problem lies. Our visual system is trained to recognize objects in three spatial dimensions. From that point of view, 'visualize' is hardly the aim! What we must do is develop a new kind of geometric intuition. And, over several decades, that's what mathematicians did. To begin with, they played little games of analogy. Like:

● A line segment has 2 end points,
● A square has 4 corners,
● A cube has 8 corners.

What comes after 2, 4, 8 . . . ? Aha! *Therefore*

● A four-dimensional hypercube has 16 corners,
● A five-dimensional supercube has 32 corners,
● A six-dimensional superdupercube has 64 corners,

and so on. It was all a wonderful game of 'let's pretend', eventually backed up by precise definitions and calculations with coordinate systems like (u, v, w, x, y, z) for 6-dimensional space. It had an internal consistency, and more to the point, it *felt like geometry*. For example, in 3-space there are five regular solids (tetrahedron, cube, octahedron, dodecahedron, icosahedron). You can prove that in 4-space there are *six* regular hypersolids! But in 5-, 6-, 7-space, there are only three. Isn't that curious? These spaces have their own individual identities. Maybe there's something here worth sorting out.

Gradually the notion of a multidimensional space became respectable, especially when it began to suggest really nice mathematics. The main architect of all this was the English mathematician Arthur Cayley. When in 1874 the Royal Society put up a portrait to the great man, James Clerk Maxwell gave a speech which ended with a poem:

> March on, symbolic host! With step sublime,
> Up to the flaming bounds of Space and Time!
> There pause, until by Dickenson depicted,
> In two dimensions, we the form may trace
> Of him whose soul, too large for vulgar space
> In *n* dimensions flourished unrestricted.

Possibly these ideas might have remained mere curiosities, but it began to dawn upon the mathematical community that they had, for centuries, been studying multidimensional spaces without realizing it – as with Molière's hero M. Jourdain, astonished to discover he'd been speaking prose all his life. For example, consider the three-body problem. What do you want to calculate there? The positions and velocities of the three bodies. Now each body has three position coordinates (it lives in ordinary 3-space) and three velocity coordinates (ditto). So you're looking at a problem involving 18 distinct quantities. You're thinking in 18-space.

A bicycle has (at a conservative estimate) five main moving parts: the handlebars, the front wheels, the crank–chain–rear-wheel assembly, and two pedals (Figure 34). Each requires one position coordinate and one velocity coordinate to describe it: an engineer would say it has 'ten degrees of freedom'. To ride a bicycle, you must gain intuition about the motion of a point in 10-space! Maybe that's what makes it so hard to learn. Oh, and that's *without* putting in variables for where the bicycle is on the road.

Pretty reformulations, however, are ten a penny. Most are also pretty useless.

This one isn't. It provides a beautiful geometric framework that makes it far, far easier to 'see' what's going on in dynamics. It takes a while to learn it, and in practice nobody really has a very good idea of what 10-space looks like; but it definitely helps. A topologist, for instance, will draw two rough circles on a blackboard, say 'consider two 7-spheres in 10-space', and not notice anything peculiar going on: neither will the audience.

Albert Einstein – and his predecessors – made respectable the idea of time

Figure 34   A bicycle has (at least) five degrees of freedom: handlebars, left pedal, right pedal, front wheel, and crank–chain–rear-wheel assembly. Mathematically, it takes ten dimensions to represent the motion of a bicycle: five of position and five of velocity.

as a fourth dimension. (Not 'the' fourth dimension: fourth dimensions are also ten a penny. On a bicycle you've got seven to choose from, once you've decided which are the first three.) But it goes much further than that. In any problem, be it physics or psychology, each distinct quantity of interest may be treated, and visualized, as a new dimension in the problem. Economists regularly seek to maximize a company's profits by juggling thousands of variables. *They are working in a space of thousands of dimensions.* (That's one reason why economics is so difficult, and I'm not joking.) A recent dramatic breakthrough in such matters, a method called Karmarkar's algorithm (Figure 35), was discovered by thinking about the problem in exactly that way: it talks of 'n-dimensional ellipsoids' without batting an eyelid.

Figure 35   Three-dimensional projection of a multidimensional polyhedron, occurring in an application of Karmarkar's algorithm. (Reprinted with permission from the AT & T Bell Laboratories, Record, © 1986 AT & T.)

Dynamics in *n*-space

What clinches the matter, though, is the way in which the idea of multidimensional spaces fits together. It's like a 999-dimensional hand in a 999-dimensional glove.

For example, the picture we derived above, of the dynamics of a pendulum, generalizes to multidimensional spaces. A system with n degrees of freedom – n different variables – can be thought of as living in n-space. The n coordinates of a single point in n-space define all n variables simultaneously. Which is easier to think about: one moving point in a notional 10-space, or all the dynamical complexity of a bicycle, wobbling about, handlebars jiggling to and fro, pedals pumping up and down?

Yes, well. Forget the 10-space bit, just think of a point. Better? Good.

How do the laws of motion come into the picture? They tell us how a given initial point moves in its multidimensional space. It traces out some curve – what Einstein calls a 'world-line'. You can now imagine a whole bunch of initial points, moving along these curves. They're like particles of some fluid, flowing along.

A particular motion of a bicycle corresponds to that of a point in a fictitious 10-space. All possible motions of a bicycle correspond to the flow of a fictitious fluid in this fictitious 10-space.

Theorem   *If the system is Hamiltonian (no friction) then the fluid is incompressible.*

I hope that brings you down to earth with the same bump that I always experience. It isn't an abstract game! This is *real!*

What I mean is, something rather deep must be going on if the geometric picture turns dynamics not just into the motion of some silly fluid in some silly space, but renders it incompressible. (That is, the 10-dimensional analogue of 'volume' *stays the same* as the fluid flows.) The incompressibility theorem was discovered by Joseph Liouville in the 19th century, and its consequences have been spectacular.

If the system isn't Hamiltonian – that is, if there's friction, say – then you can still think of a fluid, but it's no longer incompressible. You can get an idea of all this by comparing Figures 32 and 33. Imagine a blob of two-dimensional fluid filling the small circle at the bottom of the U-tube in Figure 32. (*Don't* think of fluid filling the 'inside' of the tube: only the *surface* of the tube corresponds to physical reality!) As time passes, this blob of fluid just rotates, trapped inside the little circle. Its area doesn't change. But a comparable blob of fluid in Figure 33 has to spiral down the energy levels, towards the bottom of the tube, so it must shrink. This is the basic difference between a Hamiltonian system and a non-Hamiltonian, or dissipative, system.

Incompressibility is such a natural notion that the theorem can't be coincidence. Unless you agree with Kurt Vonnegut in *Cat's Cradle*, that the Deity made the universe as an elaborate practical joke.

6

Strange Attractors

They have strange limits and one must learn to observe them. It is that surface simplicity of theirs which makes a trap for the stranger. One's first impression is that they are entirely soft. Then one comes suddenly upon something very hard, and you know that you have reached the limit and must adapt yourself to the fact.

Sir Arthur Conan Doyle, *His Last Bow*

There seem to be two main types of mathematician. Most work in terms of visual images and mental pictures; a minority thinks in formulas. Which type of thinking is used doesn't always depend on the subject-matter. There are algebraists and logicians who think in pictures, and I know that one leading topologist has real trouble visualizing three-dimensional objects. Johannes Müller, a famous biologist, said that his mental picture of a dog was like this:

DOG

There are also fashions in mathematical presentation. For decades, everyone draws lots of pictures. Then, suddenly, pictures are no longer *de rigueur* and the style becomes very formal. Laplace boasted that his *Analytical Mechanics* contained no pictures, only analysis. In times closer to the modern era (the 1950s) you find few diagrams in the works of Nicolas Bourbaki, the pseudonym used by a group of mathematicians (mostly French) who attempted to formalize the structure of mathematics. Usually the distaste for diagrams arises from some crisis of logic caused by too much sloppy thinking and free-wheeling joy-rides into new mathematical territory. But as the formulas become ever more impenetrable, visual imagery rises once more to the surface of the collective mathematical subconscious.

Poincaré's great contribution was to put geometry back into mechanics, to undo Laplace's emphasis on analytic methods and calculations. Another historical cycle, another turn around the spiral staircase. By geometry I don't mean the stilted theorem–proof–q.e.d. that used to be inflicted on innocent children in the name of Euclid: I mean *pictures*. Poincaré released visual imagination from the prison of analysis and let it roam free once more. Having recycled into formalism with Bourbaki, today's mathematics is heading back towards the geometric twist of the spiral as fast as its legs will carry it.

Let's look at some of Poincaré's ideas. I've modernized the language, but the viewpoint remains his.

Time Flies Like an Arrow

We'll start with a system having two degrees of freedom, that is, where we can draw the pictures in the plane. Unlike the pendulum, which also lives in the plane (or, at least, on a cylinder, which is much the same), this system will not be Hamiltonian. In fact, it won't correspond to any particular physical model. It will be a purely mathematical construct, intended to illustrate the typical behaviour that a system with two degrees of freedom is likely to run into.

You'll recall that, given a single differential equation, we can visualize the motion of all possible initial points by thinking of an imaginary fluid, flowing along the trajectories of the equation. If you choose a starting point, that is, a set of initial conditions for the equation, then the coordinates of its subsequent motion are the solutions to the differential equation with that initial condition.

The picture of how these flow-lines fit together is called the *phase portrait* of the equation (Figure 36). 'Portrait' seems clear enough, and it's more imaginative than many mathematical terms. The curious word 'phase' seems to have come from electrical engineering. Oscillating waveforms have an *amplitude* – how big they are – and a *phase*, whereabouts in the cycle they are. If you plot both, you get a picture in the plane. Well, that's my theory, anyway.

The flow is indicated by curved lines, corresponding to the time evolution of the coordinates of various initial points. Arrows mark the direction of motion as time flows. We've already met two phase portraits, for the simple harmonic oscillator and the pendulum, in Figures 28 and 29.

Notice how the flow fits together: the arrows on nearby curves are fairly closely aligned. This means that the notional fluid, whose flow is represented by the lines, doesn't get torn apart: the motion is *continuous*.

There are four features of this particular flow, which I'd like to draw your attention to.

First, on the left-hand side, there's a point towards which all nearby

Figure 36   Phase portrait of a flow in the plane, showing (left to right) *a sink, a saddle, a limit cycle, and a source.*

flow-lines spiral. This is known as a *sink*. It's rather like a plughole down which the fluid is gurgling, hence perhaps the name.

Over on the right-hand side is a plughole in reverse, a point from which fluid spirals away. This is called a *source*. Think of fluid bubbling up from a spring.

In between is a place where flow lines appear to cross. This is known as a *saddle*. Actually the lines don't cross; something more interesting happens, which I'll describe below. If two jets of a real fluid run into each other, you see saddles.

Finally, surrounding the source on the right is a single closed loop. This is a *limit cycle*. It resembles an eddy, where fluid goes round and round. A whirlpool.

In a few pages' time we'll see that, roughly speaking, flows in the plane possess (some or all of) these features, and typically nothing else. There can be several of each feature, but you won't find anything more complicated. I'll also explain why I use the word 'typically' here. But first, let's acquaint ourselves more closely with these four fundamental features of flows in the plane – differential equations with two degrees of freedom.

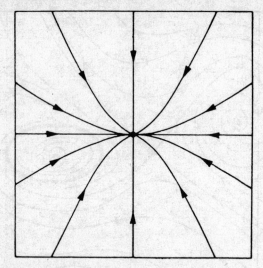

Figure 37   A sink.

Sinks

A sink (Figure 37) is a place where a flow-line degenerates to become a single point, towards which all nearby points flow. If you start the system out at the central point of a sink, nothing happens. It just sits there. So the sink itself represents a steady state of the system. For example, a lump of dough in a mixing-bowl can stay at rest at the bottom.

Meanwhile, if you start the system at some point near to the sink, it will move towards it. If you start your lump of dough a little up the side of the bowl, it will roll stickily down, until it reaches the bottom and stops. (I'm using sticky dough to introduce friction: if you used a frictionless marble, you'd have a Hamiltonian system and something rather different would happen.)

This means that the steady state at a sink is *stable*. If you take the point that represents the state of the system and move it a little way off, then it just spirals back towards where it started from. If you push the dough a little bit up the side of the bowl, it rolls back.

Sinks, then, are stable steady states.

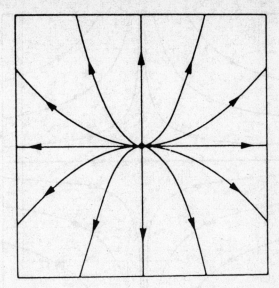

Figure 38   A source.

Sources

Sources (Figure 38) are also steady states. But now, nearby points move away. This is like a lump of dough perched on an overturned bowl. It can be made to balance at the top, if you're very careful, but if you give it a push, it rolls sideways and falls off. That is, the steady state is unstable.

Remember that the dough is only very slightly sticky: it won't stick to a slope. And think of a bowl with a rounded bottom, not a flat one. Perhaps a better analogy is trying to balance one smooth pebble on top of another. You can do it – with care – but a breath of wind, and it slides off.

Saddles

Saddles (Figure 39) are more interesting. They're also the sort of thing that only a mathematician would think of – except that Mother Nature has an even more vivid imagination. In a sense, they're steady states that are stable in some directions and unstable in others.

Imagine a rather inexperienced rider sitting on a horse, on a saddle that has been greased. If the rider moves forward or back in the saddle, he just

Figure 39   A saddle: the lines crossing at the centre are its separatrices.

slides back to the central position. But if he starts to slip sideways, he topples off. His position is stable with regard to forward or backward displacements; unstable with regard to sideways ones. It's this kind of picture that gives the name 'saddle' to such points.

The point at the middle of the 'cross', the saddlepoint proper, is – like all trajectories that reduce to single points – a steady state. Two flow lines are called the *separatrices* (singular: *separatrix*) of the saddle. They're so named because they separate the way nearby points flow. Imagine coming up a separatrix from the left of the picture. If you start just above it, you make a sharp left turn as you near the saddlepoint; if you start below you make a sharp right turn.

It rather looks as if the flow gets pulled apart at a saddlepoint. But I said above that it doesn't. This is because the separatrices don't really run into the saddlepoint, in the following sense. If you approach the saddle along its separatrix, you'll take infinitely long to hit it. So near a saddle the flow becomes infinitely slow. The fluid is stretched sideways, but not torn apart.

You might imagine that saddles are less common than sources and sinks. In fact, they're not. Here's another analogy, which helps explain why. Imagine a mountain landscape, and think of places where the ground (or at least the tangent plane) is horizontal. There are peaks, points from which every direction is down, analogous to sources. There are dips, from which every direction is up, analogous to sinks.

And there are *passes*, where some directions are up and others are down. These are analogous to saddles.

Passes are just as common as peaks and dips, in mountain country. Look at a map of the Swiss Alps. Similarly, saddles are just as common as sources and sinks. You can see them, for example, on the isobars of weather-maps, as well as the closed loops marked HIGH or LOW that surround sources and sinks of pressure. Isobars are plotted at convenient pressures – multiples of 10 millibars. Thus you seldom see the separatrices themselves, with their characteristic 'cross' shape; but you can recognize their presence by the four 'back-to-back' curves that occur nearby.

Limit Cycles

Now limit cycles are really interesting. If you start on one (Figure 40), you go round and round and round forever, repeating the same motion over and over again. The motion is periodic.

There are two basic kinds of limit cycle. The one shown is a stable limit cycle: nearby points move towards it. There is also an unstable limit cycle: nearby points move away. (To draw one, reverse all the arrows in the picture).

Limit cycles differ from sources, sinks, and saddles, in that you can't detect them by looking just near one point. You have to look at a whole region. This is what makes periodic motion harder to detect than steady states. It's also what makes it much more interesting mathematically.

In 1927 a Dutch electrical engineer called Balthasar van der Pol found an extremely important limit cycle. It occurs in a mathematical model of an electronic valve (vacuum tube in the US). Those were used in radios until the transistor was invented in 1947 by William Shockley, John Bardeen, and Walter Brattain at Bell Telephone Laboratories. A similar mathematical analysis applies to transistors, too. Van der Pol's limit cycle corresponds to a valve that is oscillating: putting out a waveform that goes up and down repeatedly. It sounds like a whistle, or a screech.

Oscillating radio waves are the basis of radio transmission. The idea is to start with a regular, very rapidly oscillating wave, and then to change the shape according to the sound that it is supposed to represent. The two standard ways to do this are amplitude modulation (AM) and frequency modulation (FM). The first changes the size of the wave; the second changes the spacing between waves. But you need a regular oscillator first, before

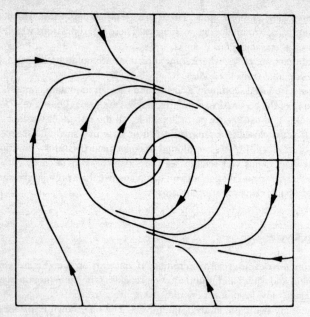

Figure 40   A stable limit cycle is a closed loop towards which nearby trajectories converge.

you've got anything to modulate. So the limit cycle in van der Pol's mathematical oscillator has important consequences for technology.

Typically, That's It

Poincaré, and a Swedish mathematician named Ivar Bendixson, proved a theorem to the effect that 'typically' only these four types of behaviour occur in a system of differential equations in the plane.

But it isn't true that *every* differential equation has only those four features. You can easily concoct more complicated things: places where three lines cross, or limit cycles that are stable on the inside and unstable on the outside.

It's here that the word 'typical' comes in. In a sense that can be made perfectly precise – but at the expense of technicalities like 'epsilon-homeomorphisms', not suitable for this book – you can show that these exceptions are infinitely rare. If sinks, sources, saddles, and limit cycles are coins landing heads or tails, then the exceptions are a coin landing on edge. Yes, it *might* happen, in theory; but no, it doesn't, in practice.

This kind of effect is fairly common in mathematics, and it litters the landscape of dynamical systems theory. If you try to list absolutely everything that can happen, you find that it's infinitely complicated and impossible to disentangle. But if you ask what is 'typical – what occurs with nonzero probability, if you like – then everything is much, much nicer. So common is this situation that dynamical systems theorists have invented (or, rather, borrowed) a technical term for it: *generic*. Behaviour is generic if it does the typical things and avoids the infinitely rare exceptional things.

I'm not suggesting that the secrets of the exceptions must be forever a mystery: sometimes you can make progress on atypical – nongeneric – systems. There's even a kind of hierarchy of typicality: typical, fairly typical, moderately typical, not-at-all typical, yugh.

For practical purposes, for mathematics that works in applications, and for satisfactory and not over-complex theories, the typical, the generic, is what you should study. Bearing in mind that *what is typical depends on what things you're talking about.* Typical Hamiltonian systems behave very differently from typical non-Hamiltonian ones. If you toss a coin in a swamp, typically it lands neither heads nor tails: it sinks. If you toss it on to a table covered in wet clay, it has a much better chance of landing on edge. If you're walking down the street, the typical person you meet is not the Chancellor of the Exchequer; if you're walking through the Houses of Parliament, it may well be.

Every interesting system is in some sense typical, in a sufficiently limited context; and if you want to understand that system, it helps a lot to find out what that context is. It's rather like George Orwell's *Animal Farm*, only here the message on the barn reads

ALL SYSTEMS ARE TYPICAL
BUT SOME ARE MORE TYPICAL THAN OTHERS

Swinging a Cat

One final type of classical motion deserves attention: *quasiperiodicity*. Here several different periodic motions, with independent frequencies, are combined together. (The *frequency* of a periodic motion is the number of periods per second. So long periods correspond to low frequencies, short periods to high frequencies.) Imagine an astronaut in lunar orbit swinging a cat round his head in a space capsule. (Yes, I *know* there isn't room to swing a cat in a space capsule. Indulge me.) The cat goes periodically round the astronaut, the astronaut goes periodically round the Moon, the Moon goes round the Earth, the Earth round the Sun, and the Sun revolves round the centre of the galaxy. That's five superposed periodic motions.

Figure 41   Topologically, quasiperiodic motions take place on a torus: (left) *combination of motions in small and large circles;* (right) *the resulting torus.*

In the topological picture a quasiperiodic motion looks like a spiral movement on a torus – a doughnut (Figure 41). You can see this for a combination of two periodic motions, because there are two directions 'round' a torus. One passes through the hole in the middle; the other, at right angles to this, runs round the 'equator'. If you start rotating round and round through the hole, and then add a little push along the equatorial belt, you'll get a spiral movement.

If you combine two periodic motions whose periods have a common measure – that is, are both integer multiples of the same thing – then the result is actually *periodic*. If one motion has period 3 seconds, say, and the other 5 seconds, then the combination will repeat every 15 seconds.

But if there's no common measure – for example, if the periods are 1 second and $\sqrt{2}$ seconds – then the motion *never* repeats exactly. It does, however, 'almost repeat', in the sense that you can find states which are as close as you like to the initial state. This is why the name '*quasi*periodic' is used.

With two periods, the criterion for the combination to be periodic is that the ratio of the periods should be a rational number – an exact fraction p/q where p and q are whole numbers. If the ratio of periods is irrational – not an exact fraction – then the two periods have no common measure and their combination never repeats. It 'almost repeats' for approximate common multiples of the period, that is, fractions which are very close to the ratio of the periods.

Quasiperiodic motion is *not* typical in a general dynamical system. Despite this, it's often found in classical dynamics. The main reason is that it's entirely typical in Hamiltonian systems, and classical dynamics concentrates on that case. Celestial mechanics is littered with superposed cycles, as the swing of the cat illustrates. Another reason is that in any system with circular symmetry, whether Hamiltonian or not, two-period motions are typical. The symmetry 'stabilizes' the combination of two periods. And circular symmetry is common.

A third reason for studying quasiperiodicity is that even though it isn't typical, quasiperiodic motion is often observed during the transition from one typical kind of motion to another. In a sense, it's a kind of motion that we understand, which can be responsible for other kinds of motion that we don't. As such, it can sometimes provide a useful starting-point for research into new kinds of motion – such as chaos.

Insight, not Eyesight

Poincaré and Bendixson could only prove their theorem for systems with two degrees of freedom. The plane has all sorts of special features, which they exploited to the hilt; but 3-space causes snags. For instance, what does the flow near a *knotted* closed loop look like? (Yes, differential equations can have knotted solutions. The Lorenz equation, in the next chapter, in an example.) There are no knots in the plane, but there are in 3-space: the mathematics has to face up to this.

In the early 1960s an American topologist, Stephen Smale, took up the qualitative theory of differential equations where Poincaré – and his successors, notably George Birkhoff – had left off. Topology had advanced a great deal in the intervening half-century: maybe the time was ripe for progress. And even if most topologists had forgotten that topology came out of problems in physics, Smale hadn't.

I must say at once that there were many important contributions to dynamics between Poincaré and Smale – I'm selecting a single thread from a rich tapestry. Liapunov introduced a set of numbers, now known as Liapunov exponents, which are currently used as one method for detecting the presence of chaos. The work of Aleksandr Andronov, Aleksandr Adol'fovich Vitt, and S. E. Khaikin on nonlinear oscillators deserves mention, together with basic topological ideas of Solomon Lefschetz. The Russian school founded by Andrei Kolmogorov has made numerous fundamental discoveries, inspired by the kinetic theory of gas dynamics. In particular it took the notion of entropy, previously a concept in thermodynamics, and defined it for an arbitrary dynamical system. The Kolmogorov–Sinai criterion, nonzero entropy, is one of the most reliable tests for chaos. An important class of chaotic systems was introduced and studied by D. V. Anosov, and Ya. G. Sinai was the first person to prove the extremely difficult result that a system of elastic particles, modelling a gas, really does behave chaotically. Vladimir Arnold has had a tremendous influence on the development of modern dynamics, especially in Hamiltonian systems, and a little of his work is described later.

Smale had a very original mind. In his Ph.D. thesis he proved a general theorem which implies, among much else, that you can turn a sphere inside out. It's allowed to pass through itself, but it has to stay smooth – no kinks anywhere at any stage. This seemed so unlikely that his supervisor didn't believe it; but it turned out that Smale was right. However, it wasn't until many years later that anyone worked out exactly how to do it. One of the people who did, the French mathematician Bernard Morin, was blind. As I said, 'visualize' isn't quite the word. Insight, not eyesight – that's what you need for topology. Smale was the leading topologist of the time, responsible for several other major breakthroughs, including the first proof – in five or more dimensions – of a problem that Poincaré had posed in 1906, and which everybody else thought was totally impenetrable.

To emphasize the new point of view, Smale used the term *dynamical system* instead of 'system of differential equations'. And he thought about dynamical systems in terms of their geometry – the topology of the phase portrait – rather than the formulas used to define them. In fact, he hardly ever wrote any formulas down. Of course, this tended to baffle the classical differential equation theorists. Smale went on to infuriate them by bombarding them with conjectures which they already knew were false. But this was just his way of getting to grips with the real problem; and soon he was bombarding them with true theorems that surprised even the experts.

One of the first questions he asked is a very natural one: what is the analogue of the Poincaré – Bendixson theorem in three (or more) dimensions? That is, what is the list of typical ways for a system of differential equations to behave?

Poincaré had made a start on this. He had found all the possible typical types of steady state. There are four. They are sources, sinks, and two different types of saddle. A source still has all nearby points moving outwards, and a sink is the opposite of a source. A saddle can either have a surface of outward-moving points and a line of inward-moving ones, or a line of outward-moving points and a surface of inward-moving ones.

You can of course get limit cycles in 3-space, but now they come in three kinds: stable, unstable, and saddle-like.

That seemed to be the lot. Nobody had found any other *typical* flow features.

Structural Stability

The first thing Smale had to do was decide on a precise meaning for 'typical'. You can't prove good theorems if you don't have a clear idea what you're talking about.

The necessary idea had been invented by Aleksandr Andronov and Lev

Figure 42   Structural instability: a saddle with three separatrices breaks up under small perturbations, forming three separate saddles and a sink.

Pontryagin in the 1930s, for systems with two degrees of freedom. They used the term 'coarse systems'. The idea is that atypical behaviour can always be 'broken up' by making very tiny changes to the equations. For instance, a place where three flow-lines cross can be broken up into a configuration of three saddlepoints (Figure 42).

On the other hand, the four typical types of behaviour in the plane *don't* change if you make sufficiently small changes to the equations. If a mountain range moves *slightly* – a few metres, say – under the influence of a tiny earthquake, then peaks remain peaks, dips dips, and passes passes. They all move around a bit, but you can't totally destroy a peak with a *tiny* earthquake.

Smale generalized Andronov and Pontryagin's idea to systems with many degrees of freedom, and coined the term *structurally stable* to mean a flow whose topology doesn't change if the equations describing it are altered by a small enough amount. *This is a quite different idea from a stable state of a given equation.* That's a solution which is stable to small changes in the initial conditions. But structural stability is a property of *the whole system*, and it is stable with respect to small changes in the entire system of equations.

Now Smale asked: does every structurally stable dynamical system in 3-space possess only sources, sinks, the two kinds of saddle, and the three kinds of limit cycle? More generally, can we make similar statements for systems with an arbitrary number of degrees of freedom?

There seemed to be no examples that disproved this conjecture: everything anyone had found, which was more complicated than sinks, sources, saddles, and limit cycles, turned out to be structurally unstable and hence not typical. On the other hand, Smale simply could not establish that these were the lot. The theorem – if there really was a theorem – resisted all efforts to prove it.

Attractors

From Smale's point of view, the most important property of a dynamical system is its long-term behaviour. This 'selects' a much simpler set of motions from among those of the entire system.

For example, in the system of Figure 36 above, an initial point either disappears off the picture (which I'll ignore), stays where it is (one of the three steady states), or converges towards the limit cycle and goes round and round. So out of all the possible motions, the long-term behaviour selects precisely those features that we decided were especially noteworthy.

Engineers have a similar view. They talk of 'transients', when the system is switched on, as opposed to what it settles down to if you wait for a while. I'm not saying that transients aren't important for some questions: when you switch on a computer, the wrong transients can blow up a circuit board. But

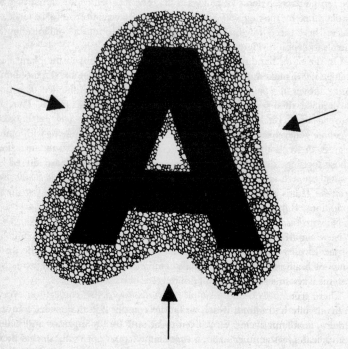

Figure 43   Schematic representation of a general attractor, here shown as a black A: nearby regions (shaded) contract towards the attractor as time passes.

for an overall view of the general nature of the system, rather than fine detail, you can ignore transients.

So what does a general dynamical system do in the long run?

It settles down to an *attractor*. An attractor is defined to be . . . whatever it settles down to! At this stage, not having proved any general theorem like Poincaré–Bendixson, we can't say in detail. But by analysing this idea we get a way to pin the concept down better. The essence of an attractor is that it is some portion of the phase space such that any point which starts nearby gets closer and closer to it (Figure 43).

We also insist that an attractor can't be broken up into two smaller subsets which each satisfy this definition. That is while we want the sink and the limit cycle in our example to be attractors, we don't want the combination 'sink + limit-cycle' to be considered a *single* attractor. This part of the definition is put in so that the attractors are the individual 'features' of the dynamics, that we've already made a fuss about, and not silly mixtures of them. You can generally forget about it except when proving theorems.

The Poincaré–Bendixson theorem tell us that for structurally stable systems in the plane – typical ones – the only attractors are

- single points
- stable limit cycles.

If you like, the only long-term motions are

- stay at rest in a steady state
- repeat some series of motions periodically.

Or, more simply,

- sit still
- go round and round.

Smale asked: is this also true in *n* dimensions, rather than just two?

Wrapping Mapping

There was a good reason why Smale couldn't prove that the only attractors in typical systems are points and limit cycles.

It isn't true.

Eventually he realized this. The first example – which goes back to the Russian mathematicians V. V. Nemytskii and V. V. Stepanov in 1949 – had four degrees of freedom, but eventually 3-space went the same way as 4-space.

I'm going to describe the basic idea first. It won't be a *bona fide* dynamical system to start with. However, once we've got the fundamental idea right, it can be prettied up to take care of the technical small print.

In a genuine dynamical system, time flows continuously from minus infinity to plus infinity, and passes through everything in between. In our stripped-down model system, time will flow in steps of a single instant, 1, 2, 3, . . . units. *There will be nothing between 1 and 2*: no time of $1\frac{1}{2}$ units, or 1.22789, or whatever. Only whole numbers: a digital clock rather than an analogue one. The system will click from one state to the next at each tick of its digital clock. The technical term for this is *discrete dynamics*; and we'll see below that there are actually close connections between discrete dynamics and genuine continuous dynamics, which mathematicians exploit to the full.

The system will be a point moving on a circle. For simplicity of description, choose units so that the circumference of the circle is exactly 1 unit. Then I can describe where the point is on this circle by a number between 0 and 1, its angular distance in these units round the circle from some chosen zero position.

In my self-appointed role as Master of the Universe, I now decree that the point shall obey the following dynamical law: if at a given instant it's at position x, then at the next instant it moves to $10x$. Geometrically, the circle is stretched to ten times its length, and wrapped ten times round itself (Figure 44). The law is applied at each instant in turn, so the point moves by iterating the mapping

$$x \rightarrow 10x$$

A *mapping* is just a rule 'x goes to something specified in terms of x', hence the little arrow. We've already found out what 'iterate' means: *repeat*.

I'm going to try to follow where the point goes as this tenfold wrapping is iterated. But I won't try to do this in too much detail. Divide the circumference of the circle into ten equal sectors labelled 0, 1, 2, . . . , 9. By the *itinerary* of a point on the circle I mean the list of sectors that it visits as the wrapping procedure is iterated.

In terms of the angular unit of measurement, sector 0 is the interval from 0 to 0.099999 . . . , sector 1 runs from 0.1 to 0.199999 . . . , and so on. Thus I might say that a point starts out at 0.25543786. This means it lives in sector 2, a little more than half-way along.

When I apply the mapping, and wrap the circle ten times round itself, its length expands by a factor of 10. So the point moves to 2.5543786. Now comes the clever footwork. One unit round the circle just gets you back to 0, and so do two units, so the result is really just the same as the angle 0.5543786. This is in sector 5. When we iterate the mapping, this is what we observe:

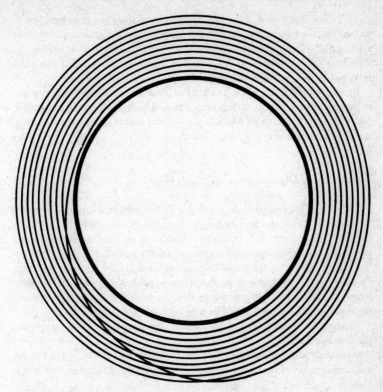

Figure 44 Stretching a circle and wrapping it ten times round itself (schematic).

time 0	0.25543786		sector 2
time 1	2.5543786	= 0.5543786	sector 5
time 2	5.543786	= 0.543786	sector 5
time 3	5.43786	= 0.43786	sector 4
time 4	4.3786	= 0.3786	sector 3
time 5	3.786	= 0.786	sector 7
time 6	7.86	= 0.86	sector 8
time 7	8.6	= 0.6	sector 6
time 8	6	= 0.0	sector 0

after which you just get 0, 0, 0, . . . At each stage you just multiply by 10 and chop off the first digit. The itinerary of such a point visits, in turn, sectors 2, 5, 5, 4, 3, 7, 8, 6, 0, 0, 0, . . . Have you seen those numbers before?

Yes, they're the decimal digits of the point we started with.

This is no accident. If you multiply by ten and drop the first digit, you're just shifting the decimal expansion one place left. So the same applies to any starting point. For instance, if I start with a point at $\pi/10 = 0.314159265\ldots$, then its itinerary visits sectors 3, 1, 4, 1, 5, 9, 2, 6, 5, . . . in turn. The dynamics recreates the successive decimal digits of π!

Be that as it may, I hope you'll agree that this stripped-down discrete dynamical system is very straightforward, and without doubt deterministic. Not only is there an exact formula for where x moves to, namely $x \rightarrow 10x$, but this formula is very easy to calculate.

Footprints of Chaos

First curiosity. Suppose that the starting point has a decimal expansion exactly the same as π, for the first billion decimal places; but thereafter goes . . . 1212121212 . . . forever. Call this new number π'. It's ever so close to π, far closer than any practical measurement could distinguish.

Under iteration of the tenfold wrapping, both π and π' have the same itinerary for the first billion steps. But after this, point π' just oscillates between sectors 1 and 2, while π goes on to visit . . . whatever the billion-onwards digits of π are. I have no idea, but they certainly aren't 121212 . . .

So two initial conditions π and π', extremely close together, eventually end up doing totally independent things.

Second curiosity. Suppose I take a die, with faces marked 1–6, and I throw it infinitely many times at random. I end up with an infinite list, something like

     1162541456522124366451432 . . .

and so on. (I got this by actually throwing a die, so it's a perfectly typical specimen, though I couldn't spare the time to produce an infinite sequence.) This is a random sequence of numbers.

And there's a point on the circle whose decimal expansion mimics this sequence, namely

    $x = 0.1162541456522124366451432 \ldots$

If I iterate the mapping starting at x, I generate the random sequence. So a deterministic mapping, applied to this particular initial point, generates a sequence as random as the throws of a die.

Third curiosity. 'Almost all' numbers in the interval 0 to 1 have decimal expansions that are random. This was proved by an American mathematician called Gregory Chaitin, who studied the limitations of computability. It's believable if you say it right: a number chosen 'at random' will have random digits. So the deterministic dynamical system that we've constructed behaves in this random fashion, not just for a few weird initial points, but for *almost all of them*!

Fourth curiosity. Let's ask when the itinerary of a point is periodic: repeats exactly over and over again. The answer is: *when its decimal expansion repeats*. There's a theorem that says such numbers are precisely those that are rational: they are exact fractions p/q with p and q whole numbers. There are infinitely many rational numbers between 0 and 1 (such as 2/3 or 199/431), and infinitely many irrational numbers between 0 and 1 (such as $\pi/10$, $\sqrt{2} - 1$). They're totally mixed up together: between any two rationals is an irrational, between any two irrationals, a rational. So the initial points that lead to periodic motions, and those that do not, are mixed up like sugar and flour in a cake. This also means that the periodic points are all unstable – if you disturb them slightly to a nearby irrational, they aren't periodic any more. In fact, *all* the possible motions are unstable!

Incidentally, don't imagine that somehow rationals and irrationals alternate along the interval – which admittedly is what the above description may suggest. On the contrary, 'most' numbers in the interval are irrational: rationals are very, very rare.

Bizarre.

Of course, you might argue that this is a pretty silly equation. Real dynamical systems don't do that kind of thing. For a start, in the above system the distinct initial points 0.42 and 0.52 both move to the same point 0.2 at the first stage; but in a genuine dynamical system different points never merge when they move. So all of the strange behaviour just described is an artefact of the ridiculously artificial recipe for the dynamics. Right?

Wrong.

Poincaré Sections

To see why, we must take another look at a fundamental idea of Poincaré's. I've mentioned it already: how to detect periodic solutions by looking at a cross-section.

Consider a system in the plane, having a stable limit cycle. Remember, that's a closed loop, and nearby points move towards it. A topologist would call it a periodic attractor. Draw a short line segment cutting across the limit cycle (Figure 45). For each point in the segment, follow its dynamical path. Eventually it hits the segment again. It may be actually on the limit cycle: if so, it comes back to where it started. Otherwise it ends up closer to the limit cycle than it was to begin with.

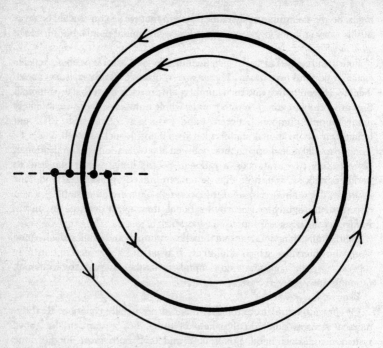

Figure 45   Poincaré section (dashed) through a limit cycle (heavy line): initial points on the Poincaré section contract towards the point representing the limit cycle on their first return.

That is, the recipe 'follow the dynamics until you first hit the segment again' determines a mapping from the segment to itself which compresses it down towards the point at which the limit cycle hits it. You've heard of the 'point of no return', but this is the point of *first* return. If you iterate the first-return mapping, you get the first return, then the second, then the third . . . You're *sampling* the full dynamics at regular intervals of time. An electronic engineer would call this 'stroboscopic sampling'. It's hi-fi turntable is moving at the right speed: the sampling there is done using a light that switches on and off at the frequency of mains electricity, and illuminates periodically placed marks on the turntable.

Now, let's take another system, which may or may not have a limit cycle. We don't know, yet. Suppose there's some line segment in phase space, with the property that every initial point in the segment eventually comes back and hits the segment again. Maybe there is, maybe there isn't: let's see what happens when there is.

I claim that *necessarily* there is at least one limit cycle running through the segment. The reason is a theorem in topology: every continuous mapping of a line segment to itself must have at least one *fixed point*: a point that maps to itself.

The idea behind the proof is something like this. The left-hand end of the segment maps to some point of the segment. If this point is also the left-hand end, there's your fixed point. If not, the left-hand end moves to the right. Similarly, the right-hand end moves to the left, so the whole segment shrinks down inside itself.

Look along the segment from left to right. Points near the left-hand end also move right; points near the right-hand end move left. Somewhere in between must be a place where the motion changes from rightwards to leftwards. The only way to change, continuously, from rightward motion to leftward motion is through zero motion. If I'm driving along a road and to start with I'm turning right, and later I'm turning left, then somewhere in between, for an instant, I must be going straight ahead. (There may be more than one such place: on a road full of Z-bends I have to straighten out, momentarily at least, between each bend and the next.)

Let me recap. *If* there's a line segment, such that every point starting on it eventually comes back to it, *then* there is at least one periodic solution passing through that segment.

Leaving aside the thorny issue of *finding* such a segment, we see that this is a rather remarkable theorem. *It doesn't depend on the detailed dynamics.*

It does use one general feature of the dynamics, though the 'fluid' doesn't get torn apart. The flow is continuous. But that's *all* it uses. What we've done is the essence of qualitative dynamics. We've used a topological fact to deduce a dynamical result. The topological fact is: 'every continuous mapping from an interval to itself has a fixed point'. The topological fact is the existence, given a suitable segment, of a periodic motion.

As already mentioned, this type of segment is called a Poincaré section. The associated mapping is its *Poincaré mapping.* There's a similar idea in three dimensions; but now the segment has to be replaced by a piece of *surface.* Typically, this is a topological disc – a small patch of surface without any holes. Mappings from a disc to itself can be very complicated (Figure 46). Despite this, there's a general theorem in topology about mappings from a disc to itself: again there *must* be a fixed point. So a flow in three dimensions that has a Poincaré section which is a disc must have a periodic trajectory passing through that disc.

In fact there's an n-dimensional version. The Poincaré section is an $(n-1)$-dimensional hyperdisc; and a rather difficult result called the Brouwer fixed-point theorem leads to the conclusion that at least one periodic trajectory must pass through it.

Topology, as I've said, is very powerful.

Figure 46   In two dimensions, a Poincaré section can be very complicated. In the Ueda attractor, illustrated here, points swirl rather like the surface of a cup of coffee being stirred. (Reproduced by permission of John Wiley & Sons Ltd., © 1986.)

It also shifts the emphasis. If I give you a dynamical system, say the motion of a prune in a bowl of porridge being stirred by Little Baby Bear, and ask 'is there a periodic solution?', then instead of trying to solve the equations and examining the result for periodicity, you end up looking for Poincaré sections instead. 'Someone's been iterating *my* Poincaré mapping,' said Mummy Bear. You can imagine that the techniques involved are rather different.

Solenoids in Suspension

What has this to do with making the tenfold circle-wrapping mapping into respectable dynamics? Smale realized that you can work a Poincaré section backwards. Given a piece of surface – say a topological disc – and

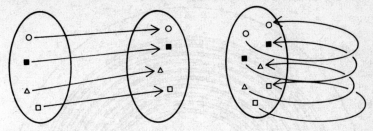

Figure 47   Suspension: a mathematical trick to turn a mapping (left) *into a flow in a space one dimension higher* (right).

a mapping from the surface to itself, you can concoct a dynamical system for which it is a Poincaré section and the 'first return' map is the one you started with.

To do this, you introduce a new 'direction' which is like a circle that cuts the disc at right angles. An initial point on the disc flows off it, round this circle, but in such a way that when it next hits the disc it does so as prescribed by the original mapping from the disc to itself. This trick is called *suspension* (Figure 47). It's the sort of thing that's natural to a topologist asking general questions about flows in *n*-space, but wouldn't occur if you were a chemist trying to understand the dynamics of a nitroglycerine explosion. However, you can write down an explicit differential equation if you want one. In science, you normally start with a physical problem and extract a differential equation. But Smale moved into the Designer Differential Equation business. The subject has never been the same since.

The upshot of all this is that anything you can see in a mapping of *n*-dimensional space can also be seen in a flow in $(n + 1)$-dimensional space. Conversely, the way to understand flows in $(n + 1)$-dimensional space is to look at mappings of *n*-dimensional space. In particular, flows in 3-space, not very well understood, reduce to mappings in 2-space, which we hope may be easier. Similarly flows in 4-space, which you have to work very hard even to think about, reduce to mappings in 3-space, where you can at least hope to draw pictures.

So instead of looking for a flow in 4-space, Smale looked for an unorthodox mapping in a 3-dimensional space which would have similar properties to our circle mapping when iterated. Here's what he found.

As Poincaré section, take the *interior of a solid torus*. A doughnut, American-style, with a hole. Dough included, this time we're not just talking of the surface of the torus. Define a mapping of the torus to itself as follows. Stretch it out to ten times its circumference and roll it thin; then put it back inside itself so that it wraps ten times round, without passing through any point

Figure 48   The tenfold wrapping applied to a solid torus to avoid self-intersections. Because the torus is three-dimensional, there is room for one winding to pass underneath the others without hitting them.

more than once (Figure 48). (Mathematicians normally use the number 2 rather than 10 here, but to see what goes on then you have to think in binary: I've rewritten history a little to make life easier for us.)

Imagine repeating this transformation of the doughnut. On the next application of the procedure it gets even thinner, and wraps 100 times round itself; then 1,000, 10,000, and so on.

Where does it go in the long run? You get something akin to an infinitely thin line wrapping infinitely many times round the torus. We'll examine this statement for hidden bugs in a moment; but it's not too far off the beam. There's an electrical gadget called a *solenoid*, in which miles of copper wire are wrapped around a metal core to make an electromagnet. Mathematicians borrowed this name for Smale's construct.

Two eminent dynamical systems theorists, colleagues of mine, were discussing all this in an American bar not long after its discovery, waving their hands graphically round and round, and chattering animatedly. 'Ah,' said the barman. 'You must be talking about *solenoids*!' This wasn't the kind of conversational gambit that they expected. Was the barman a mathematics graduate student working his way through college? It turned out he'd been in the navy, and what he was referring to was a real electrical solenoid.

At least the story shows that 'solenoid' is an appropriate name.

Anyway, we get this crazy mapping of a solid torus, in 3-space. Now we plunge our hands into the topological hat and extract a rabbit. Suspend Smale's solenoid mapping, and you get a flow in 4-space with his crazy mapping as a Poincaré section.

If you're not used to thinking in 4-space, you'll get the wrong picture at this point. You'll imagine a point starting in the middle of the dough, and wandering around through 3-space until it eventually ends up back inside the dough again. That's wrong. It moves out of 3-space altogether, immediately, without passing through the dough, wraps round in an entirely new dimension, and then hits the dough again somewhere else. As an analogy, using time as the fourth dimension, if you time-travel from *now* into the future, you leave the present 3-space *immediately*.

If you iterate the mapping from the torus to itself a large number of times, all initial points move closer and closer to the solenoid. So the solenoid is an attractor for the dynamics on the Poincaré section. The suspension of the solenoid – what you get when you whiz round in the extra dimension – is therefore an attractor for the full 4-dimensional flow.

Furthermore, it's structurally stable. To see why, imagine making a very small change to the wrapping mapping. The result will still look pretty much the same. You can't change continuously from a wrap-ten-times mapping to a wrap-nine or a wrap-eleven-times. To change continuously from ten to eleven you have to pass through ten and a half, but there's no way to wrap a torus ten and a half times without breaking it. That means the dynamics after making a small change to the mapping looks topologically the same as it did to begin with; and that's what structural stability means.

Finally, the solenoid is not a single point, and it's not a circle. So it can't be one of the traditional typical attractors. Two mathematicians, Floris Takens and David Ruelle, coined a name for this new type of attractor. A structurally stable attractor that is not one of the classical types, point or circle, is said to be a *strange attractor*. The name is a declaration of ignorance: whenever mathematicians call something 'pathological', 'abnormal', 'strange', or the like, what they mean is 'I don't understand this damned thing'. But it's also a flag, signalling a message: *I may not understand it, but it sure looks important to me.*

Cantor Cheese

The solenoid is not quite as crazy as it looks. Although it isn't a nice classical point, or circle, it has a distinguished pedigree. This is highly relevant to later developments, so I'll say a little more. The appropriate object is known as the *Cantor set* (Figure 49), because it was discovered by Henry Smith in 1875. (The founder of set theory, Georg Cantor, used Smith's invention in 1883. Let's face it, 'Smith set' isn't very impressive, is it?) The Cantor set is an interval that has been got at by mice. Infinitely many vanishingly small mice, each taking tinier and tinier bites.

Less colourfully, to build a Cantor set you start with an interval of length 1, and remove its middle third (but leaving the end points of this middle third). This leaves two smaller intervals, one-third as long: remove their middle thirds too. Repeat indefinitely. You get more and more shorter and shorter intervals: pass to the limit where the construction has been repeated infinitely many times. This is the Cantor set.

You might think that nothing at all is left. But, for example, the points 1/3 and 2/3 escape removal, and so do 1/9, 2/9, 7/9, and 8/9. *All* the end-points of removed segments remain. So do quite a lot of other points, as it turns out. The recipe involves expansion to base 3: if you like that sort of thing, see if you can describe exactly which points survive to make up the Cantor set.

The total length of the intervals removed is 1 – the original length of the interval you started with. So in some sense the 'length' of the Cantor set is zero! That's reasonable, the Cantor set consists mostly of holes. It's more like a dust than an interval.

There are other constructions which end up with something that is topologically equivalent to a Cantor set. One of the prettiest is to start with a circular disc, and remove everything except for two smaller discs (Figure 50). Like a button with two holes to put the thread through, except you keep the holes and throw away the button. Repeat this construction on each smaller disc, continue to repeat it indefinitely, and pass to the limit. Although it may not be obvious, this set is just a disguised Cantor set. I call it the *Cantor cheese*. It's been got at by mice, too.

The same is true if you make three holes at each stage – or ten. Yes, I agree that it's a surprise that these all give topologically the same result. But topology is a pretty floppy sort of thing: it leaves a lot of room for manoeuvre. You can find rigorous proofs in the topology texts – and they're nontrivial stuff.

The Cantor cheese – ten-holed variety – lives inside the solenoid. Imagine slicing through the doughnut to get a circle. When we wrap the doughnut

Figure 49   Construction of the Cantor set by repeated deletion of middle thirds.
The vertical dimension is exaggerated for clarity: ideally the line has no width.

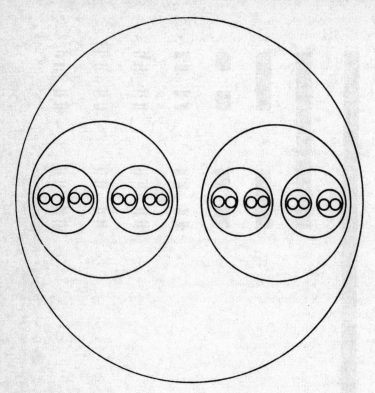

Figure 50   The Cantor cheese: alternative construction of a topological equivalent to the Cantor set, using pairs of circles.

round ten times, it meets the slice in ten smaller circles. The next stage gives a hundred smaller circles, and so on: exactly the same procedure. So the solenoid has the Cantor cheese as a cross-section. This dramatically verifies that it isn't a point or a circle!

Genuine Chaos

Equipped with the solenoid, we are now ready for an electrifying discovery. Not only does the wrap-ten-times mapping have the four curious properties noted – sensitivity to initial conditions, existence of random itineraries, common occurrence of random itineraries, and cake-mix periodicity/aperiodicity. *So do the solenoid and its corresponding differential equation.*

Philosophically, this raises a serious question. Suppose there is a physical process that is modelled by those equations. In the manner of a classical applied mathematician, I ask for a solution to the so-called *initial value problem*: given a starting point, *predict* where it will go in the long run.

The answer is 'I can only do that if you tell me the starting point to infinite precision. I want its entire decimal expansion, all the way out to infinity. Not just the first billion digits – they're irrelevant anyway after the billionth iteration. *The lot.*'

But this is a practical impossibility. Even ten-digit precision is better than most experiments can achieve.

In a sense, ten-digit precision tells us nothing about long-term behaviour. If you give me just ten digits, I can find an initial point which agrees with those ten, but thereafter does anything you like. Stays at 7 forever. Mimics π. Determines every fifth digit in $\sqrt{2}$. Runs through the sequence of primes represented in base 6. Lists the prices of all stocks and shares in the *Financial Times* Top 100, starting on 25 April 1963 and continuing indefinitely. *If you want to make a killing in the City, all you have to do is find the right initial point.*

The model predicts, to within experimental accuracy, all possible itineraries, once you've got the first ten over. The long-term behaviour is completely indeterminate.

On the other hand, what model could be more deterministic than 'shift along one digit'?

Diatribe Dialogue

If you don't like the idea of chaos, there's only one hope.

SCEPTIC: Look, this guy Smale's designer differential equations are all very well; but the real world doesn't behave like that.

CHAOSOPHER: If it can happen structurally stably in the mathematics, it can happen observably in nature.

SCEPTIC: Then why haven't I seen any equations like his?

CHAOSOPHER: Because you've been looking for regular behaviour. No physicist who ran into equations like those would dare publish them.

SCEPTIC: Well, what about experiments, then? You're bound to observe that sort of behaviour in experiments!

CHAOSOPHER: But you do, all the time. Unfortunately, though, there's a snag. Do you know any experimentalists who would publish a paper saying 'I got totally random results?'

SCEPTIC: Mmm, you've got a point there. But the fact remains, you'll never convince working scientists about this chaos stuff unless you show them it happening in nature.

CHAOSOPHER: I agree. We're working on it. It's not easy, you know. We've got to develop a totally new way of thinking about dynamics. It's *hard*. But anything that shows up as naturally as this in the mathematics has to be all over the place. If we don't find it, I'll be surprised.

SCEPTIC: Not half as surprised as I'll be if you *do* find it!

7

The Weather Factory

Let chaos storm!
Let cloud shapes swarm!
I wait for form.

Robert Frost, *Pertinax*

'Show me it happening in nature.'

That's what the sceptics wanted. To the topologists of the 1970s it seemed a tall order. But it had already been done, in 1963 – though neither the topologists nor the physicists knew it.

Glorious Failure

In 1922 Lewis Fry Richardson, an unorthodox deviser of half-baked ideas whose name floats in and out of the history of applied dynamical systems, published *Weather Prediction by Numerical Process*, a report on a glorious failure. Richardson had tried to use mathematics to predict the weather. Towards the end of this volume he described a fantastic vision, the Weather Factory. He imagined an enormous army of people, housed in a vast building rather like the Albert Hall, operating desk-calculators. (For those too young to have seen such machines, they look rather like a cash-register with a handle at the side. Oops, you won't have seen a cash-register either. Like a tin box with a rounded front. Sliding levers allow the user to set up the digits of numbers to be calculated with, and the handle is turned once to add, many times to multiply. To subtract, it's turned backwards, and division is done by repeated subtraction.) A mathematical conductor on a central podium would direct their efforts, and they would communicate with each other by telegraph, flashing lights, and pneumatic tubes. Richardson estimated that it would take 64,000 people to predict the weather at the same speed with which it actually happened – 'real time' in today's parlance.

And he said this: 'Perhaps some day in the dim future it will be possible to advance the computations faster than the weather advances at a cost less than the saving to mankind due to the information gained. But that is a dream.'

Prophetic words. The 'dim future' was a mere thirty years away. In 1950 the American ENIAC computer made the first successful calculations in weather prediction. By 1953 the Princeton MANIAC machine had made it clear that routine weather-prediction was entirely feasible.

Mind you: to predict the weather is one thing. To predict it *correctly* is another.

Climatic Chess

The game of chess involves a number of pieces and a board ruled into squares. Moves in the game take place at discrete time intervals, according to the laws of the game.

Numerical weather-prediction is like a huge game of three-dimensional chess. Imagine a fine grid of points drawn on the surface of the Earth, at several heights to track the up–down motion of the atmosphere as well as north–south and east–west. This is the chessboard. The weather *now* is described by assigning, to each grid point, several numerical values: pressure, temperature, humidity, wind-speed. These are the chess-pieces.

The weather *tomorrow* also corresponds to a position in the game – but the disposition of the pieces is different. 'Cyclone to Queen's Knight 743.' 'Blizzard to King's Lynn, Showers with Sunny Intervals to Bishop's Stortford.' We can measure today's weather using meteorological stations, ships, weather-balloons, and satellite pictures. So we know how to set up the pieces. The main question is, what are the rules of the game?

The rules are the equations of motion of the atmosphere. As we saw, these were found centuries ago by the likes of Leonhard Euler and Daniel Bernoulli. By letting time flow in tiny discrete steps, say one second long, the equations can be viewed as rules telling us how to get from the position now to the position in one second's time.

Predicting the weather one second ahead may not sound a practical contribution to the weighty problems of humankind, but that's just one move in the game. Repeat the calculation, and you have the weather two seconds into the future. After 86,400 iterations, you'll know the weather a day from now. After 8,640,000 you'll know the weather a hundred days from now. After 8,640,000,000 . . .

And in essence that's how it's done. Thousands upon thousands of repetitive calculations based on explicit and deterministic rules. Just what the computer is good at.

Twixt Zero and Infinity

There's a philosophical curiosity involved in all this. The atmosphere isn't really a perfectly divisible continuum; it's a lot of fairly solid little atoms charging around like lunatics crashing into each other. The equations of classical mechanics replace this discrete physical reality by a smooth ideal fluid. But in order to solve those equations we approximate them by something discrete again. We let time click ahead in tiny steps, rather than flow continuously, and we divide space up into a fine grid. This is forced by the structure of computers: they can only do arithmetic to some definite number of decimal places, say ten, in which case everything is an integer multiple of 0.0000000001. To represent an infinite decimal exactly requires an infinite amount of computer memory, which isn't feasible.

The philosophical point is that the discrete computer model we end up with is *not* the same as the discrete model given by atomic physics. But there's a very practical reason for this: the number of variables involved in the atomic model is far too large for a computer to handle. It can't track each individual atom of the atmosphere.

Computers can work with a small number of particles. Continuum mechanics can work with infinitely many. Zero or infinity. Mother Nature slips neatly into the gap between the two.

So we do the best we can. Mathematicians hope that this double approximation provides answers that are close to the real thing. There are no substantial theoretical proofs that this is so; but there's compelling evidence that *it works*. Until some genius develops new theoretical tools, we accept the miracle and plough ahead regardless.

It is, however, worth remembering that when you 'put the problem on the computer' you do nothing of the kind: you represent some idealization of the problem in the computer. This is one reason why the computer cannot be a universal palliative for the ills of science and society. It just isn't clever enough yet.

Megaflop

The calculations for weather-forecasting must be done at breakneck speed. The speed of a supercomputer is measured in *megaflops* – a megaflop being one million arithmetical calculations per second. In 1989 the Cray X-MP supercomputer at the European Medium-Range Weather Forecasting Centre at Reading, UK, operated at a top speed of *800 megaflops* (Figure 51). It could give a passable prediction of tomorrow's weather, for the entire northern

Figure 51   Cray X-MP supercomputer, capable of 800 million calculations every second. (Courtesy of Cray Research Inc.)

hemisphere, in about half an hour. Every day it made ten day's worth of predictions: half the world's weather a week and a half ahead. The predictions were generally fairly accurate about four days ahead; but after that they tended to drift away from the actual weather. Faster computers have not stopped this happening.

Another curiosity of the method is worth remarking. You might think that the way to get the best possible prediction is to use the most accurate possible equations. However, a fully accurate model will include not just large-scale weather movements, but sound waves in the atmosphere. Sound wave solutions to the equations play nasty tricks in the computer's discrete approximation, known as numerical instability. Errors in calculation (not mistakes by the computer, but limitation on the inherent accuracy of arithmetic when you can't tell the difference between 0.00000000001 and zero) blow up very fast and obliterate the actual weather! The solution, suggested by Jule Charney of the Massachusetts Institute of Technology in 1944, is cunning and surprising. The model is deliberately *coarsened*, to filter out the sound waves. You *don't* use the most accurate possible equations: you deliberately make them less accurate – to bring out the desired features.

Figure 52   When the weathermen got it wrong . . . Kew Gardens devastated by the 'hurricane' of 15 October 1987.

This is not a straightforward subject that we are dabbling in.

'Four days ahead,' I said. There *are* long-range forecasts, but you'll do better if you assume that this year's weather will do what it did last year. The main defect in current methods of weather-forecasting is that they're not very good at predicting *sudden* changes in weather-patterns. When I visited the European Medium-Range Weather Forecasting Centre they told me: 'We can predict the weather accurately provided it doesn't do anything unexpected.'

On Thursday 15 October 1987, Britain experienced its worst storms since 1703 (Figure 52). It would have been called a hurricane, except in Britain we don't get hurricanes. Mind you, within a few months it had become 'the hurricane' whenever anybody talked about it. The television weather service failed dismally to predict it, even at a mere twenty-four hours' notice. The following Monday the *Guardian* newspaper carried the following article by Andrew Rawnsley under the caption 'Computer under the Weather':

> The perpetrator of the worst weather forecast since records began was traced to a small town in Berkshire last night.
>
> With no apparent shame for missing the worst storms in 285 years,

it continued to pump out predictions of light showers, bright intervals and moderate winds at the rate of about a forecast a minute.

The answer to all those force 10 headlines – WHY WEREN'T WE WARNED? – is called the Control Data Cyber 205, the Meteorological Office's Bracknell-based number-cruncher and, according to a straw poll yesterday of the weathermen who rely on it, currently the most hated computer in Britain. According to its operators, the Cyber is capable of 400 million operations a second and can produce a 24-hour world forecast at 15 altitude levels in under five minutes. Unfortunately, it missed the worst storms since 1703, routing them 80 miles east into the North Sea while the real thing decided to travel via southern Britain. 'It is a pity things went wrong,' conceded a Met Office spokesman.

Nobody seemed to know why. 'It got it right at the beginning of the week,' said one forecaster at the London Weather Centre yesterday. 'It had the depression on the right track on Tuesday. Then it changed its tack.

'After the gales earlier in the week I thought we might have strong winds on Thursday,' he said, the weathermen's way of telling the computer I-told-you-so. 'We had our doubts, but we have to take the party line.'

There was an equally powerful whiff of smugness at the Reading home of the Cyber's biggest rival, Cray 1. Using the same data from satellites, ground radar, merchant ships and weather balloons, Cray predicted ferocious winds for the European Centre for Medium-Range Weather Forecasting.

The Met Office's internal investigation into Cyber's miserable performance will attempt to find out what went wrong. 'It's difficult to know,' said one of the Cyber's 10 operators yesterday. 'It's possible that a small piece of information got into the computer which shouldn't have.' Past triumphs of misforecasting by the 205 have, apparently, included predicting snow in July.

'There are plans to replace it,' said one of the cybermen. Others rallied to the Cyber's defence. 'Depressions have this habit of doing rather unexpected things,' said a forecaster. 'They can be very contrary.'

Future research may overcome such difficulties. But there are theoretical reasons for believing that there's an inherent limitation to the accuracy with which we can predict the weather. Four or five days, maybe a week – and no further.

Look up the words in a dictionary.

Mega: big.

Flop: failure.

Mathematician at Heart

But I'm getting ahead of the story. Flashback to 1963. In that year, Edward Lorenz of the Massachusetts Institute of Technology published a paper with the title *Deterministic Nonperiodic Flow*. Lorenz had set out with the idea of being a mathematician, but the Second World War intervened and he became a meteorologist instead. Or so he thought. In fact, he was still a mathematician at heart. (Mathematics is like an addiction, or a disease: you can never truly shake it off, even if you want to.) Let me quote the abstract, where Lorenz summarizes his results.

> Finite systems of deterministic ordinary nonlinear differential equations may be designed to represent forced dissipative hydrodynamic flow. Solutions of these equations can be identified with trajectories in phase space. For those systems with bounded solutions, it is found that nonperiodic solutions are ordinarily stable with respect to small modifications, so that slightly differing initial states can evolve into considerably different states. Systems with bounded solutions are shown to possess bounded numerical solutions.
>
> A simple system representing cellular convection is solved numerically. All of the solutions are found to be unstable, and almost all of them are nonperiodic.
>
> The feasibility of very long-range weather prediction is examined in the light of these results.

When I read those words I get a prickling at the back of my neck and my hair stands on end. *He knew! Thirty-four years ago, he knew!* And when I look more closely, I'm even more impressed. In a mere twelve pages Lorenz anticipated several major ideas of nonlinear dynamics, before it became fashionable, before anyone else had realized that new and baffling phenomena such as chaos existed.

Lorenz, as I've said, thought he was a meteorologist, and naturally he published his paper in *Journal of the Atmospheric Sciences*. The meteorologists, who were either non-mathematical or versed only in traditional mathematics, really didn't know what to make of it. It didn't look especially important. In fact Lorenz's equations were such a mangled, lopped-off version of the real physics, that the whole thing was probably nonsense.

There are several thousand scientific journals published per year, running on average to well over a thousand pages. If you read a lot you can just about keep up with the publications in your own field. Yes, it's just barely possible that the Spring issue of the *Goatstrangler's Gazette* might contain an idea of enormous importance in dynamical systems theory, but the same goes for a thousand other obscure journals too. With the best will in the world, the best you can do is look in the places you know about. The topologists, whose necks would doubtless have prickled like mine had they come across Lorenz's seminal opus, were not in the habit of perusing the pages of the *Journal of the Atmospheric Sciences*.

And so, for a decade, his paper languished in obscurity. Lorenz knew he was on to something big, but he was ahead of his time.

Let's take a look at what he did.

Courage of His Convections

Hot air rises.

This motion is known as *convection*, and it's responsible for many important aspects of the weather (Figure 53). Thunderclouds form as a result of convection; that's why you tend to get thunderstorms on a hot humid day. Convection can be steady, with the warmer air drifting gently upwards in a constant manner; or unsteady, with the atmosphere moving about in a much more complicated way. Unsteady convection is far more interesting, and more obviously relevant to weather. Since the simplest behaviour after being steady is to change periodically, the simplest kind of unsteady convection is some sort of periodic swirling effect.

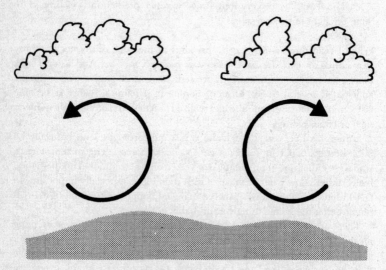

Figure 53   Convection cells, caused by hot air rising.

The study of convection has a distinguished history. In about 1900 Henri Bénard carried out a fundamental experiment, discovering that when a thin layer of fluid is heated from below it can form convection cells, looking rather like a honeycomb. Lord Rayleigh derived the basic theory of the onset of convection. But there's always more to learn. In 1962 B. Saltzman wrote down the equations for a simple type of convection. Imagine a vertical slice of atmosphere, warm the air at the bottom, keep it cool at the top, and watch it convect. What you expect to see is regularly spaced swirls, the convection cells, going round and round in a periodic fashion. In a manner typical of classical applied mathematics, Saltzman guessed an approximate form of the solution, substituted it into his equations, ignored some awkward but small terms, and took a look at the result. Even his highly truncated equations were too hard to solve by a formula, so he put them on a computer.

He noticed that the solution appeared to undergo irregular fluctuations: unsteady convection. But it didn't look at all periodic.

Lorenz was interested and decided to investigate further. Noticing that only three of Saltzman's variables played a role in this effect, Lorenz threw the rest away. This was a highly cavalier but perfectly conscious act. He obtained a system of equations that has now become a classic:

$$\frac{dx}{dt} = -10x + 10y$$
$$\frac{dy}{dt} = 28x - y - xz$$
$$\frac{dz}{dt} = -\frac{8}{3}z + xy$$

Here x, y, z are his three key variables, t is time, and d/dt is the rate of change. The constants 10 and 8/3 correspond to values chosen by Saltzman; the 28 represents the state of the system just after the onset of unsteady convection, as we'll see in a moment. These numbers can be changed, depending on the values of physical variables.

If you cross out the terms xz and xy on the right-hand sides, you get a set of equations that any mathematician worth his salt will solve with his eyes shut before breakfast. Boring, though.

But you can do something more useful along those lines. You can find the steady states of the system, where all three expressions on the right vanish, and x, y, z remain constant. There are three: one representing no convection and two others, symmetrically related, representing steady convection. You can also analyse the stability of the system near these states by a method known as *linear stability analysis*. You find that if the 28 is reduced below 24.74 then the state of steady convection is stable. At the critical value 24.74, convection starts up. Lorenz's choice of 28 occurs just after the onset of unsteady convection.

At this point linear theory abandons you. It works well *near* the steady state; but when the steady state becomes unstable, that necessarily means you have to consider what happens as the system moves away from the steady state. So linear theory can tell you where the instability occurs, but not what happens as a result. A pair of binoculars can show you where the brow of the next hill is, but not what lies beyond.

It's a start. Now you know *where the interesting behaviour occurs*. But what is it?

The Advantages of Having a Computer

There's no way out: *you have to solve the equations*. By hook, crook, cunning trickery or brute force. By far the most reliable method is brute force: compute the solution numerically.

Lorenz had a computer. In the early 1960s this was unusual. Most scientists distrusted computers and hardly anybody had one of their own. The machine on which I'm typing this paragraph is a far better computer than Lorenz had, and I'm using it for word processing. It's like using a Rolls-Royce to deliver milk. Times change. Anyway, Lorenz had a Royal McBee LGP-300 computer, a not very reliable maze of vacuum tubes and wires. So he put his equations on his Royal McBee and let it royally McBuzz away, at a speed of about one iteration per second. (My word processor is about fifty to a hundred times faster.)

Catch-22: to get out of the bind, the place, people, culture, and time must be right. Poincaré was the person, France the place – but the time and culture were wrong. Lorenz was the person, MIT the place; the culture for chaos is the computer culture, and that was well under way. When everyone has a computer, the *fact* of chaos is impossible to miss. Realizing its importance is another matter, though. For that, the time must be right too – other people have to appreciate that something really interesting is going on. The time wasn't right. More accurately, Lorenz was ahead of his time.

His paper shows the first 3,000 iterations of the value of the variable *y* (Figure 54). It wobbles periodically for the first 1,500 or so, but you can see the size of the wobble growing steadily. Lorenz knew from his linear stability analysis that this would happen: but what happened *next?*

Madness.

Violent oscillations, swinging first up, then down; but with hardly any pattern to them.

He drew plots of how various combinations of *x*, *y*, *z* varied. In the (*x*, *y*)-plane he saw a two-lobed figure like a kidney (Figure 55). Sometimes the point circled the left-hand lobe, sometimes the right.

The trajectories of his equations, he realized, lived on something rather

Figure 54 Lorenz's plots of 3,000 numerically computed steps in his equations for convection: (left) oscillations grow and become chaotic; (right) two views of the motion in phase space (American Meteorological Society, Journal of the Atmospheric Sciences, 20 (Edward N. Lorenz)).

Figure 55   The Lorenz attractor: trajectories cycle, apparently at random, round the two lobes.

like a squashed pretzel. A surface that had two layers at the back, but merged to a single layer at the front. The point that represented the state of the system would swing round one or other of these surfaces, pass through their junction, and then swing round again.

Lorenz knew that trajectories of a differential equation *can't* merge. So what looked like a single sheet at the front must really be two sheets very close together.

But that meant that each sheet at the back was double too; so there were four sheets at the back ... So four at the front, so eight at the back, so ... 'We conclude,' said Lorenz, 'that there is an infinite complex of surfaces, each extremely close to one or the other of two merging surfaces.'

It's not surprising that the meteorologists were baffled. But Lorenz was on to something big.

It's amazing what a bit of xz and xy can do for you.

The Butterfly Effect

It's not true to say that Lorenz found no pattern, that nothing was predictable. On the contrary, he found a very definite pattern. He took the peak values of the variable z, and drew a graph of how the current peak relates to the previous peak. The result was a beautifully precise curve, with a spike in the middle (Figure 56).

Lorenz's curve is a kind of poor man's Poincaré section. Instead of plotting a variable at regular periods of time, he plots z every time it hits a peak. The

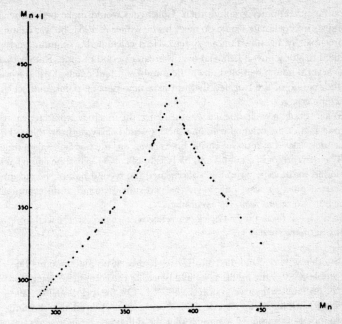

Figure 56 Order in chaos. If the size of an oscillation is plotted against that of the previous oscillation, a precise curve results (American Meteorological Society, Journal of the Atmospheric Sciences, *20 (Edward N. Lorenz)).*

time intervals are then irregular, but not badly so, because there's a definite underlying rhythm to the Lorenz attractor.

Using the curve, you can *predict* the value of the next peak in z provided you know the value of the current peak. In this sense, at least some of the dynamics is predictable.

But it's only a short-term prediction. If you try to string the short-term predictions together to get a long-term prediction, tiny errors start to build up, growing faster and faster, until the predictions become total nonsense. Indeed, Lorenz's curve has the same stretch-and-fold characteristics that we've learned to associate with chaos, and the stretch makes the errors blow up.

Lorenz noticed this too. Later it was called the 'butterfly effect'. He discovered it by accident.

He'd had his McBee for several years, since about 1960. He used to set up model weather-systems and let them run, sometimes for days on end. The computer would type out the solution trajectory as a long series of numbers

– no fancy computer graphics then. Colleagues would make bets on what Lorenz's microclimate would do next. In the winter of 1961, he was running a precursor of his now famous system. He'd calculated a solution, and he wanted to study how it behaved over a greater period of time. Rather than wait several hours, he noted down the numbers it had reached when it was in the middle of the run, fed them in as a new starting-point, and set the machine going.

What should have happened was this. First, the machine would repeat the second half of the original run, and then it would carry on from there. The repetition served as a useful check; but missing out the first half saved time.

The meteorologist went off and had a cup of coffee. When he came back, he found that the new run had *not* repeated the second half of the old one! It started out that way, but slowly the two runs diverged, until eventually they bore no resemblance to each other.

In his book *Chaos* James Gleick, a science writer who interviewed Lorenz, tells what happened next.

> Suddenly he realized the truth. There had been no malfunction. The problem lay in the numbers he had typed. In the computer's memory, six decimal places were stored: .506127. On the print-out, to save space, just three appeared: .506. Lorenz had entered the shorter, rounded-off numbers, assuming that the difference – one part in ten thousand – was inconsequential.

From the traditional way of thinking, so it should be. Lorenz realized that his equations weren't behaving the way a traditionally-minded mathematician

Figure 57   The butterfly effect: a numerical simulation of one variable in the Lorenz system. The curves represent initial conditions differing by only 0.0001. At first they appear to coincide, but soon chaotic dynamics leads to independent, widely divergent trajectories.

would expect. Lorenz introduced a striking image which became known as the 'butterfly effect' (Figure 57). The flapping of a single butterfly's wing today produces a tiny change in the state of the atmosphere. Over a period of time, what the atmosphere actually does diverges from what it would have done. So, in a month's time, a tornado that would have devastated the Indonesian coast doesn't happen. Or maybe one that wasn't going to happen, does.

There are butterflies everywhere. But who is to say that their flapping wings cancel each other out?

Lorenz, again:

> The average person, seeing that we can predict the tides pretty well a few months ahead would say, why can't we do the same thing with the atmosphere? It's just a different system, the laws are about as complicated. But I realized that *any* physical system that behaved nonperiodically would be unpredictable.

Lorenz ends his 1963 paper with some speculations about the possibility of weather-forecasting. His argument is simple and original. Imagine recording a very accurate series of measurements of the state of the atmosphere, comparable to those that you wish to use for forecasting. Collect such data for a very long time.

> The crucial point is then whether analogues must have occurred since the state of the atmosphere was first observed. By analogues we mean two or more states of the atmosphere which resemble each other so closely that the differences may be ascribed to errors in observation.

If two analogues *have* occurred, then you will make identical predictions of the future weather, starting from either of them. That is, your weather-predicting scheme must predict *periodic* variation of the weather. But this is nonsense; the whole difficulty with weather-prediction is that the weather is *not* periodic.

If analogues haven't occurred, there's still hope: the entire weather system may be quasi-periodic, almost repeating the same states over again, but with tiny variations, slowly growing. In such a case, long-term weather prediction might be possible. In fact, all you have to do is look back in the records for a close analogue of today's weather, and see what happened last time.

This line of argument fails, Lorenz notes, if 'the variety of possible atmospheric states is so immense that analogues need never occur'. And he leaves one crucial question dangling: 'How long is "very long range"?' He says that he doesn't know the answer, but 'Conceivably it could be a few days or a few centuries'. Twenty-five years later, the centuries have been ruled out, and 'a few days' looks spot on.

Swat That Butterfly

The image conjured up by the phrase 'butterfly effect' is a vivid one, and it has captured the public imagination. Perhaps too much so, for it places too much emphasis on the capricious and unpredictable nature of chaos, which is only one of its contrary aspects. The other aspect of chaos, paradoxically, is stability. That's what the word 'attractor' implies: if the system is somehow displaced from its attractor, then it rapidly homes back on to it. So chaos is a strange and beautiful combination of stability and unpredictability.

This strong element of stability means that it is wrong to think of chaos as being unpredictable in all respects. It all depends on what you want to predict. If you want to predict whereabouts on its attractor a chaotic system will lie in the distant future, and all you know is where it is now, then you've got problems. On the other hand, you can safely predict that *even after a random disturbance* the system will quickly return to its attractor – or, if it has several attractors, it will return to one of them. (Predicting which one is not as easy as you might think: see Chapter 16.)

This is all very well, but 'lying on an attractor' is a rather abstract statement. What would we actually *see*? We don't normally observe attractors directly: what we observe is observables – measurements made on the system, quantities that depend upon its state but do not directly represent that state. And as we make repeated observations, what we end up with is a sequence of numbers, not an attractor. The jargon for such a sequence of numbers is 'time-series'. What does 'lying on an attractor' tell us about a time-series? At first sight, very little, but with a bit of practice you can teach yourself to detect the underlying attractor and notice if the system moves to a different one. The key is to concentrate not on the quantitative aspects of the time-series, but on its 'texture'.

As Poincaré recognized, dynamics on an attractor is *recurrent*. That is, the state of the system repeatedly comes close to every point of the attractor, and in particular returns close to any previous state. Like a cow grazing in a field, its detailed movement is unpredictable, but in the long run every clump of grass in the pasture will be visited over and over again. (The mechanism for grazing is different, of course: if a clump is not visited for a while then the grass there will grow, and become an obvious target for a hungry cow.) Although dynamics on a chaotic attractor is recurrent, it is not periodic. That is, the time intervals between repeated visits to a given small region may be highly variable. So sometimes the chaos cow returns to the same part of the field after a few hours, but sometimes she takes several weeks. The only thing you can be sure of is that if you wait long enough the chaos cow will come past again.

Imagine a point A, moving along its attractor in phase space, and passing very close to a point B that it has previously visited. The butterfly effect tells us that the trajectories of A and B will diverge exponentially fast. However, exponentials take a little time to get going. For instance, if you start with 0.001 and keep doubling, it takes ten steps to make it bigger than 1. In contrast, a mere seven further steps will make it bigger than 100.

On any exponential curve, yesterday is invisible and tomorrow is explosive.

The early slow (though still exponential) growth implies that, on the scale of our observations, A and B appear to follow much the same course – for a while. The closer A and B are, the longer this approximate agreement continues. So the time-series for B will include a segment of ten or so steps that looks suspiciously similar to the corresponding part of the time-series for A. This means that chaotic time-series are characterized by the presence of recurrent 'motifs'. More strongly, *every* short subsequence of the time-series will recur indefinitely. So, for example, if you have a time-series that exhibits some specific pattern of wiggles – such as $\wedge\!\!\wedge\!\!\wedge\!\!\wedge$, say – once, but never again, then it does not lie on an attractor.

This recurrence of motifs is the basis of Lorenz's suggestion for forecasting weather using previous 'analogues', and it tells us that the time-series corresponding to an attractor has a characteristic texture. By becoming attuned to this texture, not only can you tell the difference between a time-series that corresponds to a chaotic attractor and one that does not: you can usually tell *by eye* whether two time-series come from the same chaotic attractor, or different ones.

What does this mean for our proverbial butterfly? Can it *really* cause a hurricane?

What the butterfly does is disturb the motion of the point in phase space that represents the Earth's weather. Assuming that this point lies on an attractor, albeit a highly complex multidimensional one, then the tiny flapping of the butterfly can divert the point off the attractor only very briefly, after which it rapidly returns to the same attractor. However, instead of returning to the point A that it would have reached if undisturbed, it returns to some nearby point B. The trajectories of A and B then diverge exponentially, but because they lie on the same attractor, they generate time-series with the same texture. In particular, a hurricane – which is a characteristic weather motif – cannot occur in the perturbed time-series unless it was (eventually) going to occur in the original one. So what the butterfly does is to alter the timing of a hurricane that – in a sense – was going to happen anyway. Don't take that too literally: the butterfly may trigger the conditions needed to make a hurricane form, or prevent one that would have formed. But most of the time it will just have a minor effect on where and when a hurricane that has been building up for global reasons – the right kind of warm, humid air in the right place – will occur.

Hurricanes are a recurrent and characteristic feature of the time-series that we call 'normal global weather-patterns', and as such they are evidence that this time-series lives on a single attractor. The butterfly does not flip the weather to a new attractor: it just displaces it a bit on the same attractor. It is true – or, at least, all available evidence strongly suggests – that if you could run the weather twice, with the only difference being the presence or absence of a flapping wing, then the hurricanes in the second run would happen at different times from those in the first run, and there might be a few more or a few less. But both runs would have the same texture – they would visibly represent the same *kind* of weather system.

Similarly, in our pastoral analogy, suppose that the cow's attention is momentarily distracted by a butterfly (sorry, but obviously it *has* to be a butterfly) fluttering away from a clump of daisies. Then she will probably wander round the field in a different path from the one that she would have taken, had the butterfly remained still. However, you will still see the same cow, ambling about in the same kind of way around the same field, and the grass will still get grazed. You probably wouldn't notice any difference. But the fate of individual blades of grass may change a lot, with some surviving for weeks instead of seconds, and others for seconds instead of weeks. It is the timing of events, not the range of possible events, that changes.

Given all this, it is an exaggeration to claim the butterfly as the *cause* of the big changes that its flapping wing sets in train. The true cause is the butterfly in conjunction with everything else. There are billions of butterflies in the world, and the lazy agitation of their wings is just one source of tiny vortices in our atmosphere. The weather is determined by the combined effect of all such influences. The proverbial butterfly is just as likely to cancel out a hurricane as to create one – and it may just raise the average temperature of India by a hundredth of a degree, or generate a small grey cloud over Basingstoke.

Philosophers long ago learned to distinguish between proximate cause and ultimate cause. The proximate cause of you burning your finger is the hot saucepan that you've just picked up; the ultimate cause is the Big Bang that formed the universe that condensed into the stars whose nuclear reactions resonated to create the carbon out of which your tender organic integument is made. These two meanings of 'cause' address two distinct issues and answer two distinct questions. Similarly, the proximate cause of a hurricane is the presence – in a particular place and at a particular time – of large circulating masses of warm, humid air. The ultimate cause tracks right back through the winds of history to the Big Bang. In there somewhere is a butterfly, but it is just one of a trillion factors (I underestimate) that have contributed, since the dawn of time and the birth of space, to the presence of that mass of humid air *now*, *here*, using *these* molecules of water and gas. It is no more sensible to blame the flapping of a butterfly's wing than it is the flipping of a quark's quantum state.

Whatever the influence of all those butterflies may be, it keeps our weather on a single attractor, the one that we recognize by its texture as 'normal weather patterns'. However, dynamical systems can possess more than one attractor. Tiny perturbations like those of the butterfly cannot switch our weather from one attractor to another, but larger changes might. The word that captures the phrase 'texture of normal weather patterns' is *climate*. Indeed there is a good case for identifying 'climate' with 'attractor', and if we do, then what we are discussing now is climate change. And then it is not the flap of a butterfly's wing that should concern us, but the massive build-up of human-made greenhouse gases.

In 1993 Tim Palmer of the European Centre for Medium-Range Weather Forecasting distinguished two types of prediction. The first was 'initial value problems' – given the weather today, tell me what it will be next December. The butterfly effect implies that this kind of prediction is possible only on short timescales. The second was how particular overall features of the weather vary as some external parameter is changed. For instance, how does a particular level of carbon dioxide (the main greenhouse gas) affect the mean summer temperature or the number of winter snowstorms? The butterfly effect is irrelevant to such questions, because they are about climate, which lives on the entire attractor, not any particular path across it. One tiny butterfly can't flap anything as big as the weather off its nice, stable attractor. Palmer cites quantitative evidence for stable planetary-scale patterns of climate, for example in the wind patterns in the northern hemisphere, to back up this assertion. All of this suggests that if we stop being obsessed by predicting how the weather will change over time, and concentrate on how the climate will change if particular global parameters – such as the levels of greenhouse gases – vary, then we can beat the butterfly and foresee the future.

Stretch and Fold

Having understood that there are more ways to foretell the weather than bashing it with a butterfly, we can safely take a closer look at the mathematics of the butterfly effect. We will see later that the butterfly effect is as much a blessing as a curse, but we will see that only by considering its nature in some depth, so it's useful to have a simple model. In fact we've already seen an example of the butterfly effect in Chapter 6: Smale's solenoid, or its simpler model, the mapping $x \rightarrow 10x$ on a circle. There the same sensitivity to initial conditions occurs. Two points π and π', agreeing to a billion decimal places, wander about independently of each other after a billion iterations.

That may not sound so bad. But two points agreeing to six decimal places evolve independently after only six iterations.

Where does this sensitivity come from?

It's a mixture of two conflicting tendencies in the dynamics.

The first is *stretching*. The mapping $x \to 10x$ expands distances *locally* by a factor of ten. Nearby points are torn apart.

The second is *folding*. The circle is a bounded space, there isn't room to stretch everything. It gets folded round itself many times, that's the only way to fit it in after you've expanded distances by ten. So, although points close together move apart, *some points far apart move close together.*

The expansion causes points that start off close together to evolve differently. At first, the difference grows regularly. But once the two points have moved far enough apart, they 'lose sight of each other'. No longer must one mimic the behaviour of the other.

The mixture of stretching and folding is also responsible for the irregular motion. Yes, some points must move closer together again. But which? *How can you tell?* Large differences now are due to very tiny differences many iterations back. You can't see what's coming in advance.

That's unpredictability.

You can see the stretch-and-fold process going on in Lorenz's system. Each half of the front of the surface winds round to the back and is stretched to double its width, before being 'reinjected' into the front part again.

It's now pretty clear that Lorenz's strange infinitely sheeted double-lobed surface must be a strange attractor – the *Lorenz attractor*. And his differential equations, while a somewhat hacked-down version of the physics, are down-to-earth equations in three variables with some kind of physical pedigree, be it ever so littered with mongrels. They aren't artificial designer differential equations, labelled 'CAREFULLY MADE BY TOPOLOGISTS' with a green dough-nut logo on the label.

And in fact you can find real physical systems which are very well modelled by Lorenz's system of equations, at least if you vary the numbers 10, 28, and 8/3. One such system is a waterwheel. Another is a dynamo. A third, at the frontiers of physical research, is a laser.

But when Lorenz wrote down his equations, nobody knew that. All they could see was the obvious: he'd got them by hacking bits off the equations for convection. Most scientists were worried about the effect of those missing bits. They didn't understand that Lorenz didn't care whether his equations made physical sense.

Lorenz had opened a door into a new world.

Nobody stepped through.

Door? What door?

8

Recipe for Chaos

When you can gather it up, start pulling it with your fingertips, allowing a spread of about 18 inches between your hands. Then fold it back on itself. Repeat the motion rhythmically. As the mass changes from a somewhat sticky, side-whiskered affair to a glistening crystal ribbon, start twisting, while folding and pulling.

Irma S. Rombauer and Marion Rombauer Becker, *Joy of Cooking*

As a child, I lived in a South Coast seaside town. My parents used to take me out regularly for walks – it was just after the war and at that time they had no car, so we got plenty of healthy exercise. Sometimes we would walk to the harbour, down the High Street. It was a steep, narrow street lined with tiny shops, cobbles underfoot, and near the top was a shop that sold homemade sweets. Naturally, this caught the child's attention. There was seaside rock, with the name of the town running right through in tiny red letters, and you could watch them assembling it, like a short log, out of red strips and white wedges, before rolling it thin and chopping in into sticks. And there was a machine that stretched and kneaded the sticky sugary mixture from which the rock was made. Two shiny steel arms rotated slowly, and simultaneously moved from side to side. A heavy strand of the sticky material hung between them, like a skein of thick knitting wool held between a pair of hands, and was repeatedly stretched and folded, stretched and folded. It fascinated me, and not just because of the end product. I didn't realize it at the time, but it was my first encounter with chaotic dynamics.

The sweetmaker didn't realize it either, but he was exploiting two character-istic features of chaos. *Mixing* – to make sure the ingredients were uniformly distributed – and *expansion*, to introduce long crystalline strands into the sugar, producing the brittle crackability of true seaside rock.

The really curious thing – so familiar that we scarcely notice it, let alone question it – is that the motion of the *machine* is perfectly regular. The toffee-pulling machine moves periodically, round and round, to and fro. But the toffee goes chaotic. Regular cause: irregular effect.

Everyone who uses a cake-mixer, egg-whisk, or food processor is performing an exercise in applied chaotic dynamics. A mechanical device, moving in a regular and predetermined fashion, is randomizing the ingredients. How is this possible?

Stretch and Fold

Stephen Smale conjectured that typical dynamics is steady or periodic. When he found out that this was wrong, he replaced the conjecture by a question: what *is* typical dynamics?

There are two main ways to make progress in mathematics.

One is 'pure thought'. Spend a lot of time thinking, in a rather general way, about what really makes the problem tick. Play around with general features. Try to dig out the fundamental ideas.

The other is to look at examples, preferably as simple as possible, and to pin down exactly how they work.

In practice you need both to get anywhere. A mathematician working on a problem will mess around with simple examples until he decides he's in a rut, and then he'll switch to a more general point of view and worry about that for a while, and then he'll go back to a slightly different set of examples and ask slightly different questions. Then he'll badger all other mathematicians within earshot. He'll telephone colleagues from Knoxville to Omsk. If he gets really stuck he'll go off and do something else: tackle another problem, change the oil in the car, build a fishpond, climb a mountain. And, often at the least appropriate moment, inspiration will strike. It seldom solves everything, but it keeps the process going. Anselm Lanturlu, a cartoon character created by the French physicist Jean-Pierre Petit, captured the feeling exactly in *Euclid Rules OK:*

> I'VE UNDERSTOOD IT! Well, that is . . . I'm not exactly sure WHAT I've understood, but I have the impression I've understood SOMETHING.

Thinking very generally about dynamics, no details, just the broadest possible picture, leads to something like this.

Traditional dynamics:

- Sit still.
- Go round and round.

The distillation of five centuries of science, into its geometric essence. What's the geometric essence of chaos?

• Stretch and fold.

The missing ingredient.

Well, not the *only* missing ingredient. Chaos is a rich mixture, full of exotic spices and strangely shaped fruits; it has its quota of nuts, too. But its basic ingredient, the flour-and-water of chaos, is stretch and fold.

Let's thumb through the cookbook.

From Radar to the Horseshoe

Right at the end of the Second World War, in 1945, two Cambridge mathematicians, Mary Lucy Cartwright and John Edensor Littlewood, were studying forced oscillators. An *oscillator* is something that wobbles repetitively, like a pendulum; and a system is *forced* when some time-varying push is given to the dynamics from outside. For example, you might imagine hanging a pendulum from a pivot which is attached to a motor and slides up and down like a piston. This example of a forced oscillator combines two distinct periodic motions: the 'natural' oscillations of the pendulum and the 'artificial' oscillations of the driving force. In general these will have different periods, that is, the natural motion will get out of step with the forcing. This leads to a complicated interaction.

Forced oscillations are everywhere. A less obvious one is the sleep–wake cycle, in which a natural biochemical rhythm is forced by the regular day–night cycle caused by the rotation of the Earth. The heartbeat is another: see Chapter 13.

Anyone brought up on classical linear theory would expect the combination of two oscillatory motions to lead to quasiperiodic motion with two superposed frequencies. However, forced oscillators don't always do what classical mathematics might lead us to expect. Nonlinear effects arise, and the result is often chaos.

The van der Pol equation, mentioned earlier in connection with radio valves, is a nonlinear oscillator. Cartwright and Littlewood proved that under suitable conditions a forced van der Pol oscillator displays complicated aperiodic motion. With hindsight, this must be counted as one of the earliest discoveries of chaos. Their work was part of the war effort. Electronics meant radar, and it is no coincidence that the van der Pol equation arose in electronics.

In the 1960s Stephen Smale was thinking about the forced van der Pol oscillator, but not about warfare. He invented a model system with similar geometry, corresponding to a simpler but less physical equation. Take a square, stretch it out into a long thin rectangle, fold it up into a horseshoe shape, and replace it roughly within its original outlines (Figure 58).

Figure 58   Smale's horseshoe mapping simulates chaotic folding. A square is stretched, folded, and replaced on top of itself. When iterated, the mapping produces an intricate multi-layered structure.

Stretch and fold.

If you think about iterations of this procedure, you'll see that the next stage produces a sort of horseshoed horseshoe, with three U-bends, the stage after has seven U-bends, the next fifteen, and so on. Each iteration doubles up the existing bends and adds an extra one. So you get, in the limit, an infinitely wiggly sort of curve. Now start again, but just think of some initial point in the square, rather than the whole square. As it is iterated, it must 'home in' on the infinitely wiggly curve – because the whole square does! So we may as well assume it's actually *on* the curve, and at each iteration, it hops around from one point on the curve to another. Because the curve is so wiggly, it turns out that the motion on it is to all intents and purposes random. This is the geometry that underlies the chaotic behaviour noticed by Cartwright and Littlewood.

The horseshoe has other important features. It has the same infinitely layered structure that Lorenz deduced must be occurring in his attractor, and which shows up in the solenoid and its closely related Cantor set.

Not only that. Inside the horseshoe is a saddle point, and one separatrix of this saddle winds away and crosses another. The result is a homoclinic tangle – dynamical spaghetti – closely resembling that which so horrified Poincaré. The main difference is that Poincaré's example arose in Hamiltonian dynamics – no friction. Smale's system can occur in dissipative systems – where friction is present – too.

So this one example bears a family resemblance to many other chaotic systems. But in several respects it's simpler. In particular you can study it using geometry and topology, rather than a computer.

By studying the horseshoe, Smale was able to make progress where Poincaré had given up, and this led to an explosion of new ideas in dynamical systems theory.

Dynamics Bolognaise

Michel Hénon is a French astronomer. In 1962 he was thinking about how stars move within a galaxy. This led him to a mathematical model, a dynamical system whose behaviour depended on its level of energy. In celestial mechanics differential equations are usually Hamiltonian: there's not much friction out in space.

The conventional wisdom at the time was that trajectories should be periodic, or more generally quasiperiodic, separable into several distinct periodic components. Classical methods, such as perturbation theory, tended to *start* with this as an assumption. Not surprisingly, all of the solutions that were obtained in this way agreed with the conventional wisdom. On the whole, few were bothered by this vicious circle, if they noticed it at all.

Hénon had been given the classical training, like everyone else, and he started out expecting quasiperiodic behaviour. With a graduate student, Carl Heiles, and armed with a new and underrated tool, the computer, he began studying what happened to the regular orbits as the energy in the system increased.

At low energies, trajectories were regular and periodic: conventional wisdom was confirmed. But at higher energies they broke up. What should have been nice closed curves in the dynamical pictures fell apart into a random smudge of dots. There were islands of regularity sitting in a complicated way in a sea of chaos (Figure 59). Meatballs of regularity in a stochastic spaghetti. Dynamics Bolognaise. Hénon and Heiles *proved* nothing rigorously, but they drew pictures of what they saw on their computer, and made some inspired guesses as to what was going on. Then, being astronomers, not mathematicians, they moved on to other problems.

Magnetic Trap

A mathematical explanation for Hénon and Heiles's discovery has been given by Jürgen Moser, in terms of what he called *twist maps*. Other scientists found the same phenomena in various applications. In 1960 the Russian physicist B. V. Chirikov was working on plasmas – gases, so hot that some of their electrons are stripped off. The ultimate aim of plasma research is to construct a working fusion reactor to supply cheap and safe electrical power. To make such a reactor, the plasma must be confined at high temperatures and pressures for sufficiently long periods of time. No ordinary material can survive the heat, so a magnetic trap is used. But plasmas and magnetic fields interact in a very complex fashion.

Chirikov was trying to understand this. He came up with a model for the dynamics of a plasma in a magnetic trap, in the form of a Poincaré mapping, now known as the *standard mapping* because it arises so often. By analysing the standard mapping, Chirikov discovered that chaos can occur in a plasma, causing instabilities which let it escape from the trap.

The standard mapping has a special feature: it is *area-preserving*. That is, if the mapping is used to transform any region of phase space, its area remains unchanged after the transformation. This reflects the fact that the full system is Hamiltonian: conservation of energy in the full system becomes preservation of area in a Poincaré section.

The standard mapping involves a numerical parameter that controls the dynamics. Chirikov found that there is a critical value of this parameter, at which the motion becomes chaotic. The mechanism whereby chaos is created in the standard mapping is an especially fundamental one, known as the 'breakdown of KAM tori'. That is, it has the same pattern of stable islands and random behaviour that Hénon and Heiles found; but now the islands start to break up. This is the *chaos border*, and it possesses an intricate and important structure. There are still many problems about area-preserving mappings which mathematicians and physicists would dearly like to solve.

Figure 59   The restricted three-body problem, in an approximation due to Hénon and Heiles: (left) *trajectories computed by classical series approximations are always regular in form;* (right) *computed trajectories reveal islands of regularity amid a sea of chaos; energy increases from top to bottom.*

Figure 60   The fine-scale structure of the Hénon attractor.

Puff Pastry

By 1976 Hénon had come into contact with dynamical systems theory, and had heard about strange attractors. He went to a lecture on the Lorenz attractor, which posed – but on the whole did not answer – questions about its geometric fine structure. He began to wonder whether this was the new mathematical idea that would explain his earlier results, and decided that a good first step would be to understand more about the Lorenz attractor.

Hénon is a scientist who, while not working directly in mathematics, has a mathematician's instincts, and a willingness to dabble with simple, non-physical, stripped-down models, in the hope of obtaining mathematical rather than physical insights. There are many such people in the annals of chaos. He invented a much simpler system of equations than Lorenz's which incorporated their main feature: stretching and folding.

Hénon got a picture very like Smale's horseshoe. The same convoluted zigzags and multiple U-bends. His computer experiments revealed the same infinitely layered structure predicted by theory (Figure 60). Hénon's attractor is U-shaped, but it isn't a curve: it comes in layers, folded over on each other like puff pastry. It's a very elegant and delicate structure. It's also rather complex: it doesn't seem possible to describe its geometry in complete detail. Yet all of this structure is implicit in the very simple equations that define it.

If you run the equations on a computer then, no matter what values you start with, successive points rapidly home in on this delicate structure, never breaking the multi-layered pattern. But, on the other hand, you can never guess whereabouts within the layers the next point will fall. *That simple equation knows something that you don't.* The interplay between regularity and randomness is baffling.

In the dynamical systems theory of the 1970s there were two distinct strands. In one, topologists exploited geometrical properties to establish rigorous results about concocted systems which they hoped had some connection with nature. In the other, physicists started from equations which they knew had connections with nature, found approximate solutions on computers, and saw similar structures to those that the topologists were seeing. But were they really the same? Or were people seeing things that weren't truly there, just because that's what they now expected to see?

The problem is that you can't completely trust a computer picture. It's a faithful representation of what the computer has calculated, all right. But the computer can't do *exact* calculations – at least, not without a completely different approach to the whole game – so its calculations are a complicated sort of approximation to the real thing. Is an accurate answer to an approximate question the same as an approximate answer to the exact question?

Sometimes. But by no means always. For instance, if you solve the equations for the motion of an aeroplane in a very slightly sticky fluid, the solution is *not* close to that for a fluid whose stickiness is zero.

Hénon's equations are so simple that it was hoped that they might provide the 'missing link' of dynamical systems theory: apply the topological results to a specific equation and confirm the numerical analysis rigorously. But dynamical systems theory is so hard that until very recently, nobody could manage to do this – even for so simple a system. There was even a respectable school of thought that Hénon results were really just something with a very long period – not truly chaotic at all. But 1987 witnessed a first-rate breakthrough. Lennard Carleson found a way to prove that the Hénon attractor is indeed chaotic – at least, for 'most' values of the numerical parameters that occur in the equations. Better still, in 1995 two mathematicians solved the long-standing problem of proving, in full rigour, that there is chaos in the Lorenz equations. Everyone who has solved Lorenz's equations (correctly) on a computer has obtained the famous 'strange attractor' picture of Figure 55. However, computers do not store numbers to infinitely many decimal places; they use 'finite precision' arithmetic which introduces round-off errors, so the fine detail of the Lorenz attractor looks different on different types of computer. So is the Lorenz attractor just an artefact of the finite-precision approximation, or is it present in the exact solution to the original equations? The answer is not at all obvious, and a few diehards have even stated that the irregular behaviour found by Lorenz is entirely due to computer error.

We now know that it is *not* computer error, thanks to Konstantin Mischaikow and Marian Mrozek at the Georgia Institute of Technology. A different proof has also been given by Brian Hassard and colleagues at the University of Buffalo. Mischaikow and Mrozek have not yet confirmed that the chaos in the Lorenz equations occurs on an attractor like the one in the computer pictures, but they have established that chaos definitely happens, which is in itself a major breakthrough. Their proof is 'computer assisted', but this does not make it unrigorous because the role of the computer is to perform certain lengthy but routine calculations which could in principle be done by hand. They set up a mathematical famework that corresponds to the finite-precision calculations of the computer, and their main effort goes into devising a theory (of 'finitely representable multivalued maps') that can link the finite-precision calculations to the infinite precision of conventional mathematics. The calculations are then fed to a topological gadget, the 'Conley index', that establishes the occurrence of chaos.

In short, the mathematicians found an angle on the problem that allowed them to parlay the computer's approximation into an exact result. There are precedents for such an approach, but this one is a virtuoso performance. In mathematics one example is enough to establish existence, but now we've got several. They tell us, firmly and undeniably, that strange attractors are not just bizarre topological confections. They're really there, in simple equations, in equations that model the real world.

Beyond Beeton

If there's a message in what we've seen, it's this:

RECIPE FOR CHAOS

12 oz phase space
1 tablespoon of initial conditions
Stretch and fold repeatedly.
Season to taste.

But this is a mathematics book, not a potted Mrs Beeton, and we seek a more formal understanding of such processes, even if the subtle genius of the born cook is overlooked. I want to end this chapter by taking a much closer look at one example of chaotic dynamics, inspired by the example of the toffee-pulling machine and its ubiquitous stretch-and-fold route to chaos.

I want to capture the essence while avoiding undue complication. *Due* complication is fair game. But let's not add complications for their own sake. I'll replace the strand of rock-mix by a line segment of unit length. I want a formula which mimics the sweetmaker's machine.

There's nothing original about the example I've chosen. It's one of the old favourites, the chimpanzee's tea-party of the chaotic zoo: the *logistic mapping*. It exemplifies not just the occurrence of chaos, but the manner is which chaos may be created.

Imagine a black box, an electronic circuit with a knob that you can turn. The box is emitting regular signals. You turn the knob slowly, and the signals change a bit, but remain regular. Then, at some critical position of the knob, the signals start to become unstructured, random. You'd be forgiven for assuming that you'd done something drastic to the black box, maybe switched on a whole new section of the circuit.

What the logistic mapping shows is that drastic changes do not have to have drastic causes. Nothing much changes in the black box circuit. Just a few fine adjustments to a variable capacitor, let's say. But it *still* can change from regularity to chaos.

The logistic mapping is also important as the place where the theory of chaos first made serious contact with experiments. And it has a close relative which has generated some of the most complex and beautiful behaviour known to mathematics from one of the simplest possible equations. But those tales must wait for later chapters. Here we familiarize ourselves with the logistic mapping and examine some of its startling properties.

Logistic Mapping

Consider a line segment of unit length. A point on this line segment is represented by a number x between 0 and 1, giving its distance from the left-hand end. The logistic mapping is

$$x \rightarrow kx(1 - x)$$

where k is a constant between 0 and 4. Iterating the mapping we get the discrete dynamical system

$$x_{t+1} = kx_t(1 - x_t)$$

We can think of t as representing time, but now time must click along in whole number steps, 0, 1, 2, 3, ... Then x_t is the value of the variable x at time t.

Geometrically, the logistic mapping stretches or compresses the line segment in a non-uniform manner, and then folds it in half. For instance, take k = 3, so that $x_t = x$ transforms to

$$x_{t+1} = 3x(1 - x)$$

The numbers between 0 and 0.5 are mapped to the numbers between 0 and 0.75. For instance, 0.5 goes to $3 \times 0.5\,(1 - 0.5) = 0.75$. The numbers between 0.5 and 1 are mapped to the numbers between 0.75 and 0: the same interval in reverse order. So the effect of the mapping is to stretch the original segment so that it covers the segment between 0 and 0.75 *twice*.

In general, for given k, the mapping folds the interval up and lays it down on top of the interval between 0 and k/4. If k is small this is a compression rather than a stretching; and we'll see a difference in the dynamics. If k is bigger than 4 the interval pokes outside itself under iteration, and some values of x shoot off rapidly towards infinity. This is not very pleasant to contemplate at this stage, which is why I've assumed k lies between 0 and 4.

To study the dynamics of the logistic mapping we must look at its long-term behaviour – its attractors. That is, we want to iterate the mapping over and over again and watch what happens to x. But there's an extra layer of structure: we wish to do this for various values of k, and see how the pattern changes as k is varied.

So k is the 'knob' on the black box, and the equation above describes the internal circuitry. You can investigate the effects of setting k to various values using a pocket calculator or a home computer; and I strongly urge you to check out everything I say. However, I'll describe what happens: partly for the benefit of those without access to such machinery, and partly to point out the main features of interest.

Steady State Regime

The range of k values between 0 and 3 is the *steady state regime*, the least interesting from the point of view of dynamics. Pick k in this range, say k = 2, and iterate the mapping. For example, take $x_0 = 0.9$. Then, by applying the formula repeatedly with $t = 0, 1, 2, \ldots$, we find a sequence of values

$x_0 = 0.9$
$x_1 = 0.18$
$x_2 = 0.2952$
$x_3 = 0.4161$
$x_4 = 0.4859$
$x_5 = 0.4996$
$x_6 = 0.4999$
$x_7 = 0.5$
$x_8 = 0.5$

and there she sits. There's a point attractor, a stable steady state, at $x = 0.5$.

Fixed Point   **Period 2**   **Chaos**

Figure 61   Graphical iteration of the logistic mapping using cobweb diagrams (left to right): *steady state, periodic point, chaos. (Reproduced by permission of John Wiley & Sons Ltd., © 1986.)*

You can check it's a steady state quite easily: if $x = 0.5$ then $2x(1 - x) = 0.5$ as well. Iteration doesn't alter the value 0.5.

The stability can also be checked by a calculation, but you can see it geometrically by drawing what mathematical economists call a *cobweb diagram* (Figure 61). This is a graphical method of iteration. First draw a graph of the formula $y = 2x(1 - x)$, obtaining an inverted parabola. Draw the diagonal line $y = x$ on the same diagram. To iterate a starting value x_0, draw a vertical cobweb from x_0 and see where it hits the parabola. Then draw a horizontal cobweb to hit the diagonal. The horizontal coordinate of this point is x_1. Repeat, forming a 'staircase' between the parabola and the diagonal line. The coordinates of successive 'risers' of the staircase are the successive iterates x_t.

When k = 2 the cobweb wanders up the diagonal and then spirals in towards the point where the parabola hits the diagonal. This is the fixed point; and the stability follows because the cobweb spirals *inwards*. If it were to spiral outwards, you'd have an unstable fixed point.

If you experiment you'll find that the cobweb spirals inwards, provided k is smaller than 3. So for k in the range of 0 to 3 you get a single stable fixed point, and the long-term dynamics is to do absolutely nothing. The position of the fixed point moves slightly as you tune up the knob k, but nothing else happens.

Period-doubling Cascade

When k is exactly 3, the fixed point is 'marginally stable': convergence to it is *extremely* slow. This is a sign that we're on the verge of something dramatic. Indeed, when k < 3, the fixed point becomes unstable, and the cobweb spirals *out*.

Whenever you know a solution to a dynamical system, and it becomes unstable, you should ask yourself 'Where does it go now?' In practice, it won't sit in an unstable state, even though that does satisfy the equations. It will wander off and do something else. Often the something else is much less obvious, and therefore more interesting, than the unstable state you started from. This is an easy way to learn rather a lot of new things: it's called *bifurcation theory*.

In which spirit: where does the steady state of the logistic mapping go when k is bigger than 3, say 3.2?

If you draw cobweb diagrams, you'll find that the outward spiral slows down and eventually converges on to a square loop. The value of x_t flips alternately between two distinct numbers. This is a *period-2 cycle*. So the steady state loses stability and becomes periodic. In other words, the system starts to wobble.

On a computer with a sound generator you can make it play a kind of rudimentary music, using the successive values of x to determine the notes to be played (Figure 62). For example, you could stretch the range [0,1] of x to cover one octave: *do–re–mi* and all that. The steady state tune is repetitive and boring: *fa–fa–fa–fa–fa* . . . forever. The period-2 tune at least has the merit of rhythm: *so–mi–so–mi–so–mi–* over and over again. Beethoven, it is not.

If you increase k to about 3.5 the period-2 attractor also goes unstable, and a period-4 cycle appears: *so–fa–la–mi–so–fa–la–mi–* . . . By 3.56 the period has doubled again to eight; by 3.567 it has reached 16, and thereafter you get a rapid sequence of doublings to periods of 32, 64, 128, . . . (If you try this out on your home computer please bear in mind the warning given in Chapter 1 about different makes of computer giving different results. The same goes for all that follows.)

So rapid is this *period-doubling cascade* that by k = 3.58 or so it is all over: the period has doubled infinitely often. At that point, having done its best to stay periodic by paying the price of longer and longer periods, the logistic mapping becomes chaotic. If you listen to it you can still hear almost-rhythms, little runs of half-familiar tunes, but nothing repetitive. It still isn't Beethoven, but it's not totally unlike the music of some modern minimalist composers.

Order amid Chaos

From this point on, the music gets ever more chaotic. At the maximum value, k = 4, the tune wanders densely through the entire octave of available notes. That is, given a trajectory – a sequence of x-values with a given starting-point

Figure 62   Schematic representation of the iterates of a logistic mapping $x \to kx(1-x)$ *in 'musical' notation. The height of the 'notes' represents the value of* x, *and the 'stave' is drawn arbitrarily. The constant k is* (top to bottom) *2, 3.2, 3.5, 3.56, 3.6, 3.8, 4.0. As k increases, the music becomes more random in quality.*

Figure 63   However, increasing k in the logistic mapping does not always increase the randomness: at k = 3.835, a period-3 cycle occurs.

– will pass as close as you wish to every point of the interval. The entire interval has become an attractor.

So it all looks pretty simple. As k runs from 0 and 4 you get a steady increase in complexity of dynamical behaviour:

steady → periodic → chaotic

with the period-doubling cascade as the mechanism whereby chaos sets in. The 'tuning knob' k just makes everything more and more complicated as you turn it.

Oh, it's not as easy as that!

Try, for example, the value k = 3.835, well into the chaotic regime. For the first fifty or so iterations, it all looks nice and chaotic, as you'd expect. But then the tune changes: *mi–so–ti–mi–so–ti–* . . . repeating indefinitely. Period *three* (Figure 63). Where did *that* come from?

According to my computer, the cycle is

$$0.1520744 \rightarrow 0.4945148 \rightarrow 0.9586346$$

If you increase k *very* gently the periods then go 6, 12, 24, 48, 96, . . . in a new period-doubling cascade!

Even more baffling is what happens at k = 3.739. Now you get a cycle of period five (Figure 64):

$$0.8411372 \rightarrow 0.4996253 \rightarrow 0.9347495 \rightarrow 0.2280524 \rightarrow 0.6582304$$

Figure 64   Period-5 cycle in the logistic mapping at k = 3.739.

repeated indefinitely. Yes, near that you'll find periods 10, 20, 40, 80, . . .

This isn't such a cosy picture. The knob k isn't just a simple 'chaos generator'. It's not true that increasing k always makes the dynamics more complicated. On the contrary, buried within the chaotic regime are little 'windows' of regular behaviour.

Where do the windows come from? It's a complicated story, but one that's now well understood. We even know in what order the periods arise. The fundamental theorem was proved by a Russian mathematician, A. N. Sharkovskii. Write the integers in the following order:

$$3 \to 5 \to 7 \to 9 \to 11 \to \ldots$$
$$\to 6 \to 10 \to 14 \to 18 \to 22 \to \ldots$$
$$\to 12 \to 20 \to 28 \to 36 \to 44 \to \ldots$$
$$\to 3.2^n \to 5.2^n \to 7.2^n \to 9.2^n \to 11.2^n \to \ldots$$
$$\to 2^m \to 2^{m-1} \to \ldots$$
$$\to 32 \to 16 \to 8 \to 4 \to 2 \to 1.$$

First, the odd numbers in *ascending* order. Then their doubles, quadruples, octuples . . . finally the powers of 2 in *descending* order. If, at a given value of k, the logistic mapping has a cycle of period p, then it must also have had cycles of period q for all q such that $p \to q$ in this ordering. So the first cycles to set in have periods 1, 2, 4, 8, . . . – the period-doubling cascade. The period 17, say, sets in *before* period 15 does; but before those, period 34 has set in, and before that periods like 44 or 52 which are odd multiples of 4, and before those 88 or 104 or 808 which are odd multiples of 8 . . .

What really boggles the mind is that this same bizarre ordering applies not just to iterations of the logistic mapping, but to iterations of *any* mapping on the unit interval that has only one hump. This result was the first hint that some of the patterns of chaos might be *universal*, that is, not specific to individual examples but representative of entire *classes* of systems.

Big Fleas, Little Fleas . . .

But there's something even more mind-boggling about the periodic windows of the logistic map.

There's a way to get an overview of the entire dynamic behaviour of the logistic mapping for all values of k in one go. It's known as a *bifurcation diagram* (Figure 65). A bifurcation is any change in the qualitative form of the attractor of a dynamical system; and the logistic mapping is just littered with bifurcations.

The way to get it is this. Draw a graph with k running horizontally and x vertically. Above each value of k, mark those x-values that lie on the attractor for that k. Then each vertical slice gives a picture, in the interval from 0 to 1, of the corresponding attractor. So, for example, when k is less than 3, there is just a point attractor, and you must mark a single value of x. This gives a curve.

Figure 65   Bifurcation diagram for the logistic mapping. The constant k increases from 2 to 4 horizontally. The vertical coordinate is the state x. Note the fig-tree of period-doublings, followed by the growth of chaotic bands. (Reproduced by permission of John Wiley & Sons Ltd., © 1986.)

Owners of home computers might like to experiment, before reading on. Imagine a graph in which k runs horizontally from 0 to 4 in stages of, say, 0.2. Plot *x* vertically, between 0 and 1. (You'll have to stretch the scales to see anything sensible.) At each value of k, iterate *x* for a few hundred steps *without* plotting any points, and then continue for another twenty or so steps, plotting the *x*-values above the chosen k.

Here's what you'll see. At k = 3 the hitherto single curve splits into two (and 'bifurcation' makes sense in English as well as Mathish), splitting again and again as k runs through the period-doubling regime. You see a beautiful tree structure. I call it the *fig-tree* (Figure 66) because it led to a wonderful discovery by the American physicist Mitchell Feigenbaum, to be described in the next chapter but one. (*Feigenbaum* is German for 'fig-tree'. I've got one further Germanic pun up my sleeve, too. Sorry about that.)

Around k = 3.58 the fig-tree culminates in infinitely many branches and the system goes chaotic. The branches of the fig-tree broaden into

Figure 66   Schematic view of the fig-tree: regular, repetitive branching, with infinitely many branches occurring in a finite space.

Figure 67   Detail of Figure 65 within a periodic window: the entire structure is repeated in miniature. And there are windows within windows (arrowed) . . . (Reproduced by permission of John Wiley & Sons Ltd., © 1986.)

bands of chaotic attractors. The bifurcation diagram is pocked with random dots.

But look more closely. Every so often, there's a thin white strip in the picture with just a few tiny dots inside it. These are the periodic windows (Figure 67).

If you look at the window around $k = 3.835$, where the basic period is three, you'll see that it contains three tiny fig-trees of its own. Choose one of them and magnify the picture to bring out the fine detail.

You'll find that this sub fig-tree also ends in bands of chaos. Within those bands there are again thin white strips with just a few tiny dots. Windows within windows. In these are tinier fig-trees, and so on.

In fact, inside any window is a precise copy of the *entire* picture. The bifurcation diagram for the logistic mapping contains *tiny copies of itself*, perfect in every detail. This is called *self-similarity*, and it's important.

The Cotswold village of Bourton-on-the-Water has a tourist attraction: a model village. In the appropriate corner of the model village is a model of the model village. In the appropriate corner of that is a model of the model of the model village. In Bourton-on-the-Water, the sequence stops there. But in Bifurcation-on-the-Logistic, it goes on forever, and each copy is a perfect replica of the original.

9

Sensitive Chaos

I like to wash,
By way of experiment,
The dust of this world
In the droplets of dew.

Bashō

The Japanese poet Bashō was born in 1644 in Ueno. He was the son of a minor samurai serving the ruling Tōdō family. At the age of forty he set out on the first of a series of journeys, recorded in *The Records of a Weather-exposed Skeleton*. His poem, quoted above, describes a spring at the hermitage of Saigyō: 'The famed spring was just as it had been when the poet described it, shedding its clear drops of water with a drip–drop sound.'

Basho was seeking to renew his identity by contemplating nature, and he found beauty in something as simple as a falling drop of water. We shall follow in his footsteps, but seeking a complementary beauty, that of the mathematician rather than the poet. The two are not unrelated: both search out simplicity within complexity.

The patterns of flowing water have fascinated many people besides Bashō. For instance, the Royal Library at Windsor contains many drawings by Leonardo da Vinci showing complicated cascades of water (Figure 68). To depict fluid motion accurately represents a challenge to any artist. His spectators have a good mental image of how water behaves, and when a painting fails to reflect this image accurately, they can see immediately that something's wrong. But the image isn't articulated consciously: they can see there's an error, but seldom have any idea what it might be. In the same way, when I look at pictures of hunting-horses in a pub, I can see that they look funny, and I even have some idea what's wrong: it's the pattern of the legs as they gallop, or maybe the height of the horse's body off the ground. But for the life of me I couldn't tell you how to draw a galloping horse.

Figure 68   Torrent *by Leonardo da Vinci (Windsor Castle. Royal Library, © Her Majesty the Queen).*

Leonardo combined the instincts of a scientist with the vision of an artist, and he took conscious steps to improve the accuracy of his work by making careful studies of animals, the human body, clouds, trees – anything that a painter or sculptor might wish to depict. And he and his contemporaries were unusually interested in water.

Water, then held to be one of the four elements out of which everything in the universe is made, was more than just a liquid. It was a symbol for the processes of life. Because, like life, water *flows*. It is born, it grows, it moves, it changes, it dies. A trickle from a spring becomes a stream, a river, a rushing torrent, an ocean. A river can meander sinuously across a flat plain, carve deep canyons in ancient rocks deposited on the sea bed a hundred million years ago, plunge in a spectacular waterfall, or clog with silt and spread into a gigantic fanlike delta at its mouth. A calm sea can become a raging monster with froth-capped breakers; a storm-swept sea can suddenly die to a flat calm. The German poet Friedrich Leopold, Freiherr von Hardenberg, who used the pen name Novalis and lived at the end of the 18th century, called water the 'sensitive chaos'.

Not a bad description.

Plumbing the Depths

We tend to take water for granted. It's something that comes out of a tap. We seldom think of the colossal feats of engineering behind that mundane fact. One day, when the Victorian tunnel that serves our particular area collapses, such questions will acquire a new dimension of urgency, but for now, as we wash our hands or fill a bucket, our thoughts are far away.

What better instrument than a humble tap to help us plumb the depths of the sensitive chaos?

Have you ever looked at how water flows from a tap? *Really* looked, I mean, not just shoved your toothbrush under it? Inspired by my own rhetoric, I did so this morning, probably for the first time in my life. I can't guarantee that your tap will do what mine did, but I recommend the experiment anyway, you'll learn a lot. Let me tell you what I saw.

The essence of scientific observation is to be systematic. I admit that many important discoveries – such as the anti-bacterial activity of penicillin – are made by chance, but they are confirmed and exploited by more systematic methods. A million monkeys bashing typewriters will eventually write *Hamlet*, but I wouldn't care to wait around for it. So I set myself a systematic task. How does the pattern of water emerging from a tap change when the rate of flow is *slowly* increased?

Open the tap just a little bit. What happens? The tap drips, of course.

If you let everything settle down to a steady motion, you'll find that the tap drips regularly, with a constant interval between each drip and the next.

Open it up a little more. The speed of the drips increases, but they remain regular. Keep increasing the flow in tiny steps: the same thing continues to happen. Patience. The life of a scientist is one of vast periods of tranquillity, punctuated by brief and sudden drama and excitement.

There comes a point at which the falling drops join together to form a steady stream. Found it? Good. But I'm forced to point out that you've missed the really interesting bit. *Before* the drops merge into a stream, several other transitions take place, rather close together. If you've been impatient, and increased the flow rate in too big steps, then go back and try again.

The first of these transitions is that the rhythm of the falling drops changes. Instead of a steady *drip–drip–drip* it becomes more like *dripdrop–dripdrop–dripdrop*, a close pair of drops, then a pause, then another pair. It's still regular, but it's different.

Perhaps with good instruments you'd be able to find further changes in the rhythm, also regular, also different. By eye and ear, I couldn't manage it. What I saw next was much more puzzling. The pattern of falling drops becomes *irregular*. They're following each other pretty fast by now, but you can still see and hear separate droplets; and the rhythmic sound has gone, replaced by something much more complex.

So there's one transition to ponder: drips that lose their rhythm.

Soon after, as I've already said, the drops merge into a steady stream. When the stream first forms, it may still break up into droplets lower down, but soon it becomes steady and smooth, a thin tapering thread from tap to washbasin. Fluid dynamicists call this *laminar* flow: the fluid moves in thin layers (*laminae*) sliding smoothly over each other like a pack of cards being spread out on a table.

Increase the flow rate to roughly normal levels. The emerging water remains laminar, although it may develop extra structure, as if the jet were trying to split into two parts, or perhaps spiral.

Now turn it on really full. The smooth laminar flow breaks up, the water hits the washbasin with enormous force, and the flow becomes frothy and irregular again. This is *turbulent* flow, and our second important transition: laminar to turbulent.

Turn off the tap, mop up the mess. The experiment is over. Now comes the mathematics.

Accumulating Wobbles

What we've seen is two versions of the transition to turbulence. The first, occurring for the rhythm of droplets, is effectively a discrete dynamical system – provided we ignore the detailed structure of the individual drops. The second, a laminar stream becoming turbulent, is a continuous system. In both cases, a regular motion suddenly becomes irregular.

Turbulence is immensely important in many branches of science, from astronomy to meteorology (Figure 69). It's also important in practical engineering problems. Turbulence can destroy a water pipe or an oil pipeline, break up a ship's screw, or cause an airliner to crash. Engineers have devised various methods, ranging from rule-of-thumb to sophisticated statistics, for dealing with practical instances of turbulence. But its true inner nature remains a problem of the highest order.

Fundamental science of that kind comes more properly into the domain of physics rather than engineering. What does a mathematical physicist in the classical mould make of the phenomenon of turbulence?

The classical equation for the flow of a viscous fluid, developed from that of Euler, is the brainchild of a Frenchman, Claude Navier, and an Irishman, Sir George Stokes. Fluid flow governed by the partial differential equation of Navier and Stokes is deterministic and predictable. Before the advent of chaos, these were considered synonymous with 'regular'. But turbulence is irregular. Conclusion: *something goes wrong with the equations.*

This is not implausible. Remember, the equations describe a highly idealized fluid, one that is infinitely divisible and homogeneous. But a real fluid is composed of atoms (take your pick among competing levels of detail, from tiny hard balls to quantum swirls of probability). Turbulence appears to involve tinier and tinier vortices. But a vortex of subatomic dimensions is a physical absurdity. If a real fluid were to obey the Navier–Stokes equations at this level of detail, it would have to shred its own atoms.

So conceivably turbulence is a macroscopic effect of atomic structure. Inaccuracies in the Navier–Stokes equations, of atomic dimensions, propagate through the physical flow, increasing in size, to be observed as turbulence. This is the Leray Theory, and it dates from 1934, a time when atomic physics was especially novel and fashionable.

Within a decade, the mathematical physicist Lev Landau had realized that there was another possibility. A paper he wrote in 1944 begins: 'Although turbulent motion has been extensively discussed in the literature, the very essence of this phenomenon still lacks sufficient clarity.' Landau then puts his finger on a key question: *where does turbulence come from?* 'In the author's opinion,

Figure 69   Turbulence in Jupiter's atmosphere near the Great Red Spot.

the problem may appear in a new light if the process of initiation of turbulence is examined thoroughly.'

Imagine a system sitting in a stable state. Sometimes, perhaps if suitable external controls are varied, this state may become unstable. For example an object resting stably on a table may slide off if the table is tilted, or a balloon may burst if it is overinflated.

When I take my car to have a tyre changed, the garage mechanic puts the wheel on a fancy piece of machinery which whirls it round and round. Guided by the numbers on the machine's screen, he hammers metal weights into the rim of the wheel, to balance it. The reason for this rigmarole is that an unbalanced wheel starts to vibrate if it revolves too rapidly, a condition known as *wheel wobble*.

In dynamics, wobbles are fundamental mathematics. One of the most basic ways for a state to lose its stability is by wobbling.

When a hitherto stable state acquires a wobble, a new periodic motion is added to its existing motion. A wheel, rotating smoothly, begins to vibrate: now there are two superposed periodic motions, the rotation and the vibration.

Landau saw the onset of turbulence as a build-up of wobbles. He theorized that in its early stages, turbulence is the superposition of three or four different periodic motions, and as it becomes fully developed, the number of periodic motions becomes infinitely large.

The basic mechanism for the creation of wobbles is called *Hopf bifurcation*, after Eberhard Hopf. A sink (steady state) becomes unstable and turns into a source, surrounded by a limit cycle representing periodic motion (Figure 70). In 1948 Hopf proposed a rather more detailed theory along the same lines as Landau's. The Dutch scientist J. M. Burgers had not long before studied a simplified version of the Navier–Stokes equations, and Hopf adopted

Figure 70   Onset of a wobble, or How a steady state becomes periodic. The mechanism is known as Hopf bifurcation: a sink loses stability and becomes a source, throwing off a limit cycle. (Reproduced by permission of John Wiley & Sons Ltd., © 1986.)

similar tactics. He came up with another approximate model which, most unusually, could be solved explicitly; and he showed that it followed the Landau scenario of accumulating wobbles.

For the next three decades the Hopf–Landau theory was widely accepted and used. It had several virtues. It was simple and comprehensible. The mechanism whereby an extra frequency was added to the motion was basic and natural. There were model equations, such as Hopf's, in which the scenario was known to occur. And it was accessible to classical techniques such as Fourier analysis, so you could do calculations with it.

Unlikely Scenario

But in 1970 this cosy picture was disturbed. Not shattered, for the proposal came from outside fluid dynamics, was highly speculative, and lacked any kind of experimental support. To make matters worse, it was not derived from the physics of fluid flow, but from topology.

David Ruelle, a Belgian mathematician working at the Institut des Hautes Études Scientifiques in Paris, and a Dutch visitor named Floris Takens, started thinking about turbulence from the point of view of topological dynamics *à la* Smale. Is there a *typical* scenario, a generic process, for the onset of turbulence?

That's not so clear. But what *is* clear, when you start thinking this way, is that the Hopf–Landau theory *cannot possibly be correct.* For while each of its accumulating wobbles appears to be mathematically and physically plausible, it isn't. Only the first one is.

Hopf's and Landau's intuition was derived to some extent from Hamiltonian dynamics. There conservation of energy imposes a constraint that makes multiple-frequency quasiperiodic motions commonplace. But this constraint does not apply to dissipative systems – systems with friction. And in the flow of a viscous fluid there is friction aplenty.

Ruelle and Takens were led to the following picture.

The first transition, from a steady state to a single wobble, is typical even in dissipative systems: it leads to a periodic motion. No difficulty here.

The second transition, adding an extra frequency, can certainly occur. It initially leads to motion which, from the topological viewpoint, is a flow on a two-dimensional torus; and this motion starts out looking like a quasiperiodic superposition of two independent periodic motions. But it can't remain that way, because such a motion is not typical, not generic. In practice small perturbations will break it up.

As it happened, the typical, generic, structurally stable flows on a torus were known; and they predicted something well known to electrical engineers,

Figure 71 Frequency-locking: (left) *two independent periodic oscillations combined by superposing them:* (right) *the flow breaks up to form one stable periodic cycle (heavy line) and one unstable periodic cycle. For clarity, the torus on which the motion takes place has been cut and opened up to form a square.*

called *frequency-locking* (Figure 71). The two originally independent periodic motions will interact and become entrained, yielding a combined motion that is periodic, with a single combined period.

With three superposed frequencies, something even more dramatic goes wrong. Typically the three frequencies need not even lock: instead they can combine to create a new object – which Ruelle and Takens called a *strange attractor*. The solenoid is a strange attractor, and so (it is conjectured) is the Lorenz attractor. Strange attractors have strange geometries.

The foundation of the Ruelle–Takens theory is that the Hopf–Landau scenario is, in a topologist's world-view, as likely as a pin balancing on end. The pin is unstable: the Hopf–Landau theory is *structurally* unstable. If you move the pin it will topple and crash to the table: if you make slight changes to the equations of motion the Hopf–Landau scenario will fall apart and crash into a strange attractor.

Falsifiability

Not everyone in fluid dynamics was overjoyed by Ruelle and Takens's proposal. It was, in fact, somewhat controversial. But a few people – as it turned out, enough – took inspiration from it, and began the next phase. *It's pretty, but is it right?*

In science, there's a time-honoured way to find out whether a theory is right.

Experiment.

More accurately, an experiment can tell you whether a theory is *wrong*, for you can never be absolutely certain that it's right. You can prove a theorem in mathematics, but you can't prove a theory. As the philosopher Karl Popper emphasized, testing a scientific theory is a matter of *falsification*, not *verification*. The more a theory fails to be falsified when confronted by experiment, the more likely it is to be true; or at least, the broader the range of conditions under which it works. But you can never be certain the theory is absolutely correct, even if it survives a million experimental tests; for – who knows? – it may fail at the million and first.

Thus, as the third millennium AD approaches, do scientists abandon the pursuit of Truth.

Having said which, they try very hard not to make mistakes. But we no longer live in an era of absolutes. We're learning, dreadfully slowly, not to take ourselves too seriously.

To count as scientific, a theory must in principle be falsifiable. On the island of Corfu, there's a superstition that if you see a praying mantis, it either brings you good luck – or bad luck, depending on what happens. This belief doesn't amount to a scientific theory; not because you can't measure 'luck', but because it's hard to see how an experiment could disprove the theory, even if you could.

None of this means that the inhabitants of Corfu are *wrong*. What we're discussing is limits on scientific knowledge. There may be true things in the universe that cannot be known in the scientific sense. However, it's going to be hard to resolve disputes about them.

Laboratory Classic

Is the strange attractor theory falsifiable?

As originally proposed, it certainly wasn't directly falsifiable. You can't go out and look for a strange attractor. So you can't fail to find one either. The reason is that the mathematical description of such an attractor, in the Ruelle–Takens theory, is not related to any physically measurable variables. So, as a falsifiable theory, it looks little better than one that claims turbulence to be the wake of invisible monsters swimming in the fluid, monsters undetectable by any physical apparatus.

There are several ways to get round this. One is to improve the contact between the mathematics and the physics. That seems to be very hard for turbulence – which is *not* to say it isn't important. Another is to sidestep the issue. Perhaps the strange attractor can be made to reveal itself indirectly.

The Hopf–Landau theory is much more obviously falsifiable. All you have to do is measure the component frequencies of the motion, and watch whether the wobbles pile up in the prescribed manner. If not, Hopf–Landau's a dead duck.

So, instead of trying to show that Ruelle and Takens are right, you can start off by trying to show that Hopf and Landau are wrong. Historically, it didn't quite happen that way. Instead, the experimentalists set out to show that Hopf and Landau were *right*.

But surely, you would imagine, this had been done already? After all, the Hopf–Landau theory had been widely accepted for several decades.

Not entirely. The first few stages had been observed. But, as the wobbles piled up, it became harder and harder to get accurate enough measurements.

Further progress needed a new idea.

Harry Swinney, a physicist at the University of Texas, Austin, began his experimental career working on phase transitions. When water boils, metal melts, or magnets become magnetized, that's a phase transition: a macroscopic change of state due to reorganization at a molecular level. In a sense, the transition to turbulence is a kind of phase transition in a fluid. Some of the great fluid dynamicists, such as Osborne Reynolds and Lord Rayleigh, had even thought of it that way. But the analogy seemed too loose, too inexact, to be mathematically useful.

Nevertheless, it set Swinney thinking. Could the methods he had used to study delicate phenomena in phase transitions be applied to fluids?

There are many ways for a fluid to become turbulent. The first stage in designing an experiment is to sort out which system to use. Basic science is not aimed at a specific goal like 'find the best shape for a wing-flap on a jumbo-jet', and it has the luxury of choosing which system to work with. For laboratory experiments in basic science, the important thing is that the system should be 'clean'. I don't mean that it shouldn't be covered in sticky finger marks, I mean that it should be easy to set up and run, give precise results, and give reproducible effects on repeated runs.

There is a classic laboratory system in fluid dynamics, originally invented by the French hydrodynamicist Maurice Couette. He wanted to study 'shear flows' where the fluid is wrenched apart, and he came up with an arrangement of two cylinders, one inside the other (Figure 72). With the outer cylinder fixed, and the inner one rotating, there is a constant and controllable shear.

What you'd expect to happen in such a system is that the fluid goes round and round with the cylinder, fast in the middle and slow at the outside. And that's what Couette found.

In 1923 the English applied mathematician Geoffrey Ingram Taylor experimented with speeding up the inner cylinder, and he made a puzzling discovery. If the speed is high enough, then the fluid stops going round and round smoothly, and breaks up into pairs of vortices, like a tube of Polo mints with the wrapping removed. In fact this is a beautiful example of the Hopf–

Figure 72   Apparatus for the Taylor–Couette experiment (schematic). The space between two cylinders is filled with fluid, and the cylinders are rotated. The gap between the cylinders is here exaggerated for clarity: it is usually 10–20 per cent of the radius of the outer cylinder.

Figure 73   Wavy vortices in the Taylor–Couette experiment. Note the dislocation two thirds of the way down, where the number of waves is in the process of changing.

Landau type of instability, where a new periodic motion is created. But it's only the first stage of the Hopf–Landau scenario.

Subsequently, experimentalists and theorists studied the Couette–Taylor system (or Taylor–Couette system, as non-francophiles often call it) in enormous detail. It may well be the most studied of all fluid flows. They found a tremendous variety of pattern-formation effects. The vortices can become wavy (Figure 73). The waves can go up and down like horses on a roundabout, giving modulated wavy vortices. There are twisted vortices and braided vortices. There are spiral patterns like a barber-pole, wavy spirals, modulated wavy spirals, and interpenetrating spirals.

And, at high speeds, the system goes turbulent.

All this richness of behaviour is produced by a piece of apparatus the size and shape of a vacuum flask, in a precisely reproducible fashion. So Swinney and his collaborator Jerry Gollub decided to perform their experiments on this laboratory classic.

Illumination from the Laser

At that time, fluid dynamics usually made measurements on flowing fluids by inserting probes or injecting streams of dye. These methods tended to interfere with the flow, and weren't very sensitive or accurate, but people in the field had become inured to such problems and expected little better. Swinney had his eye on a much more sensitive device: the laser.

Today lasers are commonplace. If you've got a compact disc player then you've got a laser. As every fan of *Star Wars* knows, lasers are what you zap the imperial guards with. Lasers produce a beam of coherent light – light in which all the waves are in step, and reinforce each other instead of cancelling out. What you've got is a very precise and accurate torch.

If you listen to the siren of a fire-engine as it goes past, you'll notice that the pitch of the siren appears to change, becoming lower once the fire-engine has gone by. This is the Doppler effect, named for the Austrian scientist Christian Doppler who first noted it in 1842. In effect, the sound waves are speeded up when the fire-engine approaches, and slowed down as it departs.

The same effect works with light, only now it's the colour, the frequency, that changes. If you shine a laser at a fire-engine, and compare the colour of the returning light with the colour of the light you originally sent out, you can tell how fast the fire-engine is going.

More to the point, if you suspend tiny flakes of aluminium powder in a fluid, you can use a laser to tell how fast the flakes – and presumably the fluid – are moving. This technique is known as Laser Doppler Velocimetry.

If you have a complicated signal that is a mixture of waves of different frequencies, then it's possible to analyse the signal mathematically and to extract the individual components. You can also find how strong each component is – how much it contributes to the total. The method is basically Fourier analysis: representing a curve as a sum of sine and cosine curves.

The result of this analysis can be summarized as a *power spectrum*, a graph showing the strength of each component frequency (Figure 74). The figure shows five series of observations (the graphs on the left) together with their power spectra (on the right). The time scale for the observations (in

Figure 74   Time-series of observations in a convection experiment, and the corresponding sequence of power spectra, showing how the strength of component frequencies changes. Spikes indicate well-defined frequencies in a periodic or quasiperiodic motion; broad bands indicate chaos.

seconds, s) and the frequency scale (in hertz: 1 Hz = 1 oscillation per second) are at the bottom.

For example, the top left picture shows a very regular rhythm, with about one oscillation every ten seconds. This is picked out in the corresponding power spectrum on the right as a single spike, marked f_2, close to 0.1 Hz. The second series of observations is much less regular, and its power spectrum has several spikes. A trained eye can see that they are all built up by adding together multiples of two basic frequencies f_1 and f_2 at around 0.03 Hz and 0.1 Hz.

These spikes on the power spectrum correspond to sharply defined component frequencies that are much stronger than any nearby frequency.

A quasiperiodic signal has a power spectrum consisting mostly of sharp spikes, like the top three pictures in Figure 74. A noisy, 'random' signal has a broad-band spectrum, whose component frequencies are smeared out, like the bottom picture. A mixture of the two is also possible, as in the fourth picture.

The power spectrum is a kind of 'frequency fingerprint' of a series of observations, and it can be used to detect the presence of certain types of behaviour.

Swinney and Gollub used a computer to extract, from their laser data, the power spectrum of the fluid velocity. This is exactly what you need to observe the successive creation of new frequencies as predicted by Hopf and Landau.

This was their expectation.

They looked for the first transition, and found it. They repeated the experiment many times, getting very clean and accurate data. So clean and accurate, in fact, that the fluid dynamicists didn't believe them. Nobody wanted to publish their results. Their application for a research grant was turned down. Some said the results weren't new; others didn't believe them at all.

Undaunted, they continued to the next transition – and failed to find it. There was no clean creation of a new frequency. Instead, there was the gradual emergence of a broad band of frequencies (Figure 75). 'What we found was, it became chaotic.'

Contact

Science is big. It's impossible to know everything that's going on. The way people find out things that they need to know is through personal contacts. Swinney and Gollub had tested the Hopf–Landau theory – and found it wanting. But at that point they were unaware that Ruelle and Takens had proposed an alternative.

But others were. The scientific grapevine went to work. In 1974, a Belgian mathematician appeared in Swinney's laboratory – David Ruelle. Ruelle had a theory that predicted chaos; Swinney had chaos but no theory. It remained to see whether what Ruelle had predicted, and what Swinney had found, matched up.

There was indirect evidence. For example, computer calculations showed that a broad-band power spectrum is to be expected when a strange attractor is present.

By now, the pace was hotting up. More and more scientists were becoming aware of chaos, more and more mathematicians were developing its theoretical aspects. A series of experiments – at first performed by Swinney and his

Figure 75   Power spectra for the Taylor–Couette system. Initially only one frequency ω_1 is observed (periodic oscillation). Then a second frequency ω_2 appears (together with other spikes representing combinations of ω_1 and ω_2). Finally, broad-band chaos is seen. (Reproduced by permission of John Wiley & Sons Ltd., © 1986.)

colleagues, but soon by others – suggested in no uncertain terms that strange attractors are implicated in a whole range of turbulent flows.

The results only applied to the onset of turbulence, but at least in some particular laboratory systems, the strange attractor theory of turbulence was holding up well, and the Hopf–Landau theory was dead in the water. Ironically, most of the mathematical fine detail proposed by Ruelle and Takens was turning out to be irrelevant, or even wrong – not mathematically, so much as in its interpretation for experiments. But the main idea . . . There it seemed they had struck gold.

However, it still wasn't certain.

There *might* be other explanations of the observations. Something more direct was needed, something that would make the strange attractor hypothesis falsifiable in an experiment.

And that required another idea.

Fake Observables

The 1970 paper of Ruelle and Takens is not so much a theory of turbulence, as a starting-point for such a theory. The main missing ingredient is any connection between the topology and the physics. If there is, for example, some quantity that you can measure and plot, and look for a strange attractor in the results, then the theory becomes falsifiable. If you carry out such an experiment, and don't find the strange attractor, you know you're wrong.

What is an experimental observable? It's a quantity that depends on the state of the system being observed. What we're missing, in the topological theory of turbulence, is any knowledge of how it so depends. At first blush it's hard to see how to get round this, except by establishing such a connection. So one possible research programme to put the Ruelle–Takens theory on a testable basis is: *derive a strange attractor from the Navier–Stokes equations* for fluid flow. This is a problem that requires mathematical, rather than experimental, advances, and it hasn't been carried out yet. The Lorenz attractor doesn't count, because of the approximations involved.

But there's another way. Suppose you can somehow reconstruct the shape of the attractor from a series of observations in a way that's *independent* of what precise quantity is being observed. Then the connection doesn't matter.

It's a neat trick. David Ruelle and Norman Packard thought it could be made to work, and Floris Takens found a way to prove that it did.

In its simplest form, a sequence of experimental observations produces a *time-series*: a list of numbers representing the value of the observed quantity at regular intervals of time. (It can be irregular, but let's keep the discussion simple.) For example, the temperature at a given place at noon every day forms a time-series, perhaps something like

17.3, 19.2, 16.7, 12.4, 18.3, 15.6, 11.1, 12.5, . . .

in degrees Celsius.

Suppose you want to fit such data to a strange attractor. The problem is that you're contemplating, say, an attractor in three-dimensional space; but your observations only give one quantity. For instance, the Laser Doppler

Figure 76   Computer experiments in the reconstruction of attractors by the Ruelle–Takens method, in a two-dimensional plot: (a) the periodic time-series sin t + sin 2t *yields a closed loop; (b) the two-frequency time-series* sin t + sin $\sqrt{2}$t *yields (a projection of) a torus; (c) the three-frequency time-series* sin t + sin $\sqrt{2}$t + sin $\sqrt{3}$t *has no clear structure in a two-dimensional plot. A third coordinate must be plotted to reveal its quasiperiodic nature.*

Velocimetry technique just gives you the frequency of the reflected light – the speed of the fluid at one particular point, where the laser light gets reflected back. So you've squashed the attractor down into one dimension. You're seeing it in silhouette, so to speak.

If you could look at the attractor from other directions, you might build up a complete three-dimensional picture, much as an architect can convey the shape of a building by a plan, a front elevation, and side elevation. To reconstruct a three-dimensional attractor you need information from three different directions.

But there's no chance of finding those extra directions in a time-series of a single observable, is there? You need two other observables.

What Ruelle realized is that you can concoct two more fake observables from this same time-series, by displacing the time value (Figure 76). Instead of the single time-series, you compare three of them; the original and two copies, shifted one and two places along:

series 1   17.3, 19.2, 16.7, 12.4, 18.3, 15.6, 11.1, 12.5, ...

          ✓   ✓   ✓   ✓   ✓   ✓   ✓

series 2   19.2, 16.7, 12.4, 18.3, 15.6, 11.1, 12.5, ...

          ✓   ✓   ✓   ✓   ✓   ✓

series 3   16.7, 12.4, 18.3, 15.6, 11.1, 12.5, ...

In this way you get a mathematical confection: a time-series of three-dimension observations, built out of the original time-series of one-dimensional observations. Just read successive columns of triples. So here the first of these fake observations is the triple (17.3, 19.2, 16.7), representing a point in three-dimensional space which, relative to a chosen origin, lies 17.3 units to the east, 19.2 north, and 16.7 up. The next is (19.2, 16.7, 12.4), and so on. As time evolves these triples move in space. Ruelle conjectured, and Takens proved, that the paths these triples trace is a topological approximation to the shape of the attractor (Figure 77). Packard was among the first to use this method in an experiment.

For an attractor with more dimensions, you need more of these displaced time-series, but the same general idea works. There is a computational method to reconstruct the topology of the attractor from a single time-series – and *it doesn't matter which observable you use* to do it.

In practice there are other considerations, to do with the efficiency of the method. Some observables are better than others, and the method

Figure 77   Reconstructing a strange attractor (here the Lorenz attractor) by the Ruelle–Takens method (compare Figure 54).

gets bells and whistles added to it. But the idea very neatly circumvents the need to identify any physical variables whatsoever in the mathematical theory!

Strange Chemistry

Chemical reactions can oscillate. The effect was first reported in 1921 by William Bray, in the decomposition of hydrogen peroxide into water and oxygen with an iodine catalyst. But chemists then believed – wrongly – that the laws of thermodynamics forbid oscillations. Instead of following up Bray's discovery, they concentrated on explaining it away, on the grounds that his experimental method must have been at fault.

That attitude set them back nearly forty years. In 1958 the Russian chemist B. P. Belousov observed periodic oscillations in the colour of a mixture of citric and sulphuric acid, potassium bromate, and a cerium salt. Ilya Prigogine had by then shown that far from thermodynamic equilibrium, the usual laws of thermodynamics don't hold, and people were more prepared to take the results seriously. In 1963 A. M. Zhabotinskii modified Belousov's recipe, using iron salts instead of cerium, getting a dramatic red-blue colour change. He showed that circular and spiral waves can form if the chemical mixture is spread in a thin layer. Today many oscillating chemical reactions are known; and dynamical effects more complex than periodicity are commonplace.

As a sample of recent work, I'll describe a paper published in 1983 by Swinney and his collaborators J.-C. Roux and Reuben Simoyi, in the journal *Physica*. It concerns not fluid turbulence, but chemical turbulence – chemical chaos – in the Belousov–Zhabotinskii reaction.

The experiments measured the way the concentration of bromide ion varies in time. The data were subjected to various kinds of mathematical analysis. They found its power spectrum, and thereby determined the component frequencies of oscillations. They reconstructed the corresponding dynamical attractors (Figure 78, left-hand picture) by forming a second 'fake' time-series. The typical geometry of a strange attractor is clearly visible. By plotting the variables every time the motion passed through the dashed line marked on the left-hand picture in Figure 78 they obtained a Poincaré mapping, shown in the right-hand figure. The points cluster near a humped curve, showing that the underlying dynamics, although chaotic, is really quite simple, and not unlike the logistic mapping.

The results are highly detailed and are consistent with all the known mathematical properties of strange attractors. In any case, the pictures are immediately convincing. They could have come off a computer-graphics

Figure 78   A strange attractor reconstructed from experimental data on chaotic chemical oscillations in the Belousov–Zhabotinskii reaction, and a Poincaré mapping for the Poincaré section marked by a dashed line. (Reproduced by permission of John Wiley & Sons Ltd., ©1986.)

terminal plotting some analogue of the Lorenz attractor. In fact they resemble very closely a variation on the Lorenz attractor proposed in 1976 by Otto Rössler (Figure 79).

Chaos *does* occur in nature. In fact, I find it amazing how much Nature seems to know about the mathematics of chaos. And presumably knew it long before the mathematicians did. Not only does the idea of chaotic dynamics work – it works far better than anyone could have hoped. Somehow, the very delicate effects predicted by continuum models of fluids – a model that we know must be *wrong* at the atomic level – survive the approximations involved in replacing a sea of atoms by an infinitely divisible continuum. It's easy to dismiss this as something obvious, but I think that's wishful thinking. We'd *like* it to be true – and in defiance of all experience, it is. *'Whatever can go wrong, will.'* But in this case, the celebrated law doesn't apply. There's a mystery here.

But not one that has to be resolved before we can take advantage of the wonderful miracle that *it works*.

Bashō Revisited

I began this chapter with a quotation from Bashō about the poetic fascination of liquid drops. It's fitting to end by evoking some of their mathematical beauty. A dripping tap usually calls for a plumber rather than an exclamation of admiration, but we've seen that there's more to a dripping tap than just water in the wrong place. It's chaos in microcosm.

Figure 79   The Rössler attractor.

Moreover, the chaotic dripping of a tap is a discrete dynamical system, easier both to observe and to analyse than a continuous one. Instead of a laser, a microphone suffices.

Let's consider the formation of droplets in more detail.

With a gentle flow of water, a tap normally drips in a regular rhythm. The water slowly builds up on the rim, forming a drop that bulges and bloats until surface tension can no longer sustain it against the pull of gravity. Its sides begin to shrink, forming a narrowing neck; then the drop detaches itself and the process starts over again. It's hardly surprising that the drops are repetitive and rhythmic.

But if the flow of water is a little greater, something more complicated can happen. As the drop forms, it also oscillates. It doesn't get a chance to settle down to a steady, slowly-growing state. As a result, the precise moment at which it detaches depends not just on how much water has entered the drop, but also on how fast it's moving in its oscillation. In these circumstances, the drops can be produced at irregular, aperiodic intervals.

There's a clear analogy. A fluid at low speeds flows smoothly, but at higher speeds it makes a transition to turbulence. At low speeds, drops form regularly, but at higher speeds, they become irregular. Might a similar mathematical mechanism control both phenomena?

It might not. Perhaps, when the flow becomes irregular, it's because random influences such as air currents affect the formation of the drops. Bashō has an example of this, too:

> Tonight, the wind blowing
> Through the Bashō tree,
> I hear the leaking rain
> Drop against a basin.

(The Bashō tree is a species of banana, and the poet was so attached to one that grew outside his house that he took it as his pen name.) The random motion of the leaves is here responsible for any irregularity, not the delicate dynamics of the formation of a single drop.

Deterministic chaos? Or randomness?

Robert Shaw, Packard, and colleagues at the University of California, Santa Cruz, tested this idea experimentally. They let a tap drip on to a microphone. The signals from the microphone were recorded so that each falling drop produced a well-defined blip.

The blips filter away much of the detailed dynamics. They don't show the motion of the droplet while it's growing: only the instant at which it detaches. They're like a series of discrete snapshots of the dynamics. In other words, they form something very much like a Poincaré mapping, which is also a series of snapshots. Mathematically, they can be treated the same way.

The Santa Cruz mathematicians had to process the experimental data to extract the dynamics. To do this, they measured the intervals of time between successive blips. This gave them a time-series of intervals, roughly 5,000 observations long. Then, just as described above, they used Takens's method of reconstruction. They formed two further 'fake' time-series, displacing the original by one and two units, and plotted the resulting sequence of 5,000 triples using a computer.

In this way they were able to reconstruct the topology of an attractor in the dynamics of the dripping tap (Figure 80). As they report in the December 1986 edition of *Scientific American*:

> The exciting result of the experiment was that chaotic attractors were indeed found in the nonperiodic regime of the dripping faucet. It could have been the case that the randomness of the drops was due to unseen influences, such as small vibrations or air currents. If that was so, there would be no particular relation between one interval and the next, and the plot of the data would have shown only a featureless blob. The fact that any structure at all appears in the plots shows that the randomness has a deterministic underpinning. Many of the data sets

Figure 80 The dripping tap experiment: (top left) *apparatus;* (top right) *part of a time-series;* (lower left) *a three-dimensional plot of the observed data;* (lower right) *a simple mathematical model.*

show the horseshoe-like shape that is the signature of the simple stretching and folding procedure.

A strange attractor is indeed responsible. In fact, the data resemble an attractor like Hénon's.

At higher flow rates, the experimental attractor becomes very complicated, and its structure is still not understood. Nor is any very direct link known between the physics of droplet formation and this empirical model. There's plenty left to do.

A trickling tap is only one kind of turbulence, and a rather special one, but chaos has since been found in many other turbulent flows. In 1989 Tom

Mullin of Oxford University hit the front page of the journal *Nature* with a beautiful experimental observation of a strange attractor in turbulent Taylor–Couette flow. This particular experiment used a short cylinder, and in order to impose a strong rotational forcing on the flow the end plates were made to rotate with the inner cylinder. The apparatus had to be maintained at a uniform temperature in a vibration-free environment, otherwise the delicate signature of chaos would have been smeared away. Finding chaos isn't easy. Theoretical computer analyses made by Mullin in collaboration with Andrew Cliffe at AERE Harwell, using a numerical package called ENTWIFE, had already suggested that this system was likely to generate chaos. The experimental data were reconstructed using the method of Ruelle and Takens, and out popped a very tidy strange attractor. In fact it was one that the mathematicians recognized: it had turned up earlier in some equations devised by Bill Langford at Guelph, associated with the so-called Silnikov bifurcation – a standard route to chaos. So on this occasion the experimental reconstruction of a chaotic attractor led to new information about the mathematical properties of the fluid flow, and it even offered clues about what was causing the chaos.

Thus the idea that chaotic dynamics of strange attractors is responsible for at least *some* turbulent phenomena is well established. But much of turbulence remains a mystery. Fully developed turbulence, if it involves strange attractors at all, may require attractors of enormous dimensions – a thousand, a million. At the moment we can say nothing worth knowing about these. Many turbulent effects seem to be caused by boundaries – the walls of pipes, for example – and strange attractor theories haven't yet been related to the influence of boundaries.

And we shouldn't be obsessed with chaos as the only likely explanation. Recently the Russian physicist V. P. Maslov has found evidence for a kind of non-uniqueness in the Navier–Stokes equations. The equations may not actually determine the flow in all details: for given initial conditions they may have more than one solution, at least in a certain approximate sense. Maslov says the effect 'can be described figuratively. In Pushkin's famous tale *The Priest and his Worker Balda*, Balda stirs the water with a rope, calling up demons. Thus, when he spins the rope fast enough, the demons start to rave in a nondeterministic manner, causing turbulence.'

Perhaps the invisible monster theory isn't so daft, after all.

10

Fig-trees and Feigenvalues

A fool sees not the same tree that a wise man sees.

William Blake, *Proverbs of Hell*

A new mathematical technique: chaos. An old problem: turbulence. A new tool, an old task: what could be more natural than to wield the tool and see whether it fits the task? And they did, and it did.

But science doesn't always move in the direction you most expect. The herd may be stampeding towards the distant horizon, but there are always a few mavericks perversely heading the opposite way. One of those mavericks was responsible for a fundamental breakthrough. But it was a breakthrough in mathematics, leading only later to serious pay-off in the theory of turbulence. It imported a new idea into mathematics from the physics of phase transitions – a powerful technique known as *renormalization*. This in turn showed that some features of chaos are *universal* – they don't depend upon the precise equations, only on the qualitative type of strange attractor that is present. And that made it possible to perform simple experiments to test for the presence of certain kinds of chaos. But to lead – rather tangentially – into all this, I want to pick up an earlier topic: the *Voyager* spacecraft.

Bottle in a Cosmic Ocean

The *Voyagers*' Grand Tour of the Solar System did not end at Uranus. Like their *Pioneer* predecessors, they will continue out into interstellar space. In 40,000 years' time they will come within a light year of the star AC +79 3888. Over millions of years they will drift through the galaxy, perhaps encountering other planetary systems.

On the off-chance that one of these might support intelligent life, the *Voyagers* carry with them a 12-inch gold-plated copper disc – a gramophone record (Figure 81). Encoded in its grooves are 115 photographs, from a diagram of continental drift to a supermarket, and a variety of sounds ranging from 'hello' in Akkadian to Beethoven's Fifth. 'The spacecraft will be encountered and the record played only if there are advanced spacefaring civilizations in interstellar space,' says Carl Sagan. 'But the launching of this bottle into the cosmic ocean says something very hopeful about life on this planet.' I can't decide whether I think this particular cosmic gesture is a heartwarming manifestation of the indomitable human spirit, a dangerous betrayal of our galactic coordinates to a potential enemy, or a pointless conceit. I do wonder what the aliens who find this treasure will make of it: in particular, the photograph of Jane Goodall and her chimpanzees might lead to some misconceptions. But it's too late now to go and get it back.

The third photograph on *Voyager*'s record consists of mathematical definitions. There's a long-standing human tradition that the best way to make contact with alien races is through mathematics – presumably because it appears to be a universal medium of thought. Carl Friedrich Gauss suggested that the diagram for Pythagoras's theorem might be drawn in the Sahara desert for Martians and the like to observe through their telescopes. Other schemes involve transmitting the sequence of primes, or the digits of π, on the assumption that no civilized and intelligent race could fail to recognize these, and hence the intelligence and civilization of the beings transmitting them.

Where I suspect these schemes fall down is through parochialism. I *think* that π is likely to remain important in terrestrial mathematics – but I wouldn't bet too hard on it surviving another 10,000 years as an object of fundamental importance, let alone a million. I have no idea what the green-tentacled mathematoids of the Greater Magellanic Cloud think of as fundamental knowledge. In James Blish's science fiction novel *A Clash of Cymbals* the mathematics of the dirigible planet He bears a superficial resemblance to the terrestrial, but there are pitfalls: 'Here, for instance, Retma was using the *d* which in Amalfi's experience was an increment in calculus, as simply an expression for a constant.' Be warned!

Suppose that in the summer of 1975 an astronomer had recorded what might or might not be a message from a source that might or might not be natural, a series of binary blips which when translated into decimal turned out to be the number 4.669201609 . . . repeated over and over again. The scientific world would have expressed some disappointment that the signal wasn't 3.141592653 . . . because it would have stretched the imagination to argue that π was just a coincidence. But might it be some other significant number? They would hunt through their tables of basic mathematical constants, such as the base e of natural logarithms, the golden number, Euler's constant, and the square root of two: no joy. In growing disappointment they

Figure 81   Technicians mounting the gramophone record on Voyager 2.

would dig out more recondite numbers, such as Catalan's constant or the volume of the smallest hyperbolic 3-manifold . . .

No, there's nothing significant about 4.669201609. The astronomers must have found a natural source, a periodic vibration of some distant neutron star, the radiation from a black hole.

However, had the same signal been received in 1976 . . .

Don't Perturb – Renormalize!

Mitchell Feigenbaum is a physicist. In the early 1970s he was working at the Los Alamos Laboratory. Some of his colleagues would have objected to that word – working – because nobody knew quite what it was that Feigenbaum was working *on*. Including Feigenbaum himself.

He was interested in nonlinear systems. At that time, the main methods for handling nonlinearity were the perturbation techniques of particle physics, especially things called Feynman diagrams – named after Richard Feynman, the Nobel-winning physicist who invented them. As a student, Feigenbaum had learned how to do such calculations, decided they were the wrong way to think about nonlinearity, and got bored with them.

A different area of physics deals with *phase transitions* – changes in state of matter, such as liquid turning to gas. The mathematics of phase transitions is also nonlinear. When Kenneth Wilson at Cornell came up with a new idea about phase transitions, a method known as *renormalization*, Feigenbaum fell in love with it. Wilson's method was based upon the idea of self-similarity, the tendency of identical mathematical structure to recur on many levels. Now the classical picture of turbulence involves just this structure: an endless cascade of ever-smaller vortices. As Lewis Richardson wrote, in an intentional parody of Jonathan Swift:

> Big whorls have little whorls
> Which feed on their velocity,
> And little whorls have lesser whorls,
> And so on to viscosity.

Feigenbaum was not alone in thinking that Wilson's renormalization method might apply to turbulence. The onset of turbulence, mathematically and physically, looks just like a phase transition; the only difference from the usual idea of phase transition is that turbulence is a transition in a flow-pattern rather than in the physical structure of a substance. So several physicists were working on this idea. However, the evidence that it might apply was slim, and even if it did, nobody could see exactly how.

Feigenbaum, like any sensible research scientist, made no attempt to beat his brains out on the full complexity of real turbulent flow. Instead, like Smale, he wondered what the general phenomena in nonlinear differential equations might be. He decided that the textbooks didn't contain anything very useful: it was going to be a bare-hands job. So he began with the simplest nonlinear equation he could think of – our old friend the logistic mapping.

The logistic mapping had already been studied by a number of people. The ecologist Robert May had worked on it in 1971 and used it as a vehicle to point out the curious nature of nonlinear population models. In the same year Nicholas Metropolis, Paul Stein, and Myron Stein had discovered that it was, if anything, even more complicated than anyone had imagined. Paul Stein warned Feigenbaum of this, and for a time the problem went on the back burner. If the *simplest* nonlinear mapping is virtually incomprehensible, what hope for *realistic* nonlinear dynamics?

In 1975, though, Feigenbaum attended a conference, and heard Smale talking about dynamical systems. Smale mentioned the logistic map, and its period-doubling cascade to chaos. He raised the possibility that something of real mathematical interest might be going on at the point where all the period-doublings accumulated – the place where the cascade stopped and the chaos started. Feigenbaum, inspired once more, took his problem off the back burner and turned up the gas.

The Advantages of *Not* Having a Computer

You'll remember that the logistic mapping has the form

$$x \rightarrow kx(1 - x)$$

where x lies between 0 and 1, and k is a parameter between 0 and 4. Of its many features, the one that concerns us is the period-doubling cascade, which I earlier dubbed the *fig-tree* in Feigenbaum's honour.

We've seen that the fig-tree occurs as the value of the parameter k is increased from 3 to about 3.58. For k between 0 and 3 there is a unique steady state. At k = a period-2 cycle appears; at k = 3.5 the period changes to 4; at k = 3.56 it doubles again to 8, and so on. The successive doublings accumulate faster and faster; and the picture of how the attractor varies with k is like a tree with infinitely many shorter and shorter boughs, branches, twigs, twiglets, . . . , splitting in two at each stage. Smale asked what happened at the very tips of the fig-tree's utmost twiglets, when k is about 3.57, and Feigenbaum looked for an answer.

His first step was routine: calculate the exact sequence of values of the parameter k at which the various doublings occur. Today you'd automatically reach for your desktop personal computer. In those days using a computer was a lengthy process, with jobs being submitted in batches on punched cards, and results appearing days later. If you made the slightest mistake, as was common, you might get little more than a single sheet of paper with – if you were lucky – a laconic error-message. So Feigenbaum used a Hewlett-Packard HP 65 programmable calculator instead.

This turned out to be a stroke of luck, because the calculator was so slow that its operator had time to think about the results as they emerged. Indeed, *before*. The calculation began with an approximation to the required number, and then improved it step by step. Now, the better the initial approximation, the less time the calculation took. So to save time – an important consideration when you're using a calculator – Feigenbaum started trying to guess roughly what the next number in the cascade might be. Soon he saw a pattern. The differences between successive numbers were in a constant ratio, each about four times as big as the next one. More exactly, the ratio was about 4.669.

A mathematician would call this *geometric convergence*, and probably think little more of it. But to a physicist, especially one with knowledge of phase transitions, constant ratios mean *scaling*. Features of the physics must be recurring on ever-smaller scales. Little whorls within big whorls – like turbulence. Within a given structure, there must be smaller copies of the same structure, their sizes being determined by the scaling factor.

Feigenbaum had discovered evidence that, at the utmost tips of the fig-tree, there must be some mathematical structure that remains the same when its size is changed by a scaling factor of 4.669. This structure is *the shape of the fig-tree* itself. The steady attractor forms the trunk. The period-2 attractors form two shorter boughs. From these sprout even shorter period-4 branches, then period-8 twigs, period-16 twiglets, and so on. The size-ratios of trunk to bough, bough to branch, branch to twig, twig to twiglet, get closer and closer to 4.669, the nearer you get to the top of the tree.

Indeed, if you break off a bough, you get an approximate copy of the entire fig-tree (Figure 82). The same holds if you break off a twiglet. The copy is smaller, and the sizes decrease in a scaling ratio that tends to 4.669. And the further along you go, the closer the similarity in shape becomes. This is *self-similarity*. It's what you need to apply Wilson's renormalization method. Feigenbaum still couldn't see how to go about it, but he knew he was on the right track.

Snakes and Bears

Metropolis, Stein, and Stein had found some intriguing patterns in the logistic mapping; and they'd found identical patterns in at least one other mapping, the *trigonometric mapping*

$$x \rightarrow k \sin(x)$$

Figure 82   Self-similarity in the fig-tree: in the ideal case each twiglet has the same shape as the original, but reduced in size.

Inspired by this, Feigenbaum repeated his calculations, but using the trigono-metric mapping. Again he found a period-doubling cascade (Figure 83). Again the convergence was geometric: the scaling ratio of the fig-tree's branches tended to a constant.

Figure 83   The trig-tree: a period-doubling cascade in the trigonometric mapping (compare Figure 65).

That wasn't really very surprising; after all, there's got to be *some* pattern to the numbers, and they have to shrink fast enough to cram infinitely many branches into a finite space. Constant scaling is probably the simplest way to achieve this.

But there *was* a surprise, all the same. The *value* of the scaling ratio.

It was 4.669 again.

This was astonishing. There seemed to be no good reason why the two mappings, with completely different formulas, should throw up the same number. But the calculator said they did.

Perhaps it was just coincidence. Maybe the numbers differed in the next decimal place. The easiest way to decide was to do the calculations more accurately – and *now* Feigenbaum felt it was time to learn to use the computer. 'Think first, compute later.' It's a motto that should be engraved on every scientist's computer terminal.

For the logistic mapping, Feigenbaum quickly found a more accurate value for the scaling ratio: 4.6692016090.

He repeated the calculation for the trigonometric mapping. To ten decimal places, the two numbers remained the same.

This couldn't possibly be a coincidence. But why on earth was it happening? Feigenbaum's bewilderment is captured in an analogy offered by James Gleick in his book *Chaos*:

Imagine that a prehistoric zoologist decides that some things are heavier than other things – they have some abstract quality that he calls *weight* – and he wants to investigate this idea scientifically. He has never actually measured weight, but he thinks he has some understanding of the idea. He looks at big snakes and little snakes, big bears and little bears, and he guesses that the weight of these animals might bear some relationship to their size. He builds a scale and starts weighing snakes. To his astonishment, every snake weighs the same. To his consternation, every bear weighs the same too. And to his further amazement, bears weigh the same as snakes. They all weigh 4.6692016090. Clearly *weight* is not what he supposed.

It was a puzzle all right. But now Feigenbaum had caught a glimpse of the pattern he was hunting, he was hot on its trail.

However, it was a different trail from the one he'd anticipated.

The traditional view of physics and applied mathematics is that the most important thing in the world is the equation that describes the system under investigation. To study the flow of water in a bath *write down the equations*. Then you can throw out the bathwater and concentrate on the mathematics. In the same way that a baby grows into a mature adult, everything you want will grow from the equation.

Feigenbaum had followed this time-honoured practice and thrown out the bathwater. Apparently the baby had gone with it. *The scaling ratio didn't depend on the equation*. Logistic or trigonometric, *it made no difference*.

He'd found a pattern, all right.

But it made no sense at all.

Renormalization

Renormalization was a well-established technique, so plenty of lines of attack were open. Feigenbaum tried them all. He circulated his results informally, and talked to a lot of people. Gradually light began to filter through the mathematical murk. By the time he was ready to publish his ideas, he had a fairly complete picture of what was going on. Wilson's renormalization method was indeed behind it all, as he'd guessed at the outset: not in its usual technical form, perhaps, but in its underlying philosophy. Feigenbaum wrote two papers, the first outlining the mathematical phenomena involved, and the second sketching the reasons why many different mappings all had the same scaling ratio. His reasoning was still some way short of rigorous proof, but it carried conviction and explained that the miracle was no miracle at all, but a logical consequence of the mathematical structure. The final pieces of the puzzle were supplied by Pierre Collet, Jean-Pierre Eckmann, and Oscar Lanford, who found rigorous proofs that Feigenbaum's scenario is correct.

The basic idea is very beautiful, and I'll try to describe it, but I must warn you that you'll get only a tiny fragment of the picture and you'll have to take a lot for granted.

I'll start with an analogy that conveys some idea of what renormalization does. Recall that a process or object is self-similar if you can pick out a small part, blow it up, and recreate something very closely resembling the whole. Like the windows of the logistic mapping. Or the way that, in a turbulent fluid, you can blow up a small vortex to get a bigger one. There's a scaling ratio here too: the amount of magnification you need.

If you select smaller and smaller pieces, and magnify them to full size, the resulting picture may *stabilize* in the sense that successive versions, at higher and higher magnification, start to look almost identical. If so you can pass to the limit, ending up with a kind of finite-sized picture of the infinitesimal geometry. This procedure is called *renormalizing* the system. It has the advantage that in the renormalized version, the self-similarity is exact, not just approximate. And any property of the original that depends only on this infinitesimal geometry can be read off from the finite geometry of the renormalized object.

So renormalization is a mathematical trick which functions rather like a microscope, zooming in on the self-similar structure, removing any approximations, and filtering out everything else.

I'll give you an analogy that captures the main mathematical features: the geometry of small pieces of large circles. A circle has *approximate* self-similarity. A small enough piece of a circle is a slightly bent smooth curve. If magnified, it doesn't change its shape very much: it remains a slightly bent smooth curve. The self-similarity is not exact, however. If you blow up a piece of circle, its curvature does in fact change, albeit by a small amount.

However, a straight line has exact self-similarity: if you take a small segment, and blow it up, you reproduce the original precisely.

What does a large circle look like to an ant? Approximately, it's straight. In the same way, the large sphere that we inhabit looks flat to us diminutive apes. What would an infinitely large circle look like to an infinitesimal ant? Presumably, *exactly* straight. But you've got to be careful with words like 'infinite' and 'infinitesimal'. How can we make rigorous sense of this kind of statement?

By renormalization. To renormalize the circle, select tinier and tinier arcs, blow them all up to the same length, and compare the results. What you'll see is a sequence of straighter and straighter curves, approaching a straight line *as a limit* (Figure 84). This limit captures the 'infinitesimal'

Figure 84   Renormalizing a circle reveals that 'infinitesimally' it is just a straight line.

flatness of the circle, and converts approximate self-similarity into exact self-similarity.

The straight line possesses a certain degree of universality, too. If you repeat the renormalization, but start with an ellipse, you again end up with a straight line. Indeed, the same goes for any smooth curve. No matter which smooth curve you start with, the process of renormalization turns it into a straight line. The line is a kind of 'universal attractor' for the renormalization procedure.

On the other hand, if you start with something that has a corner, and perform the renormalization so that the corner always stays in the picture, then the limiting curve will be two straight lines meeting at an angle. So the straight line is universal only for a suitable class of initial curves, the smooth ones.

The physicists who studied phase transitions had discovered a similar phenomenon of universality. Certain physical quantities, known as critical exponents, tend to take the same values, independently of the exact mathematical model. The reason is that upon renormalization, all of the various models look the same; and the critical exponents depend only on the renormalized model.

Feigenbaum's Mapping

Feigenbaum realized that he could apply the same trick to the fig-tree. The scaling ratio of the fig-tree is analogous to a critical exponent, so the universality observed in phase transitions must be responsible for the same scaling ratio always turning up in fig-trees, no matter what the mapping might be.

Recall that the fig-tree is a diagram that shows the successive creation of periodic cycles, of periods 1, 2, 4, 8, 16, . . . as the parameter k is varied.

The basic idea is that each successive doubling of the period happens by the identical mechanism. A period-2^n cycle becomes unstable and creates a period-2^{n+1} cycle. The way this happens is that each point of the 2^n cycle, pulls apart into two. If you peer myopically at the 2^{n+1} cycle, just after it has appeared, the pairs of points will blur and you'll just see the old 2^n cycle.

There's a mathematical trick that lets you select just one point of the 2^n cycle and watch how it splits in two. Now you're looking down a mathematical microscope at a tiny part of the interval between 0 and 1. Apart from the size of this tiny interval, the geometry by which the splitting occurs is almost identical. Indeed, if you make a photograph of what you see through your mathematical microscope, and blow it up to a standard size, the successive pictures at successive period-doublings look more and more alike. So, as the size of the period tends to infinity and you approach the very tip of the fig-tree, the successive photographs resemble more and more closely some limiting picture.

The analogy with renormalization is now clear. Mathematically, the procedure is identical. Which means that we can push the analogy further, by asking what the limiting picture is – and which mapping it corresponds to.

First, we would expect a similar picture to hold good, whatever the original mapping might be – logistic, trigonometric, or anything else that has just one hump. The crucial observation is that the shape of the limiting picture is *the same* in all these cases – just as circles or ellipses both produce a straight line when they're renormalized.

To find the mapping that corresponds to the universal limiting picture, we start by observing that – in the 'circle analogy' – the straight line has a special property which makes it stand out as being unusual. It stays exactly the same when renormalized – precisely self-similar. Suppose you could find one very special mapping, for which the process of blowing up a microscope photograph doesn't just approach a limiting form, but reproduces the identical form at each step. That is, its bifurcation diagram is the archetypal Figure 82, *precisely* self-similar. Then this special mapping ought to play an analogous role to that played by the straight line. Let's call it the *Feigenbaum mapping*. Like the straight line, it's *unchanged* by the renormalization process. Feigenbaum argued that, no matter what mapping you started with, it would approach this special mapping upon renormalization – just as an arbitrary smooth curve approaches the straight line.

For the Feigenbaum mapping, the fact that the successive twiglets of the fig-tree shrink at a constant rate is an immediate consequence of its definition: the constant rate is the ratio by which successive photographs have to be blown up to repeat the identical shape. Which you can calculate, once and for all, by working out what the Feigenbaum mapping must look like. *You get only one number because there's only one Feigenbaum mapping.* As it happens, that number is 4.6692016090. Well, it's got to be *something*.

For any other mapping, however, the successive blow-ups don't just resemble each other more closely – they resemble the picture for the Feigenbaum mapping. So their fig-trees shrink at a rate that gets closer and closer to that for the Feigenbaum mapping. Thus, in the limit, you find *the same* ratio 4.6692016090.

Ellipses and circles both renormalize to a straight line, and the straight line can be characterized by a property of self-similarity. In the same way, logistic and trigonometric mappings both renormalize to the Feigenbaum mapping, and this can also be characterized by a property of self-similarity.

Feigenbaum had a more sophisticated image of the whole process. There's a kind of dynamical system going on, but it uses *mappings*, not numbers. It's a discrete system, and at each step a given mapping is transformed into the next by looking down a microscope and taking a blow-up photograph. The Feigenbaum mapping is an attractor for that system. No matter what mapping you start with – logistic, trigonometric, whatever – the dynamic takes it ever closer to the Feigenbaum mapping. So those of its properties that depend only on the late stages of the blow-up procedure come to resemble more and more closely those of the Feigenbaum mapping.

In particular you only get one *number*, 4.6692016090, because there's only one *attractor* in this dynamical system of mappings. Feigenbaum's magic number, like π, is a natural and fundamental mathematical constant. If the green-tentacled mathematoids of the Greater Magellanic Clouds are heavily into dynamics, they might think it's just the gadget to send a signal to the rest of the intelligent universe.

Feigenvalues

The physicists studying phase transitions had got used to this kind of universality, the tendency of different mathematical models to lead to the same numerical answers. They couldn't prove it was always so, but they learned to exploit it anyway. If lots of models gave the same answer, you could choose whichever one made the calculations easiest.

Once the mathematicians had sorted out the fine print, Feigenbaum was in a rather better position. He could *prove* that different mappings always give the same scaling ratio. In the rigorous version of his theory, the number 4.669 arises as an *eigenvalue* of an operator. An eigenvalue measures the amount of stretching in a special direction. So pun-loving physicists call 4.669 a *Feigenvalue*.

The universality of Feigenvalues is relative, not absolute. The scaling ratio

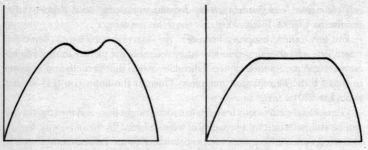

Figure 85   Mappings with multiple or flat humps lead to different Feigenvalues.

is always 4.669 for a one-humped mapping whose hump resembles a parabola. For multiple humps, or markedly different shapes of hump – flat humps, say, or pointed ones – the scaling ratio is different (Figure 85). But then there's another whole range of mappings that have the new number as their scaling ratio. The enormously varied range of mappings gets lumped together into universality classes; and within each class the scaling ratio is always the same.

And there are other numbers, too, associated with the dynamics of nonlinear mappings, which are similarly universal. For example, the scaling ratio 4.669 for a fig-tree is the ratio of the lengths of the twigs – or rather, their horizontal shadow, as measured by the parameter k. If you look at the picture of a fig-tree you'll notice that the smaller branches don't open out as quickly as the larger ones. The rate at which the branches open also scales by a universal constant, but a different one: this time it's 2.5029078750957.

Two-edged Sword

All this has some important, though curious, implications for experimental tests of chaotic models. Many real systems appear to undergo a series of period-doublings – we'll encounter one in a moment. A natural model is then a dynamical system along the lines of the logistic mapping. By Feigenbaum's universality result, two experimental predictions can be made. The size-ratio of the intervals between successive doublings should be about 4.669; the rate at which the branches open should produce a ratio of about 2.502.

To test these predictions is perfectly straightforward. You make the observations and calculate the numbers. The theory is thus falsifiable: if it's wrong, you'll get numbers like 6.221 and 0.074 instead. It would be a remarkable coincidence indeed to get numbers close to the predicted Feigenvalues, unless the theory were basically correct.

Notice that we're getting quantitative, numerical predictions from a purely qualitative model. Miraculous!

But the miracle bears a price-tag. The very phenomenon that makes it possible – universality – also means that the experiments can't distinguish between the mappings in that universality class. The trigonometric mapping will pass the same experimental test as the logistic mapping; and so will any mapping with a single hump.

Assuming that the experiment does indeed produce numbers around 4.669 and 2.502, as predicted, we can therefore be pretty sure that the behaviour is indeed describable by a discrete dynamical system, which is climbing the fig-tree to chaos. Precisely *which* system, though, is another matter. The test is for a whole *class* of equations, not a particular one.

This procedure is very different from the traditional view of experiment, in which the predictions of a single specific model equation are compared with reality.

Another thing. Suppose you don't know that the Feigenvalue 4.669 is universal. Suppose the logistic mapping is the only single-humped mapping you know about. By repeating the calculations that led Feigenbaum to his theory, you'd be able to extract the number 4.669 from that specific equation. When experiment confirms it, you'll imagine you've obtained strong evidence in favour of the logistic mapping model. You wouldn't realize that any other model of similar qualitative type will also give exactly the same number!

For instance, imagine that in another incarnation in an alternative universe you've been reborn as Galileo. You develop a theory that an object thrown in the air describes a parabola. You calculate a few numbers, do the experiment, and get good agreement. You conclude, reasonably enough, that the parabola is correct. It never occurs to you that perhaps lots of other theories will give the same numbers, that maybe you haven't confirmed a parabola at all.

So Feigenbaum's discovery of universality is a two-edged sword. It makes it relatively easy to test a particular class of chaotic models by experiment; but it doesn't distinguish between the different models in that class.

One way out is to look for more sensitive tests: the detailed structure of the period-doubling sequence, say, and not just its behaviour close to the accumulation point, the outermost fringe of the fig-tree.

But another would be to accept that, for some purposes (such as, what *is* the behaviour close to the outermost fringe of the fig-tree?) the distinction between the different models doesn't matter. Not just qualitatively, but quantitatively. For those purposes, *any theory in the same universality class will do just as well.*

Turbulent Daydreams

As I've said, Feigenbaum started out thinking about turbulence, which involves a very specific and complicated system of equations for fluid motion, the Navier–Stokes equations. Instead of studying those, though, he worked on a simplified, artificial equation, the logistic mapping. He thereby made a priceless discovery: universality. He would never have got that out of complicated equations – even though they would have been more realistic. Sometimes realism can be a pain.

The mathematical techniques for studying differential equations include an extensive repertoire of tricks to turn one problem into an apparently different one. There are changes of variable, which alter the form of the equations without changing the underlying model; there are reduction methods that eliminate a lot of variables from consideration altogether. It's technically hard to apply these to the Navier–Stokes equations; but you can daydream about the possibility without having to face up to the snags.

Now, it's asking an awful lot to expect any type of mathematical analysis to extract a genuine logistic mapping from the Navier–Stokes equations. Without universality, an analysis of the logistic mapping would be just a single example, probably not characteristic of anything else: an isolated, useless calculation. But the essence of chaos, stretch-and-fold, is much more likely to be seen in turbulent flows. And the *simplest* systems that exhibit stretch-and-fold are those qualitatively similar to the logistic mapping. By universality, any of these will yield the same Feigenvalues.

Conclusion: if it so happens that, buried away in the Navier–Stokes equations, there's a mathematical process involving a single-humped mapping, then a period-doubling cascade with scaling ratio 4.669 is going to occur. *You don't have to extract the mapping to make this prediction.* All you need to do is guess that such a mapping might be in there somewhere. It's a prediction with all the advantages of theft over honest toil.

But it's a perfectly good prediction, whatever its ethical status: you can go away, do an experiment, and see if the number 4.669 turns up. And if it does, you've obtained strong evidence that there is indeed some chaotic dynamics, a strange attractor, a single-humped mapping, buried in the Navier–Stokes equations. Experimental evidence in favour of a mathematical theorem!

Bizarre.

Having thought along these lines, Feigenbaum proposed a new route to turbulence. Not the accumulation of additional, independent wobbles favoured by Hopf and Landau. Not the one, two, and three-makes-chaos route proposed by Ruelle and Takens. Instead, a route of accumulating period-doublings, happening faster and faster, climbing the fig-tree, to pluck the fruit of chaos from the tips of its branches.

It was all very speculative. Not many people were willing to buy the leap from a simple, artificial mapping to the time-honoured partial differential equation for a fluid. Nor did they like the total absence of physical content in Feigenbaum's theory. 'It's a chaotic dynamical system, but it doesn't matter much which, and even if the experiment works, it won't help you work out which.' Discomfiting.

But Feigenbaum's leap was not a speculative leap to an unwarranted conclusion. It was a leap of imagination to an entirely warranted one. It had a better chance of being right than most people were willing to concede.

The first evidence that there was more to the idea than met the eye came from computer calculations with more realistic fluid equations. Sometimes these could be persuaded to produce a period-doubling cascade. When they did, the scaling ratio could be calculated. Numbers close to 4.669 had a remarkable habit of cropping up.

What was missing was a real experiment, on a real fluid, giving that self-same number.

By another quirk of fate, the groping in the dark that is so characteristic of basic science, such an experiment had already been performed. But neither Feigenbaum, nor the experimentalist who had already tested his theory, knew their results had anything in common.

Cold and Silence

Liquid helium is one of the weirdest substances on earth. Cooled to a temperature close to absolute zero, it can climb out of a beaker of its own accord, a macroscopic manifestation of quantum uncertainty. In quantum theory, you can't be absolutely sure that the liquid is in the beaker at all; helium escapes through this quantum loophole. You won't find liquid helium lying around in the street: not because it gets away, but because it has to be made, in a laboratory, using sophisticated methods to produce very low temperatures, around −270°C. But to Albert Libchaber, a low-temperature physicist, liquid helium was an old friend. And what made it worth all the effort of producing it was that it was very pure, and experiments with liquid helium were very 'clean'.

At room temperatures, the atoms of a liquid are all rushing randomly about, propelled by thermal agitation. What looks like a motionless beaker of water is actually, on the atomic scale, a seething ocean racked by tempests. These thermal effects produce 'noise' – not in the usual sense, but in the sense of random perturbations of experimental data. If you want to approach atomic-scale accuracy, the noise spoils your results. It's like trying to listen to a nightingale in the middle of a cocktail party: the signal is swamped by the surrounding random chatter.

To get rid of the noise, you have to shut up the revellers, that is, slow down the thermal agitation. In other words, lower the temperature. The lowest possible temperature is absolute zero, $-273°C$. At absolute zero, there's no thermal noise at all: even the atoms are frozen.

But you can't do experiments on fluid flow with a fluid that has frozen solid. You need a substance that remains fluid even at temperatures close to absolute zero. Helium is unique in this regard. It selects itself as the only substance on which these highly accurate experiments can be performed. So, willy-nilly, if you want fluid flows plus high precision, you're a low-temperature physicist and you're working with liquid helium. If you're interested in classical, rather than quantum effects, helium is very accommodating: it behaves like a classical fluid once it warms up to a comfortable $-269°C$.

Helium Rolls

In 1977, like many researchers in physics and fluid dynamics, Libchaber was interested in convection. He knew that other experimentalists, such as Swinney and Gollub, had cast doubt on the Hopf–Landau theory of accumulating wobbles. If Libchaber had been a painter he would have painted miniatures; if he'd been an engineer he would have made Swiss watches. He liked things that were small, neat, and precise; it was exactly those attributes that had attracted him to low-temperature physics in the first place. Where others might study fluid flow in a wind-tunnel 30 metres long, Libchaber's apparatus could be carried in your pocket. And the quantity of fluid that he set flowing was no bigger than a grain of sand.

Libchaber had fashioned a tiny, precise, stainless steel chamber; and filled it with liquid helium. At a few selected places the temperature of the fluid was monitored using diminutive devices made from sapphire. There was room for only one or two. Then the bottom of the cell was heated a fraction of a degree warmer than the top of the cell, creating a temperature inversion that set the warmer fluid rising, the colder falling. Within his tiny cell, Libchaber could create almost noise-free convective flows, and measure their behaviour.

Long ago, the great physicist Lord Rayleigh worked out what happens in such a cell when convection first sets in. The fluid forms cylindrical rolls, lying like felled tree-trunks, stacked side by side, with neighbouring rolls rotating in alternate directions (Figure 86). This is also the system that Lorenz studied, but Libchaber was working with a real system, not an approximate mathematical model.

Figure 86   Parallel rolls in fluid convection: neighbouring rolls rotate in opposite directions.

So small and carefully designed was Libchaber's cell that there was room for precisely two rolls. If the bottom of the cell was made slightly warmer, then the rolls developed wobbles, shimmying like a pair of belly-dancers, keeping step with each other. Again this was in accordance with classical expectations.

What happened next was not. A new oscillation appeared, but unlike the wobbles of Hopf–Landau, its period was not independent of the existing wobbles. Instead, it oscillated with exactly twice the previous period. And just above that temperature, oscillations at four, eight, and perhaps sixteen times the period could dimly be discerned. Beyond that, the earsplitting thermal noise of atoms at −267°C swamped the measurements.

Libchaber detected these oscillations using power spectra (Figure 87) computed from his observations. Recall that spikes in a power spectrum represent strong component frequencies. Running along the sequence of pictures you see first a single spike, then several spaced closer together, and so on. The spacing halves each time, which means that the period – inversely proportional to the frequency – doubles each time. The final power spectrum shows broad bands, indicative of chaos.

Libchaber had found a period-doubling sequence. A physical fig-tree. To him, it was a new and puzzling phenomenon.

By 1979, however, he had made contact with Feigenbaum. Now he knew what his observations were, and what to do with them. Like a magician, Feigenbaum had extracted from the top-hat of chaos the rabbit of universality. Libchaber had merely to calculate the scaling ratio for his sequence of period-doublings, and see whether it was close to 4.669.

It was. Close enough to make further, more accurate experiments worth doing.

Within a few more years, a whole range of experiments, made by scientists

Figure 87   Experimental evidence for a fig-tree in convection. Each new set of spikes occurs half-way between the previous ones, indicating period-doubling. A sequence of four period-doublings can be seen, followed by chaos.

the world over, had confirmed Feigenbaum's prediction to the hilt. Not just in turbulent fluids, but in all kinds of physical systems: electronic, optical, even biological. The people, the places, the culture – and now the time too was ripe. It all came together.

Chaos was a fact, not a theory.

Big science from little fig-trees grows.

11

The Texture of Reality

> We have
> a map of the universe
> for microbes,
> we have
> a map of a microbe
> for the universe.
>
> Miroslav Holub, *Wings*

A farmer, it is said, hired a team of scientists to advise him on improving his dairy production. (Stop me if you've heard this one before.) After six months' work they prepared their report. The farmer began to read, only to encounter the opening sentence: 'Consider a spherical cow.'

There's an important message behind this hoary tale. The shapes that we see in nature, and the traditional geometric shapes of mathematics, do not always bear much resemblance to one another.

Sometimes they do. In 1610 Galileo said that the language of nature is mathematics, and 'its characters are triangles, circles, and other geometrical figures'. His dramatic successes in dynamics explain his viewpoint. But by 1726 Jonathan Swift was ridiculing such a philosophy in Gulliver's *Voyage to Laputa*: 'If they would praise the beauty of a woman, or any other animal, they describe it by rhombs, circles, parallelograms, ellipses, and other geometrical terms.'

These quotations find a modern echo in a much-quoted statement of Benoît Mandelbrot in *The Fractal Geometry of Nature*: 'Clouds are not spheres, mountains are not cones, coastlines are not circles, and bark is not smooth, nor does lightning travel in a straight line.' Unlike his predecessors, Mandelbrot – then an IBM Research Fellow at Yorktown Heights but who has since held positions at Yale and Harvard – decided to do something about it. Between the late 1950s and early 1970s he evolved a new type of mathematics, capable of describing and analysing the structured irregularity of the natural world, and coined a name for the new geometric forms involved: *fractals*.

During the 1970s, when both were in their infancy, chaos and fractals appeared unrelated. But they are mathematical cousins. Both grapple with the structure of irregularity. In both, geometric imagination is paramount. But in chaos, the geometry is subservient to the dynamics, whereas in fractals the geometry dominates. Fractals present us with a new language in which to describe the shape of chaos.

Scales of Measurement

Physical phenomena usually take place on some characteristic scale of measurement. The structure of the universe, for example, is best described on length scales of millions of light years. The structure of a microbe involves scales closer to a micrometre. I suspect that this interplay between phenomena and scales of measurement is really an artefact of the limitations of the human mind, rather than a genuine truth about nature. Our minds just can't grasp something as big as the universe on a level of fine detail. So we dissect it up into large-scale structures, like galactic superclusters, and then dissect these into their component galactic clusters, and dissect those into galaxies, and galaxies into individual stars, and so on. Nature, in contrast, operates on all scales simultaneously. Be that as it may, our attempts to understand nature necessarily introduce scales of measurement that to us seem 'natural'.

This approach works well for phenomena that involve only a small range of scales. It works less well for phenomena for which a large range of scales is essential. For example the mechanism of phase transitions, where a mass of billions of atoms suddenly changes its gross physical characteristics, tends to spread itself across a rather large range of scales, mixing up the microscopic and the macroscopic. This is one reason why the mathematics of phase transitions has proved very difficult.

One of the newer techniques for dealing with this kind of problem has just made an entrance: renormalization. As we've seen, this is a method for finding the limiting infinitesimal structure of a self-similar object or process, by repeatedly magnifying smaller and smaller parts of the whole. Self-similar objects, by definition, don't have characteristic length scales: they look much the same on many different scales of measurement.

The orthodox shapes of geometry – triangles, circles, spheres, cylinders – lose their structure when magnified. We've seen how a circle becomes a featureless straight line when viewed on a large enough scale. People who think the Earth is flat do so because that's the way it looks to a tiny human. Mandelbrot invented the term 'fractal' to describe a very different type of

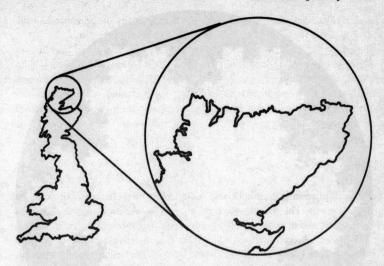

Figure 88   Fractal structure of a coastline: when magnified, new bays and promontories appear, and it continues to resemble a realistic coastline.

geometrical object: one that continues to exhibit detailed structure over a large range of scales. Indeed, an ideal mathematical fractal has structure on an infinite range of scales.

Snowflakes and Coastlines

A coastline is a good example of a naturally occurring fractal (Figure 88). Maps of coastlines, drawn on different scales, all show a similar distribution of bays and headlands. Each bay has its own smaller bays and headlands; so do these; and so on. The same general structure is visible in the magnificent sweep of the Gulf of Mexico, the Baie de la Seine, the Pendower Coves near Land's End, the gap between two rocks on the foreshore at Acapulco, or even the individual indentations of a single rock. Swift's doggerel, which inspired Richardson's parody quoted earlier, is a cliché within the fractal fraternity, but so apt that it can't be left out:

> So, Nat'ralists observe, a flea
> Hath smaller fleas that on him prey,
> And these have smaller fleas to bite 'em,
> And so proceed *ad infinitum.*

Figure 89   A mathematical fractal, the snowflake curve.

A mathematical curve with the same general features is the 'snowflake curve' of Helge von Koch, dating from 1904 (Figure 89). Here the bays and headlands are successively diminishing equilateral triangles. You wouldn't model a natural coastline by a Koch snowflake, because nature doesn't sculpt coastlines from equilateral triangles. But the snowflake curve does capture one important feature of coastlines very well, namely their *scaling behaviour*. Not only do both the natural and mathematical fractal have structure on all scales: they have – within reason – the *same* structure on all scales.

A tiny piece of coastline, magnified ten times, still looks like a coastline; the same goes for a segment of the snowflake curve. We've met this idea already: *self-similarity*. In the one case the similarity is only statistical: the average proportions of bays and headlands remains the same under scaling, although their precise arrangement may change. In the other, it's mathematically exact.

Not all natural objects scale in this manner: for example, Swift's flea. A flea can jump a metre or so. If it were to be magnified a thousand times, becoming as big as an elephant, it would not be able to jump a thousand metres. On the contrary, its legs would break under its own weight. Fleas have a natural length scale: coastlines don't.

One-and-a-quarter Dimensions

'Qualitative,' said the great physicist Ernest Rutherford, 'is poor quantitative.' But to make quantitative measurements of all the individual details of a fractal is almost impossible. Fortunately, a numerical measure of the degree of roughness of a fractal is readily available. Originally it was known as Hausdorff–Besicovitch dimension, after Felix Hausdorff and A. S. Besicovitch, two mathematicians who invented and developed it. Nowadays it's usually referred to as *fractal dimension*.

We're used to the idea that a line is one-dimensional, a plane two-dimensional, a solid three-dimensional. But in the world of fractals, dimension acquires a broader meaning, and need not be a whole number. The fractal dimension of a coastline is usually between 1.15 and 1.25, and that of the snowflake curve is close to 1.26. Thus coastlines and Koch snowflakes are equally rough.

At first, this idea may seem bizarre. How can it make sense to say that something has one and a quarter dimensions? But the snowflake is obviously more crinkly – better at filling up space – than a smooth curve, which has dimension one. And it is less good at filling up space than a surface, of dimension two. A dimension somewhere between 1 and 2 makes good sense. The Hausdorff–Besicovitch dimension is defined to capture this idea, while agreeing with the usual dimension on the usual spaces. Its precise definition is complicated and wouldn't be very illuminating, but the basic idea is to define the 'd-dimensional volume' of a shape for arbitrary (non-integer) d. Then the Hausdorff–Besicovitch dimension of the shape is the value of d for which the d-dimensional volume changes from infinity to zero.

Every shape has a specific value of d at which the d-dimensional volume makes such a switch. For the Cantor set, for example, it can be proved that d is log 2/log 3, which is roughly 0.6309; and for the snowflake it is log 4/log 3 = 1.2619.

Koch's snowflake, and Hausdorff–Besicovitch dimension, were invented to show the limitations of mathematics. Their inventors would have laughed if it had been suggested that their artificial concoctions had any bearing on the natural world. But Mother Nature knew better.

'Avoid Geometry'

The young Benoît Mandelbrot wanted to become a mathematician. His uncle, Szolem Mandelbrojt, already was one. And he had some sound advice for his nephew. 'Avoid geometry.' The mathematical fashions of his uncle's time placed great value on rigorous analysis and very little on visual imagery. The uncle recommended the young man to study and emulate a piece of mathematical research that captured the approach perfectly, a 300-page article by the French mathematician Gaston Julia on complex analysis – the calculus of $\sqrt{-1}$. Julia showed that simple mappings of the complex numbers could give rise to monstrously complicated shapes. A rival of Julia's, Pierre Fatou, worked on the same questions at much the same time; and between the two of them they polished off the whole area. At least, so it seemed in the 1940s. Julia and Fatou drew only very crude diagrams of their shapes: Mandelbrot was unimpressed. Like many youngsters before and after him, he ignored the advice of his elders.

In 1958 he joined the staff of IBM, working on a variety of apparently unrelated problems: word-frequencies in linguistics, error-bursts in the transmission of messages, turbulence, galaxy clusters, fluctuations of the stockmarket, the level of the river Nile . . . But by the early 1960s he began to realize that all of his work was somehow interrelated: it was about the geometric structure of irregular phenomena.

Mandelbrot encapsulated his ideas in a single word – 'fractal' – in 1975. He used it in the title of a remarkable book, *The Fractal Geometry of Nature*, published the same year. The book is highly geometric, in the pictorial sense, crammed with vivid and beautiful computer graphics (Figure 90). So much for Uncle Szolem.

The descriptive power of fractals was immediately evident. 'Fractal forgeries' – artificial computer-generated representations – of mountains, coastlines, lunar landscapes, and even music, bear an uncanny resemblance to the real thing. But could the theory of fractals surpass mere description, and acquire a deeper, more operational significance for science? Could it be used to predict new phenomena, and broaden our understanding of nature? Or was it merely descriptive?

And what was its proper place in mathematics?

In the mid-1970s the theory of chaos was known only to a few specialists. Mandelbrot's book doesn't mention chaotic dynamics as such. But it contains many topics that have a direct bearing on chaos, such as fluid turbulence and the large-scale structure of the universe. And perhaps the most basic fractal of them all, the Cantor set, is exactly the object that shows up in the geometry of strange attractors.

Figure 90   The Menger sponge, a fractal of dimension log 20 / log 3 = 2.7268.

Nowadays many things are much clearer. In particular, the geometric distinction between smooth forms such as circles and spheres – that is, manifolds – and rough forms, such as fractals, turns out to be precisely the distinction between the familiar attractors of classical mathematics, and the strange attractors of chaos. Indeed, it's now customary to define a strange attractor to be one that is fractal.

Moreover, the fractal dimension – that weird fractional number, invented by Hausdorff and Besicovitch but neglected by applied scientists until resurrected, polished, and exploited by Mandelbrot – turns out to be a key property of the attractor, governing various quantitative features of the dynamics.

So today, fractals appear in science in two different ways. They may occur as the primary object, a descriptive tool for studying irregular processes and forms. Or they may be a mathematical deduction from an underlying chaotic dynamic. To show the differences, and the scope of the concepts, let's take a look at both kinds of fractal modelling.

Silicon Valley

Many of the direct applications of fractals are to the physics of surfaces. Surfaces are the places where interesting things happen. Look out of the window: the regal complexity that we call life holds court inside a thin skin at the surface of the Earth. Surfaces are the boundaries between competing regimes, the places where alien worlds make contact with each other. The topography of surfaces is significant throughout science. When antibodies bind to a virus, or enzymes to a DNA macromolecule, they do so because of some affinity for the particular shape of surface involved. The surface of the polio virus (Figure 91) is fractal, and this affects the way that different chemical molecules interact with it. Chemical catalysts, so important for industry, function by causing reactions to occur at surfaces. Metallurgists worry about the form of fracture surfaces, while geologists do the same about mountain ranges. The same morphology may occur on many scales: scanning tunnelling microscopic photographs of the surface of silicon look not unlike the Grand Canyon.

Other kinds of topography are also important. Ores are seldom distributed uniformly in rocks. Clay has a highly complex structure of loosely packed molecular layers, and an apparently solid piece of ground can suddenly become a sea of mud if this molecular house of cards collapses, as occurred in a Mexican earthquake a few years back. The ultimate fate of the universe depends on the distribution of matter within it.

In 1980 Harvey Stapleton investigated the magnetic properties of iron-bearing protein molecules. If a crystal is placed in a magnetic field, which is then removed, it loses its magnetization in a characteristic fashion. This 'relaxation rate' can be quantified, and for crystals is always equal to 3. This is because a crystal is a three-dimensional object. But for proteins, Stapleton obtained values such as 1.7. He showed that this could be explained by their geometry. A typical protein molecule is folded and crumpled in a very irregular way. The crumpling resembles a fractal, and the number 1.7 can be explained as its fractal dimension.

More recently Douglas Rees and Mitchell Lewis have shown that protein surfaces – for example, that of haemoglobin, which transports oxygen in the blood – are fractal. Using computer analysis of X-ray diffraction data, they found that protein surfaces have a fractal dimension around 2.4. This suggests that the surfaces are very rough – in fact much like a crumpled paper ball, whose fractal dimension is about 2.5. Rees and Lewis also found that some regions of a protein's surface are smoother – this is, have smaller fractal dimension – than others. Like Velcro, proteins stick together best where their surfaces are roughest. Smooth regions seem to be active sites for enzymes,

Figure 91   Computer-generated model of the surface of a polio virus, showing its rough irregular structure: a fractal model is more appropriate than a smooth surface. (Arthur J. Olsen, Research Institute, Scripps Clinic, La Jolla, CA, © 1987.)

which bind more loosely to the protein. So fractal geometry allows biologists to quantify the surface structure of important biological molecules, and relate it to their function.

Aggregation and Percolation

We used to live in a village, and we had a fireplace that could burn the felled corpses of beetle-stricken elm-trees. We still own a sweep's brush: it was cheaper to buy a brush than to hire a sweep. I never enjoyed sweeping the chimney, because I always had visions of a cascade of soot going all over the furniture.

Figure 92   DLA cluster of particles, grown on a computer. (Reprinted by permission from Nature, *322, p. 791, © Macmillan Magazines Ltd.)*

Soot gets everywhere because it's soft and crumbly. It's soft and crumbly because it consists of a loosely knit aggregation of carbon particles. Similar processes occur in the electrolytic deposition of metals (electroplating) and in corrosion. In 1983 T. A. Witten and Leonard Sander devised an influential model of such processes, known as Diffusion Limited Aggregation, or DLA for short. In DLA, single particles diffuse randomly until they collide with the growing aggregate, and then stick at the collision site (Figure 92). Computer simulations of this process on a flat surface produce loose branching structures, like very irregular ferns, with fractal dimension 1.7. Similar processes in three-dimensional space lead to fractal clusters of dimension about 2.5.

When gold is deposited on a surface it at first beads in clusters, like water left in a bath after taking a shower, or dew on a spider's web. The growth of these clusters corresponds well to the DLA model. Gold colloid deposited on flat surfaces produces clusters with dimension about 1.75, close to the simulated value. There's also an interesting fractal phase transition in the deposition of gold. As more and more gold is added, the branching clusters start to join up, until at a sharply defined critical state they all join together into a single mass. This *percolation transition* is of considerable importance, and versions of it occur in many different physical systems. Percolation itself can also be modelled using fractals.

How Oil and Water Don't Mix

A very similar branching process, which has been studied much longer, is known as *viscous fingering*, a topic of some importance to the oil industry (Figure 93). In order to extract oil from a well, water is pumped in under pressure. Since oil and water don't mix, the oil is pushed out through production wells. However, the manner in which the water flows through the oil is surprisingly complicated, and the amount of oil extracted is not as great as one would wish. A better understanding of the process holds out the hope of more efficient production.

Figure 93   Viscous fingering of oil pumped into water. (Reprinted by permission from Nature, *321, p. 668, © Macmillan Magazines Ltd.)*

The standard experimental set-up for studying this problem is known as a *Hele-Shaw cell*: two flat glass plates between which is a thin sandwich of oil. Water is fed in through a hole in the middle of one plate. At first it spreads in a circular disc, but if the oil/water interface becomes too straight, it becomes unstable and develops bumps, which grow into 'fingers' that penetrate the oil in a star-like pattern. These fingers repeatedly undergo the same kind of instability, causing them to split at their tips when they become too wide. The result is a repeated branching growth, not unlike that of a developing plant. According to the experiments of J. Nittman, H. Eugene Stanley, and colleagues, the dimension is about 1.7. This is remarkably close to that for DLA, and there is now growing evidence that the two processes are mathematically related.

In practice oil does not occur in large free spaces, but mixed with particles of rock or sand. Jens Feder and others have investigated viscous fingering in a porous medium, finding that the fractal dimension is reduced to about 1.62. This means that pumping water is less effective when the oil is dispersed in porous rock strata. This kind of mathematical analysis should help oil companies to extract the precious liquid more efficiently.

The Universe and Everything

'When a young man in my laboratory uses the word "universe",' said Rutherford, 'I tell him it is time for him to leave.' But the Great Question of Life, The Universe, and Everything, has a fatal charm. Fractologists are not immune.

Astronomers used to think that on large scales the structure of the universe was the same everywhere – a homogeneous, evenly stirred mix of galaxies and vacuum. In fact, this belief gave rise to a paradox. In 1826 Wilhelm Olbers remarked that since both the diameter of a star and its light output diminish proportionately as its distance increases, the night sky should be uniformly bright, which it manifestly is not. Proposed resolutions of this paradox usually concentrate on mechanisms that screen distant starlight, such as dust clouds between the galaxies. According to a recent proposal the night sky looks the way it does because the universe has not been in existence for an infinite period of time, so much of the distant light has yet to reach us. If we wait long enough, this theory asserts, then Olbers will be proved right. He's just ahead of his time by a few billion years.

In the 1960s, Mandelbrot made a different proposal. The structure of the universe can be homogeneous, he argued, without this implying a *uniform* distribution of matter – provided the distribution is fractal. The final resolution of Olbers's Paradox remains unclear, but the universe does indeed have a complex structure that resembles a fractal much more closely than it does anything uniform (Figure 94).

The position of a galaxy can be measured very accurately, but in order to plot three-dimensional maps of galaxy distribution, its distance must also be estimated. The standard method is to exploit an empirical hypothesis known as *Hubble's Law*, proposed in 1929 by Edwin Hubble, an American astronomer. Astronomers can measure the different colours of light emitted by a star or a galaxy, thereby obtaining its *spectrum*. Hubble's Law says that the more distant a galaxy is, the further its spectrum shifts into the red. This is exactly the same Doppler effect that lets physicists use lasers to measure fluid velocities: the idea is that the universe is expanding, so more distant galaxies are moving faster, hence the red shift.

New instruments and photographic emulsions have made it easier to

Figure 94   The distribution of galaxies within a thousand light years of the Earth. Is the distribution fractal?

measure the red shift of faint, distant galaxies, and a much more detailed picture of the universe is emerging. Galaxies are not uniformly distributed. Instead they form a sponge-like network with huge voids and twisted, spindly threads of galactic matter between. The distribution is clumpy on all scales, with a measured fractal dimension of 1.2.

Margaret Geller and John Huchra have used fractal models in a rather different way, to investigate the statistics of galaxy distribution. A number of factors, such as the obscuring of clusters by interstellar dust, distort the observations; and the problem is to develop techniques to take these into account. Geller and Huchra started with a simulated fractal model of galaxy distribution for which the 'true' positions are known to the investigator. Distorting effects can also be simulated. Then methods for removing the distortion are tested on the simulated data to see how well they reconstruct the original distribution.

A fractal distribution of matter in the universe is a source of embarrassment for a lot of cosmological theorists, because most models of the universe are based upon Einstein's general theory of relativity. All such models tacitly assume that – at least on a sufficiently large scale – matter is smoothly distributed. The reason is that general relativity uses differential equations to describe changes in the curvature of space-time. The curvature of space-time determines the distribution of mass, and anything differentiable must vary smoothly. However, around 1990 the best available observations of the distribution of very distant galaxies suggested that on scales of, say, a billion light years, the clumpiness of the universe might start to even out, and cosmological theorists began to breathe a little more easily.

Not any more. In 1987 a group of astronomers known as the 'seven samurai', headed by Alan Dressler, had discovered that the Milky Way and its neighbouring galaxies all seem to be rushing headlong in the same direction (relative to the mean motion of distant galaxies), towards the constellation Leo. They suggested that there must be some unimaginably huge conglomeration of matter that was pulling them all the same way, and dubbed this hypothetical superclump the 'Great Attractor'. Tod Lauer of the National Optical Astronomy Observatories in Tucson and Marc Postman of the Space Telescope Science Institute in Baltimore carried out what they jocularly called a 'sanity check' to make sure that the Great Attractor really exists. They studied a group of galaxies thirty times as large as those observed by the seven samurai, and to their surprise they found that the whole lot were moving at about 700 kilometres per second towards the constellation Virgo – next door to Leo. They concluded that not only does the Great Attractor exist, but that it is part of an even huger clump of matter. Even on billion-light-year scales, the universe is clumpy.

Cosmologists admit that they simply do not understand how structures so large can have formed during the 15-billion-year lifetime of the universe. That's not long enough for gravitational instabilities to create such big irregularities, and the cosmic background radiation indicates that the early universe, soon after the Big Bang, was too smooth to generate the irregularities directly. As Geller says: 'It is a tough, tough problem, much harder than people realized when I was starting out. Answers are not just around the corner.'

Fractal Forgeries

One of the earliest 'applications' of fractals was computer graphics (Figure 95). To store in a computer the exact data needed to reconstruct the cratered surface of the Moon requires absolutely vast amounts of memory: reasonable enough for a catalogue of lunar geography, but pointless if the purpose is to produce a convincing background for a TV science-fiction drama. The answer is 'fractal forgeries', which mimic the desired forms without worrying about precise details.

In fact fractals and computers are a marriage made in heaven. One of the most powerful techniques in programming is *recursion*, whereby a procedure is broken down into a sequence of repetitions of itself. (Example: to build a brick wall, lay one course of bricks, then build a brick wall on top of it. The procedure 'build a brick wall' is defined in terms of itself. In practice you must also specify when the procedure stops. In this case it should stop when the wall is sufficiently high.) Fractals also break up into copies of themselves: they are recursive geometry. For fractals, unlike walls, the recursive process goes on forever.

Some years ago, Loren Carpenter made a computer movie of a flight over a fractal landscape, and was hired by Pixar, the computer graphics division of Lucasfilms. Fractals have been used in the movie *Star Trek II: The Wrath of Khan*, for the landscape of the Genesis planet; and in *Return of the Jedi* to create the geography of the moons of Endor and the outlines of the Death Star. Peter Oppenheimer has used fractal branching processes on a computer to produce abstract works of art (Figure 96) and lifelike and stylish trees and plants (Figure 97). Richard Voss, who started the whole field, continues to be active: one triumph of his is the computer generation of *convincing* clouds.

Clouds and Rain

Talking of clouds . . . Shaun Lovejoy has analysed genuine clouds using data from the *Geosat* satellite, with the remarkable conclusion that not only are clouds fractal, but they have the same fractal dimension over seven orders of

Figure 95   Fractal forgery by Richard Voss: Planetrise over Labelgraph Hill.

Figure 96 The Kiss *(New York Institute of Technology (Peter Oppenheimer)).*

Figure 97 *Fractal forgery of a tree (New York Institute of Technology (Peter Oppenheimer)).*

magnitude (Figure 98). This degree of uniformity is almost unprecedented in natural phenomena, and it means that clouds have no natural length scale. This is a surprise. The atmosphere is about 10 km high, and clouds are a convective phenomenon, so one would expect a distinguished length scale of around 10 km to make itself evident. It still may, but it doesn't show up in the *shapes* of clouds.

Lovejoy has also studied rainfall, finding that the boundaries of areas of rain are fractal. Moreover, rain tends to fall in irregular bursts, and the variations over short and long timescales are similar, so the temporal structure of rain is also fractal. Harold Hastings has made similar analyses of acid rain, aiming to improve forecasting of the stresses to which an ecosystem may be

Figure 98   Shaun Lovejoy's data on the scaling properties of clouds show a constant fractal dimension (represented by the constant slope of the line) over a surprisingly wide range of scales. The graph plots of the logarithm of the area of a cloud patch against the logarithm of its perimeter. (Solid squares show satellite data, open squares show radar data.)

subjected. He is also hoping to identify good indicator species, which could act as 'early warning devices' for acid rain damage.

Sisters under Their Skins

Fractals are novel in so many ways that it is easy to make the mistake of seeing them as a totally new world, isolated from existing mathematics. That this is not true is shown by the increasing contact between fractals and chaotic dynamics. One place where fractals and chaos come together in the study of turbulent flow. We've seen that the classical approach to turbulence, due to Lewis Richardson in 1922, is to see it as a cascade, in which the energy of fluid motion is progressively passed to smaller and smaller vortices. Such a process is clearly fractal.

As we've also seen, turbulence is an attractive topic for the devotees of chaotic dynamics. Like the Colonel's Lady and Judy O'Grady, these two theories of turbulence are 'sisters under their skins'. Strange attractors *are* fractals. The same complexity of structure that lets fractals model the irregular geometry of the natural world is what leads to random behaviour in deterministic dynamics. Itamar Procaccia has made extensive studies of the connections between fractals and turbulence, including turbulent diffusion, with applications to Lovejoy's observations of cloud shapes mentioned earlier. I've already described how Harry Swinney and his group reconstructed strange attractors from experimental data on turbulent convection. They've also computed their fractal dimension, to confirm that the attractors really are strange and to quantify the strangeness.

In 1986 K. R. Sreenivasan and C. Meneveau published a spectacular experimental study of turbulence from the fractal viewpoint. They looked at turbulent jets, surrounded by still fluid. The surface of the jet is known to have a very complicated structure. They asked whether the jet's surface is a self-similar fractal, and if so, what its fractal dimension is. Their experiments show that the answer is 'yes'. For a turbulent layer developing on a flat plate, the measured dimension is 1.37. This suggests that for a flow in a three-dimensional fluid the turbulent/non-turbulent interface should have dimension one higher, about 2.37. 'The overwhelming conclusion of this work,' they say in summary, 'is that several aspects of turbulence can be roughly described by fractals, and their fractal dimensions can be measured.' However, they warn that much more work is needed before the statement 'turbulence is fractal' can be asserted without qualification. A similar warning must be voiced for the strange attractor theories: they work best at the *onset* of turbulence, and may not be so useful for fully developed turbulence.

The Gingerbread Man

There are many ironies in the history of science. A striking one is that the work of Fatou and Julia, which put the young Mandelbrot off doing pure mathematics because of its lack of geometric content, has re-emerged as a central application of fractals to mainstream mathematics, widely hailed for its outstanding pictorial beauty. I need hardly say that Mandelbrot himself is responsible for this twist of fate.

Gaston Julia, a student of Poincaré, studied the iteration of mappings of the complex plane. Today you can't even write such a sentence down without immediately leaping to a conclusion: 'Aha! Discrete dynamics!' But in Julia's day the idea that the iteration of a mapping had anything to do with dynamics was unheard of. Dynamics was continuous; iteration was discrete – as alike as syrup and sand.

A complex number is a number of the form $z = x + y\sqrt{-1}$, where x and y are ordinary real numbers. The word 'complex' is used in the sense of 'having several components' rather than 'complicated': two real numbers x and y correspond to a single complex number z. But we know that two real coordinates define a point in the plane. Thus, just as we visualize real numbers as being spread out along a number line, we can speak of the complex numbers as living in the *complex plane*. Complex numbers have their own arithmetic, algebra, and analysis; they are among the most important and beautiful ideas in the whole of mathematics. They rely for their existence on an act of purely mathematical imagination: to agree that -1 is allowed to have a square root, and enlarge the number concept to embrace this enormous surmise.

Julia's theory is about complex mappings, for example $z \rightarrow z^2 + c$ where c is a constant. With a little harmless mathematical juggling, this can be thought of as the complex analogue of the logistic mapping. The idea is to fix a value of c, and ask what happens to any given initial value z as this formula is iterated.

At the coarsest level, there's one major distinction to be observed. Some starting values z move rapidly off to infinity; the rest do not. Imagine taking a paintbrush and painting the points of the complex plane. If they move off to infinity under iteration of the mapping, paint them black; if not, paint them white. You're delineating the *basin of attraction* of the point at infinity. The *Julia set* is its boundary.

As Julia and Fatou observed, the resulting shapes can be incredibly complicated. With modern computers, we can draw them with ease: they are also incredibly beautiful. Shapes like seahorses and rabbits, stardust and pinwheels, an endless variety (Figure 99).

To keep our ideas straight, I'm going to employ an analogy between the

Figure 99   Julia sets: a simple idea leads to intricate beauty and endless variety.

complex mapping $z \rightarrow z^2 + c$ and our old friend the logistic mapping $x \rightarrow kx(1 - x)$. Then x and z play similar roles, as do k and c. Each c has its own Julia set; this is analogous to the fact that each k has its own attractor. (Here I've bent the analogy so much that it's close to breaking-point. The 'filled in' Julia set is the basin of attraction of the point at infinity, the set of initial conditions that move towards it under iteration. The Julia set is the boundary of this basin. The attractor itself is just the point at infinity. Bear with me, it makes life easier if we ignore this distinction.)

For the logistic mapping, we invented a picture that conveys not just what the attractor for a given k is, but how it changes with k. This is the bifurcation diagram, and it led us to a wonderful discovery, the fig-tree. There's a similar object that gives an overview of how the Julia set for a given c changes as c ranges over the complex plane; but instead of the fig-tree we obtain the *gingerbread man*. More properly, it's called the *Mandelbrot set* (Figure 100). But we'll shortly see that it looks much like a gingerbread man, with a dumpy body and a round head; and 'Mandelbrot' is 'almond bread', making the pun irresistible. (That's the second Germanic pun. I promise there are no more.)

The variety of shapes for Julia sets is vast. We focus on a single, crude but distinctive feature. Some Julia sets are all in one piece; some fall apart. That is, they're either connected, or disconnected. The disconnected ones look like hundreds of specks of dust; the connected ones look like curves, or intricate designs.

To construct the gingerbread man, take your paintbrush again. Pick a

Figure 100   The Mandelbrot set or 'gingerbread man'. Mandelbrot's original picture is reproduced top left.

point c in the complex plane. Iterate the mapping $z \rightarrow z^2 + c$ for all possible z, to find the Julia set for c. See whether or not that's connected. If it is, paint c black. If not, paint it white. Do this for every c.

The result, remarkable for its intricate and curious geometry, and a total surprise, is the gingerbread man. Another way to obtain it is, for each c, to iterate the mapping $z \rightarrow z^2 + c$ starting at c = 0. If the iterates do not escape to infinity, colour c black: if they do, colour it white.

The best way to grasp the intricacy and beauty of the gingerbread man's structure is to beg, borrow, steal, or (I recommend) buy *The Beauty of Fractals* by Heinz-Otto Peitgen and Peter Richter. This is a unique object, the first mathematical coffee-table book in the world. But its striking pictures are not computer simulations of psychedelic art: they're snapshots of a deep, natural, and wonderful object, the gingerbread man. It's rightly been described as the most complex mathematical shape ever invented. (Not that that's stopped people inventing even more complex ones.) Yet you can persuade a computer to draw it with perhaps ten lines of program code. It puts the word 'complexity' in a new light.

The most startling feature of the Mandelbrot set is the way it retains its highly complicated structure if you zoom in on it at ever higher levels of magnification (Figure 101). Such a journey into the gingerbread man is an experience not to be missed; but you need a very fast computer to make the journey in speed and comfort. Each new level of detail reveals new and ever-surprising structures. Whirlpools, scrolls, seahorses, lumps, sprouts, burgeoning cacti, thin snakes, coils, insect-like blobs, zigzag lightning.

And every so often, buried deep within the gingerbread man, perhaps a millionth of the size (Figure 102), you can find . . .

Tiny gingerbread men.

Complete in every detail, including having their own sub-gingerbread men. Just as the bifurcation set of the logistic mapping has windows containing perfect replicas of itself, so does the gingerbread man.

Big fleas, little fleas . . .

Big gingerbread men, little gingerbread men.

This self-similarity of the Mandelbrot set is just one of its remarkable features. Here's another. Choose a point c on the edge of the Mandelbrot set and renormalize its shape near c by magnifying ever tinier pieces nearby to an ever greater extent. What shape do you get?

The Julia set corresponding to that value of c.

Inside the Mandelbrot set are *all possible Julia sets*, each on an infinitesimal scale, merging comfortably into each other, and each sits precisely upon its own value of the constant c. This is currently still a conjecture, but it has been proved in many cases and it is so pretty that it *must* be true.

This is only the beginning of the tale. A whole new subject, complex

Figure 101 Zooming in on the Mandelbrot set . . .

dynamics, is coming into being. ('Complex' is used here in the sense of 'complex number', not 'complicated'. It *is* complicated, though. But beautiful with it.) Among its applications are the methods whereby numerical analysis solves equations by successive approximations. For what is a successive approximation but the iteration of some mapping? It's an old idea, it goes back to Sir Isaac Newton or earlier. But fractals and chaos have breathed new life into the ancient bones.

Figure 102   . . . to reveal a tiny sub-gingerbread man, perfect in every detail.

The Fractal Cow

From the simplicity of the snowflake to the complexity of the Mandelbrot set: a natural mathematical progression, but a vastly different perspective.

Koch's snowflake curve interests mathematicians because it has infinite length but encloses finite area, and it's continuous but has no well-defined direction at any point. It, and many similar objects, were invented at the turn of the century to dramatize these and other pathologies. There were curves that filled space and curves that crossed themselves at every point. Voss says:

> Minds conceived of strange monsters without counterpart in nature. Having once discovered these monsters (and congratulated themselves on a creativity superior to nature), mathematicians banished the pathological beasts, mostly sight unseen, to a mathematical zoo. They could imagine no use for, nor interest in, their creations by natural scientists. Nature, however, was not so easily outdone.

These early concoctions of pure mathematicians, and various apparently unrelated investigations in other fields of science, fused together in the

Figure 103   Encoding a fern by fractal rules. The fern (left) is built from four transformed copies of itself (right); the fourth being the stem which is a fern squashed flat to form a line. In order to reconstruct the fern all you need are the rules.

imagination of Benoît Mandelbrot to create a new kind of mathematical model for nature. Almost all of the current work on fractals, theory and applications, can be traced back to his 1975 book. It was a spectacular exercise of mathematical imagination.

Fractals are having a commercial impact as well as a scientific one. Michael Barnsley discovered that simple fractal rules could generate highly intricate 'forgeries' of objects such as trees and ferns. The idea is that such objects are self-similar: parts of them have the same form as the whole, but shrunk (Figure 103). The list of mathematical transformations – fractal rules – that describe how to shrink the object to produce its various self-similar parts is a 'coded' representation of its shape. The object itself can be reconstructed from these fractal rules. Musing about the significance of this result, Barnsley realized that it might be a way to reduce the amount of data needed to specify pictures in a computer. If you describe Barnsley's fern in the usual way, as a list of the coordinates of those screen pixels that the fern occupies, then the amount of information required is huge. But the fractal rule that generates it is simple and requires only a few numbers. In the same way, one simple mathematical rule captures all of the complexity of the Mandelbrot set.

Of course a typical picture does not consist of precisely self-similar ferns, but when drawn on a computer screen or TV screen it does consist of hundreds of thousands of tiny elements – pixels. Barnsley reasoned that at the level of small clumps of pixels there is a great deal of self-similarity in any image. Clumps of, say, a dozen pixels might contain smaller clumps of three or four that resembled their parent clump. If so, the same idea of encoding the image by fractal rules would apply, and while the list of transformations would be a lot bigger than it is for a fern, it seemed likely that it would still be much more economical than describing the image pixel by pixel. Barnsley tried to get various companies to develop this idea, but nobody was interested in those silly overrated fractal things, so he set up his own company. It took some time to get the ideas working, but today he is the owner of a multimillion-dollar company, and his data-compression methods are widely used. For example, the widely distributed Encarta® encyclopaedia that Microsoft™ supplies on CD-ROM contains thousands of multicoloured graphic images – and every single one of them is compressed using Barnsley's method, because without it, the encyclopaedia wouldn't fit on to a single CD.

But now the theory of fractals is moving on. The early speculations have served their purpose, stimulating new and deeper investigations. As with any developing research field, the attractive early simplicities are running up against the stubborn complexities of nature. For example, the appropriate concept of fractal dimension seems to vary from one application to another. An important mathematical problem is to understand how all these various dimensions relate to each other. Much is still not understood.

The applicability of fractals is wide, but not universal. The fractal cow is not of necessity more realistic than the spherical one. It should also be said, in warning, that not all applications make essential use of the concept of a fractal. Work that twenty years ago would have been presented as a power law derived from a log–log plot of data, is now presented as the measurement of a fractal dimension. There are fashions in science, and they follow the buzzwords as well as the important breakthroughs.

But there's much more to fractals than just a few buzzwords. 'No one will be considered scientifically literate tomorrow who is not familiar with fractals,' says the physicist John Wheeler. Fractals reveal a new regime of nature susceptible to mathematical modelling. They open our eyes to patterns that might otherwise be considered formless. They raise new questions and provide new kinds of answers. 'Fractals,' says the science writer Jeanne McDermott, 'capture the texture of reality.'

12

Return to Hyperion

Blazing Hyperion on his orbed fire
Still sits, still snuffs the incense teeming up
From man to the Sun's God: yet unsecure.
For as upon the earth dire prodigies
Fright and perplex, so also shudders he:
Nor at dog's howl or gloom-bird's hated screech,
Or the familiar visitings of one
Upon the first toll of his passing bell:
But horrors, portioned to a giant nerve,
Make great Hyperion ache.

John Keats, *Hyperion*

There are two constant threads running through the history of dynamics. *Up there*, and *down here*. Thales, with his eyes on the heavens and his nose in the ditch. Galileo with the moons of Jupiter and a church lamp swinging in the draught. The grand unification of Newtonian gravitation: the planets, and the path of a cannonball. Astronomical observations provided a major spur towards the creation of statistics; but so did the heights of children. Poincaré first saw his homoclinic tangles in the mathematics of a dust-particle in the gravity-wells of Jupiter and Saturn, but Smale's understanding of them was indirectly inspired by a problem about radar.

Until now our discussion of chaos has largely been earthbound, indeed for the most part confined to the laboratory. But up there is chaos on the grandest of scales. The motion of satellites, the long-term behaviour of Pluto, the structure of the universe itself.

In the opening chapter I mentioned the strange behaviour of Hyperion, a satellite of Saturn: celestial chaos. Let's start with that.

Cosmic Potato

The most familiar shape for celestial bodies is a sphere, or more accurately
a spheroid: the Earth, for example, is flattened at its poles by a few per cent.
Hyperion, in contrast, is an ellipsoid whose principal axes (length, breadth,
and height, so to speak) are 190 km, 145 km, and 114 km. A cosmic potato
(Figure 104).

In accordance with the discoveries of Kepler and Newton, Hyperion's
orbit around Saturn is approximately elliptical. The extent to which an ellipse
deviates from circular form is measured by a quantity known as *eccentricity*:
Hyperion's orbit has an eccentricity of about 10 per cent. This is unusually
large for the planets and satellites of the Solar System, but it just means that
the orbit is a slightly flattened circle.

Hyperion's *position* in orbit is regular and predictable. You could tabulate
it decades ahead and be accurate to a fraction of a second in timing. What
makes Hyperion virtually unique among the moons and planets of our Sun
is its *attitude* in orbit: the directions in which its three axes point. Most planets
roll along like soccer balls on a flat pitch: Hyperion looks more like a rugby
football bouncing over a battlefield. If you could freeze the position of its
central point, and just watch the way it moves relative to that, you'd see it
swinging almost randomly in every possible direction.

Both its position in orbit, and its attitude, are determined by the identical
physical laws, the same mathematical equations. Its position corresponds to
a regular solution of those equations; but its attitude corresponds to an irregular
solution. Hyperion's tumbling is due not to random external influences, but
to dynamical chaos.

Why is Hyperion chaotic? For that matter, why are all the other bodies
regular? Is it the potato-like shape? Are all potatoes chaotic?

Not at all. The reasons are more subtle, more complicated, and much
more interesting. Hyperion's chaotic motion is a cosmic coincidence. At
various times in the history of the Solar System, other bodies have evolved
into, and back out of, a period of dynamical chaos. But it so happens that
Hyperion is undergoing this process at precisely the time when the human
race has become interested in it.

Vampire *Doppelgänger*

The motion of a rigid body is a classical problem first attacked by Euler. A
number of important principles emerge from Euler's analysis. First, we can
pretend that the centre of gravity of the body is fixed, and deal only with the

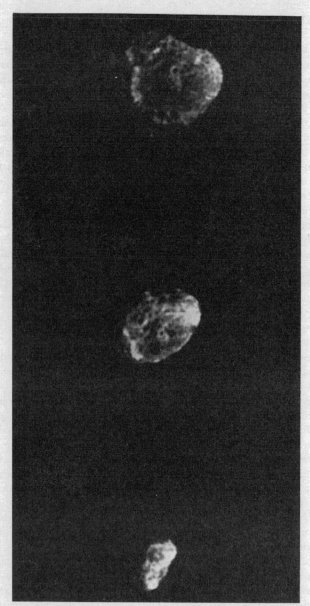

Figure 104   Three Voyager views of Saturn's unruly satellite Hyperion.

motion relative to that. Second: the shape of the body is largely irrelevant. What determines the motion is its axes of inertia. To every solid body, no matter how irregular its shape or density, there corresponds an *ellipsoid of inertia*. This is a ghostly companion, rigidly attached to the body but having no mass, and as its name suggests, it's ellipsoidal in shape. The length of each axis of the inertial ellipsoid is proportional to the inertia of the body when spun about that axis, so long axes correspond to greater inertia.

As the body moves, so does the ghost: it's a *Doppelgänger*. If the body rotates regularly, so does the ghost; if the body tumbles, the ghost tumbles too. But now comes a transmogrification. Let the ghost, vampire-like, absorb the material essence of the body, so that we have a solid ghost and an eerie spectral body, still attached to it like a living husk. How does the motion change? *Not at all*. The body and its ghost have the same inertial properties; therefore their motion is identical.

In other words, when thinking about the motion of solid bodies, you can confine your attention to uniform ellipsoids. The fact that Hyperion looks like a potato is irrelevant; but the fact that a potato's ghost ellipsoid has three *unequal* axes is crucial.

Despite all this, Euler was unable to solve the equations for a rigid body in full generality. Classical discoveries, *tours de force* of analysis, managed to solve a few very special cases, such as the motion of a circularly symmetric top. But the mathematicians found some general principles. For example, one of the simplest types of motion is when the body is spinning about one of its inertial axes. When is such a motion stable? Answer: when the axis is either the longest or the shortest, but not when it's the one in the middle.

You can easily check this experimentally. A book is an accessible example of a body with three unequal inertial axes. They run through the central point of the book, buried deep in its pages. The longest inertial axis runs from the middle of the back cover to the middle of the front cover. The shortest runs from the middle of the top edge to the middle of the bottom edge. The third, the in-between one, runs from the middle of the spine to the middle of the vertical edge (Figure 105).

You'll have noticed that the longest axis of inertia is the shortest axis of the book, and vice versa. That's not a mistake: inertia is greatest where the mass is moving fastest. If you spin the book at a given speed around its shortest physical axis, then points at the corners of the book and a long way from the axis and so move faster. On the other hand, if you spin it at the same speed about its longest physical axis, the points of the book are closer to the axis and hence move more slowly. Incidentally, my ghostly metaphor rather slid over this problem – the ghost isn't really the ellipsoid of inertia itself, but a uniform ellipsoidal body that *has the same ellipsoid of inertia* as the original body. It's fat where the inertial ellipsoid is thin, thin where it's fat.

Figure 105   A book, and its inertial ellipsoid. Note that the shortest axis (S) of the book corresponds to the longest axis (L) of the ellipsoid, and vice versa, and the two middle axes (M) correspond.

Anyway, get yourself a book. Something heavy (in the physical, not the metaphorical sense) is best: *War and Peace*, or a dictionary. Hold it between the palms of your hands with the title on the spine facing you, and spin it about its shortest axis. You'll have no trouble doing this. Now hold it by its top and bottom edges, with the spine horizontal, and spin it about its longest axis. Again, no trouble. Finally, however, hold it at the middle of the spine and the middle of its vertical edge, and try to flip it about its middle axis. You'll find that it refuses to spin properly and instead begins to twist and tumble. This is because rotations about the middle-sized axis are unstable. Next time you visit a stony beach, select a (roughly) ellipsoidal stone with unequal axes and try to spin it about the middle one. You'll find that it's very difficult to stop it wobbling.

Spin-orbit Geometry

In 1984 Jack Wisdom, an astronomer at the Massachusetts Institute of Technology, and his colleagues Stanton Peale and Francois Mignard, wrote a paper in the journal *Icarus* with the title 'The Chaotic Rotation of Hyperion'. In it, they *predicted* that Hyperion ought to be tumbling chaotically. Their analysis, in somewhat simplified form, goes like this.

Hyperion's orbit is an ellipse, but it changes slowly. Ignoring this, we can model the satellite's orbital motion by a fixed ellipse. The approximation is acceptable because Hyperion tumbles much faster than its orbit varies. Model Hyperion itself by a suitable ellipsoid, spinning about its longest axis; and assume that this axis is perpendicular to the plane of the orbit. We'll see why below. Its tumbling can then be captured by the *spin-orbit* geometry, described as follows. Because we've fixed the direction of the longest axis of inertia, just one further angle will tell us exactly what the attitude of Hyperion is. Namely, we need to know in which direction the smallest axis points. (The middle axis, at right angles to both of these, is thus determined too.) Call this the *spin angle*. One extra number will tell us whereabouts in its orbit Hyperion is: namely, the angle between its position and some fixed point of the orbit. For convenience the periapse – the nearest point to Saturn – is chosen as this fixed point, and the corresponding angle is the *orbit angle*, or 'true anomaly' in more conventional parlance. The gravitational pull that Saturn exerts on Hyperion depends on this orbit angle, which in turn depends on time; hence the gravity of Saturn can be represented as a time-varying gravitational field of a particular kind.

Anyway, you can write down the equations for all this, and you end up with a simplified mathematical model with three ingredients. One is the spin angle, the second is the rate of change of the spin angle, and the third is time – or equivalently, the orbit angle.

The gravitational pull of Saturn enters as a *time-varying* force. If instead Saturn's pull were constant in time, the equations would be a 'one degree of freedom system', and it would be possible to solve them explicitly. That would mean no chaos. But the time-variability of the gravitational term turns the equations into a 'one and a half degrees of freedom system', in which chaos is a viable option. (The extra half a degree of freedom is time. Conventionally, a Hamiltonian system with n variables has $n/2$ degrees of freedom, because variables usually come in position–momentum pairs. Here the spin angle and its rate of change form such a pair. Time doesn't, hence the curious terminology.)

The equation can be put on a computer, and solved numerically. To display the result, it's simplest to plot a Poincaré section (Figure 106). This shows the spin angle, and its rate of change, at regular intervals of time. From one interval to the next, the point representing the state of the satellite hops from one position in the Poincaré section to another. The Poincaré section doesn't show where the point goes in between, but we don't have to worry about that to distinguish regularity from chaos.

The Poincaré section shows a series of closed curves, plus a large X-shaped stippled region. The curves represent regular periodic or quasiperiodic motion: at each interval the representative point hops regularly round one of the closed curves. The stippled region represents chaotic motion: at successive intervals, the representative point hops around 'at random' *over the entire stippled region*. Hyperion might in principle be behaving in either of these ways. But the energy of its motion determines which, and chaos is the winner.

Each dot on the picture represents the state of Hyperion. The horizontal coordinate is its spin angle, and the vertical is the rate at which that angle is

Figure 106   A Poincaré section for Hyperion. The stippled region indicates chaotic motion: the points all belong to a single trajectory. The closed loops show regions of regular quasiperiodic motion.

changing. Between one orbital revolution and the next, the dot hops from one position on the picture to another. In quasiperiodic motion, this representative dot hops round and round one of the closed curves in steps of much the same size. All highly regular.

In chaotic motion, it hops rather randomly all over the stippled region which dominates most of the picture. This *whole region* is traced out by a single trajectory, if you watch for long enough.

The sharper-eyed among you will have noticed a second chaotic zone, much smaller, shaped like a thin *X* with long trailing arms, just above the big chaotic zone. This is a different chaotic motion, and because it covers such a small region, it's not a very important one.

Tidal Friction

Saturn's gravitational field also exerts a more subtle influence on Hyperion. Because the force of gravity falls off at increasing distance, Saturn attracts the near side of Hyperion more strongly than it attracts the far side. Among other things, this 'tidal' attraction causes Hyperion to revolve around its longest inertial axis, rather than its shortest, even though both would be stable in the absence of Saturn's gravity. Imagine Hyperion in a horizontal orbit, sitting at a tilt, with one side bulging towards Saturn, the other away. Suppose for definiteness that the side nearest Saturn is tilted below the horizontal. (In space, there's no distinction between 'up' and 'down', so to use that kind of descriptive language, you have to specify which is which.) Then Saturn pulls a little bit harder on the nearer bulge. This causes the satellite to tip upwards a trifle, bringing the spin axis more nearly vertical. Over long periods of time the effect of the tidal force is to make the spin axis perpendicular to the orbital plane. This is true of all bodies, not just Hyperion. The process takes a long time, however, because the difference in the forces on the two bulges is very very small; and other phenomena can militate against the effect.

Wisdom describes an experimental analogy: 'This process is nicely illustrated by tossing a partially filled bottle of Liquid Paper which is initially spinning about the longest axis.' Try it. (Make sure the top is firmly screwed on first.) Recall that the longest physical axis – the symmetry axis, along the middle of the bottle from top to bottom – is the shortest inertial axis. You'll find that the bottle refuses to spin about its longest axis (even though a totally full bottle will do so quite happily, just like *War and Peace* did). Instead it twists until it's rotating about the shortest physical axis – the longest inertial axis. The motion of the liquid within the bottle introduces a kind of tidal friction, similar in its effect to the tidal forces that Saturn imposes upon Hyperion.

This is why the model assumes that the spin axis is perpendicular to the orbital plane. There are other assumptions too. Fortunately, it can be shown by more careful analysis that once the system is in the chaotic zone, then the chaos persists even if the assumptions of the model are relaxed. However, in the chaotic zone, the orientation in which the spin axis is at right angles to the orbital plane – which we've just seen arises from tidal effects – becomes *unstable* in this more detailed model.

How It Came About

This complicates the picture, but we're finally in a position to see how Hyperion came to be in its current chaotic state.

In the distant past, Hyperion's rotational period ('day') was much faster than its orbital period ('year'). Its motion was then regular and quasiperiodic. Over aeons of time, the tidal forces from Saturn slowed its rotations and (as we saw with the Liquid Paper experiment) stood Hyperion up on end, so that its spin axis was the longest inertial axis and this was perpendicular to the orbital plane. However, once Hyperion lost enough energy to bring it into the chaotic zone, the work of millions of years was undone in only a few days. Within three or four orbits, Hyperion began to tumble in all directions.

This tumbling motion of Hyperion was predicted theoretically before *Voyager* got close enough to observe it. The initial *Voyager* pictures looked consistent with chaotic tumbling (as reported in the first edition of this book), but as more of them came in it started to look as though the motion was pretty much periodic. Disappointment turned to triumph when further pictures showed that the periodic motion had ceased and full-blooded chaos had taken over. There are many forms of chaos that resemble periodic motion for a while but then switch – somewhat erratically – to more complex motions. Hyperion happened to be in just such a state, and later this turned out to make good theoretical sense too. The moral is: don't give up too soon. God plays other games than dice.

Hyperion is the only satellite in the Solar System that, right now, is expected to tumble in this way. But the same analysis suggests that *all* irregularly shaped satellites must at some stage in their evolution pass through a period of chaotic tumbling. Phobos and Deimos, the two moons of Mars, must have tumbled chaotically at some time in the distant past. So must Neptune's smaller moon Nereid.

Resonance

There's more to the picture than just chaos. At the lower left and right, towards the edge of the chaotic zone, you can see an 'island' of regular motion. This corresponds to synchronous motion, in which Hyperion always turns the same face towards Saturn (as the moon does towards the Earth). Hyperion might eventually emerge from chaos into synchrony. Other islands can be seen too; for example, the small one at the top of the chaotic zone corresponds to Hyperion rotating twice in each orbital period. These are similar to the islands discovered by Hénon and Heiles, and Chirikov: see Chapter 8. The islands correspond to *resonances*, where different aspects of the motion occur with periods that are in some simple numerical relationship such as 1:1, 2:1, 3:2, and so on. Thus Titan, another satellite of Saturn, has an orbital period that is close to 4:3 resonance with that of Hyperion. Specifically, Hyperion takes 21.26 days to complete one orbit, and Titan takes 15.94. The ratio of these is 1.3337, convincingly close to the ratio 4:3.

In ordinary language, a resonance is a rich sound. In the imagery of Bashō:

> Breaking the silence
> Of an ancient pond
> A frog jumped into water –
> A deep resonance.

The mathematical idea of resonance is not unrelated – the rich sound heard by the poet is caused by the parts of a vibrating object (here the water) moving in step with each other.

Resonances are important in Hamiltonian dynamics, and often have chaos associated with them. To see how this occurs, we consider first the classical picture of a Hamiltonian system near a periodic orbit. In a Poincaré section, it consists just of a series of concentric circles (Figure 107). The central point represents the periodic orbit; each surrounding circle introduces a second period, independent of the first, on which the motion is quasiperiodic.

This picture has the virtue of simplicity – but the vice of being wrong. Indeed, for those who can read them, there are clear signs that something more delicate must be going on. I just said that the extra period is independent of the first. Actually, that's not always true. The second period varies continuously from one circle to the next. Consider the ratio for the two periods. If it's irrational, then the periods are independent. But if it's rational, they combine to give a genuinely periodic motion. They're in resonance. Now the rational numbers are *dense*: any interval, however small, contains a rational number. And the classical analysis fails near resonances, for the sort of reasons

Figure 107   Classical picture of a Poincaré section near a periodic trajectory. Each circle represents a quasiperiodic motion with two distinct periods.

that Poincaré discovered. So, near a dense set of classical circles, the resonant ones, you expect trouble.

Despite this worry, the classical picture does hold good for some very unusual systems, those said to be *integrable*. By a wicked irony of fate, integrable systems are the ones that can be solved explicitly by a formula. So the classical emphasis on explicit solutions leads us to study systems that are not truly representative. But by following Poincaré's and Birkhoff's lead, we can work out what the true, typical picture is.

It's almost unbelievably complex. An evocative description was given a few years ago by the physicist Michael Berry:

> Imagine winding cable starting from a 'primary' single loop of thin wire. Cover it with concentric sheaths of plastic. Interrupt this sheathing to find a secondary sheathed loop in a spiral about the primary, to close after a few windings. On this secondary loop are tertiary, quaternary, . . . windings. Continue the interrupted primary sheathings to surround the secondaries. Repeat *ad infinitum*. When this process has been completed, there will be some vacant spaces. Fill each with an infinitely long, tangled wire.

The plastic windings represent regular, quasiperiodic motion. Secondary sheathings are resonances; tertiary sheathings and the like are more delicate multiple resonances. The tangled wires are chaotic trajectories.

This isn't a computer experiment: it's a theorem. A very difficult theorem. Andrei Kolmogorov first realized that such a result might be true, and he sketched out a plan of attack. Vladimir Arnold, a student of Kolmogorov's who has become one of the world's leading mathematicians and an authority on dynamics, devised a rigorous proof, overcoming serious technical difficulties in the process. The results were then extended by Jürgen Moser. Their combined efforts led to what is now called the KAM theorem (short for Kolmogorov–Arnold–Moser). The regular quasiperiodic trajectories predicted by this theorem are known as KAM tori. Chirikov's work, described in Chapter 8, places limits on the existence of KAM tori and hence on the validity of the KAM theorem.

Ralph Abraham and Jerry Marsden, two American mathematicians who wrote one of the bibles of dynamical systems theory, call this picture the VAK (Figure 108). This stands for 'Vague Attractor of Kolmogorov', and is also the name of the goddess of vibration in the Rig-Veda, which is appropriate.

The VAK has the same disturbing quality that Mandelbrot's fractals and Feigenbaum's fig-tree have: self-similarity. The tiny islands within the VAK look, at first sight, like the classical picture of concentric loops. But that's just a result of the limitations of drawings. Each island has the same complexity,

Figure 108   What really happens near a typical periodic trajectory: the Vague Attractor of Kolmogorov. Only some of the classical quasiperiodic motions survive. Elsewhere, chaotic trajectories wind between resonance islands. (Ralph H. Abraham and Jerrold E. Marsden, Foundations of Mathematics, © 1978 Addison-Wesley Publishing Company Inc.)

indeed the same qualititive form, as the entire VAK itself. And while the simple, classical picture is atypical and misleading, the complicated self-similar structure of the VAK is not some mad mathematician's nightmare: it's what really happens.

Kirkwood Gaps and Hilda Clumps

Resonances feature prominently in another astronomical conundrum, the gaps in the asteroid belt. The largest asteroid, Ceres, was discovered in 1801 by Giuseppe Piazzi and is about 750 km in diameter. The smallest known asteroids are little more than huge rocks. There are tens of thousands of them. Most asteroids circle between the orbits of Mars and Jupiter, although a few come much closer to the Sun.

The asteroid orbits are not spread uniformly between Mars and Jupiter. Their radii tend to cluster around some values and stay away from others (Figure 109). Daniel Kirkwood, an American astronomer who called attention to this lack of uniformity in about 1860, also noticed where the most prominent gaps occur. If a body were to encircle the Sun in one of these Kirkwood gaps, then its orbital period would resonate with that of Jupiter. Conclusion: resonance with Jupiter somehow perturbs any bodies in such orbits, and causes some kind of instability which sweeps them away to distances at which resonance no longer occurs. The special role of Jupiter is to be expected: it's so massive in comparison to the other planets.

The gaps are obvious in recent data, especially at resonances 2:1, 3:1, 4:1, 5:2, and 7:2. On the other hand, at the 3:2 resonance, there is a *clump* of asteroids, the Hilda group.

Resonances have been used by astronomers as something of a catch-all.

Figure 109   The asteroids form clumps at some distances from the Sun, and leave gaps at others. Resonances with Jupiter appear to be responsible. The graph plots the proportion of asteroids against the ratio (period of Jupiter: period of asteroid).

The Moon always faces the Earth, a 1:1 resonance between its orbital and rotational periods. Mercury takes 88 days to revolve once round the Sun, and 59 days to rotate once on its axis. Two thirds of 88 is very close to 59, so Mercury's orbital and rotational periods are in a 2:3 resonance. These resonances are presumably stable (or else the bodies concerned would never have got into such a relationship). So the stability of the resonances 'explains' the observed phenomena.

But for the asteroids, apart from the Hilda group at 3:2, the explanation appears to be the *instability* of resonances! Clearly the only way to resolve this difficulty is to work out the mechanism of instability: presumably it is different in each context. Further, there must be something unusual about the 3:2 resonance, which explains the Hilda group.

Spikes of High Eccentricity

Until recently neither analytic nor numerical methods were capable of performing a sufficiently long-term analysis of any of these resonances. But advances in computing methods and the introduction of new theoretical principles is beginning to shed some light. The 3:1 resonance, in particular, is pretty well understood nowadays.

The computer calculations show that an asteroid, orbiting at a distance that would suffer 3:1 resonance with Jupiter, can follow a very irregular path. Indeed, the eccentricity of its orbit can change violently and almost at random (Figure 110). This is another astronomical example of dynamical chaos. The irregularities happen on a timescale that's short by cosmic standard, but long by computational standards: about 10,000 years.

To see what's *really* happening requires much larger timescales, covering millions of years. A typical chaotic trajectory then exhibits bursts of high eccentricity, interrupted by periods of low eccentricity, with occasional high-eccentricity 'spikes'. A body in such an orbit will follow a roughly circular path when the eccentricity is low, but a much longer and thinner elliptical path when the eccentricity is high.

A numerically computed Poincaré section (Figure 111) helps to explain these results. It shows two distinct chaotic bands. In one band eccentricity is low; in the other, it's high. Now the Poincaré section shows successive 'snapshots' of the motion of an orbiting body. The body hops around this picture, sometimes in one band, sometimes in the other. More detailed analysis shows that most of the time, the body circles round the low eccentricity band. Occasionally it gets trapped in the high-eccentricity band. Motion there is fairly quick, so it doesn't stay there long. So you see a brief high-eccentricity spike.

Figure 110   The eccentricity e *of the orbit of an asteroid in 3:1 resonance with Jupiter. The spikes correspond to sudden, large changes in eccentricity. Horizontal timescale* t *is in millions of years.*

Martian Sweeper

How does this account for the 3:1 Kirkwood gap?

In a burst or at a spike, the asteroid's eccentricity increases. It turns out than an asteroid whose orbit has eccentricity 0.3 or more becomes *Mars-crossing*; as you'd guess, this means that its orbit crosses that of Mars. Every time it does so, there's a chance that it will come sufficiently close to Mars for its orbit to be severely perturbed. An asteroid that crosses the orbit of Mars often enough will eventually come too close, and be hurled off into some totally different orbit.

Until it was realized that chaos could generate high eccentricity, Mars-crossing was not a plausible mechanism. Asteroids around the 3:1 Kirkwood gap were expected to stay well clear of Mars: there was no reason to expect a sudden change of eccentricity. But now there is such a reason, the mathematics of chaos. So it looks as if the 3:1 Kirkwood gap is there because Mars sweeps it clean, rather than being due to some action of Jupiter. What Jupiter

Figure 111   Poincaré section for an asteroid in 3:1 resonance with Jupiter has two distinct chaotic bands, explaining the spikes in eccentricity.

does is create the resonance that causes the asteroid to become a Mars-crosser; then Mars kicks it away into the cold and dark. Jupiter creates the opening; Mars scores.

A comparison between the boundary of this 3:1 chaotic zone, and the actual distribution of asteroids, is strikingly good (Figure 112). It turns out that some quasiperiodic trajectories, as well as the chaotic ones, lead to Mars-crossing: this has been taken into account when drawing the boundary.

The same mechanism that causes asteroids to be swept up by Mars, can also cause meteorites to reach the orbit of the Earth. The 3:1 resonanc with Jupiter thus appears to be responsible for transporting meteorites from the asteroid belt into Earth orbit, to burn up in our planet's atmosphere if they hit it. It would be hard to find a more dramatic example of the essential unity of the entire Solar System, or a better example of the ubiquity of chaos.

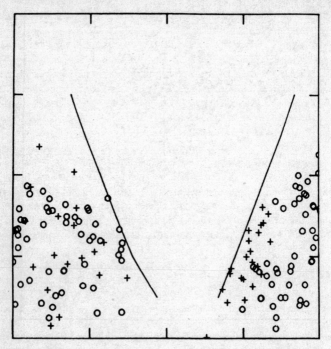

Figure 112 The boundary of the 3:1 chaotic zone: theory and observation. In theory the region between the two lines should contain no asteroids. Dots and crosses, representing observed values, confirm this prediction.

Digital Orrery

What of the Hilda Group, clumping together at a 3:2 resonance? What of other resonances?

Even a supercomputer has a hard time with long-timescale celestial mechanics. Wisdom, together with several colleagues, including James Applegate, Michael Douglas, Yekta Gürsel, and Gerald Sussman, decided there was only one answer. Build their own computer. It was to be a highly specialist machine, whose sole aim in life was to compute the behaviour of a small number of bodies moving in roughly circular orbits under Newtonian gravity. Custom-built machines can exploit loopholes that aren't available to off-the-shelf computers: if you've only got one job to do, you can find short cuts.

They called their custom-built celestial computer the *Digital Orrery*. An orrery is an old-fashioned mechanical device that simulates the orbital motion of the planets, using gears and cogs. Not unlike the Antikythera mechanism, except that the Greeks got there 2,000 years earlier.

The Digital Orrery was a parallel computer: it did several jobs at the same time. This was just one of the tricks used to speed it up. Whereas a conventional computer must fetch instructions from its memory at each stage of running a program, the Digital Orrery did a lot of its calculations in hardware. The mathematics was permanently wired in.

The Digital Orrery was used to study the motion of the Solar System for about 110 million years into the future and 100 million years into the past, a total span of more than 200 million years. Pluto has long puzzled astronomers. Its orbit is much more eccentric than those of the other planets, and much more tilted. Very recently Wisdom and Sussman have found yet another instance of Pluto's perversity: they used the Digital Orrery to show that (in their mathematical model) its orbit is chaotic. To demonstrate this they ran the Orrery twice, with Pluto in very slightly different initial positions. After several hundred million years the two predicted orbits place Pluto on opposite sides of the sun, a cosmic case of the butterfly effect.

The Digital Orrery found that for the 2:1 resonance (where a gap appears in the asteroid belt) there's a sizeable chaotic zone. But for the 3:2 resonance, the one at which the Hildas clump, *there is no chaotic zone.*

Mathematically, each resonance is a unique beast with its own special features. There's no reason why the 3:2 resonance should behave like the 3:1 or the 2:1, any more than the number 3/2 should be the same as 3 or 2. Apparently, the effective absence of chaos is one of the more striking aspects of the 3:2 resonance. Without chaos, there's no reason for orbits to gain eccentricity; without increased eccentricity, there's no reason why another planet, such as Mars, should sweep them up. The Hildas appear to have found an 'ecological niche' in the universe of chaos.

Saturn's rings are another happy hunting ground for resonances. The rings are full of gaps, which are either caused by resonances with one of Saturn's many moons, or are actually swept out by tiny moons (which is a 1:1 resonance since the moons are *in* the gaps). This sweeping-out process is chaotic, which has led to a way of predicting not just the presence of new, unobserved moons, but their masses too. Carl Murray (Queen Mary and Westfield College) and other theoretical astronomers have shown – first by computer simulation and then analytically – that the width of the gap swept out by a moon varies as the 2/7 power of the moon's mass. Apparently moonless gaps detected by the orbiting Hubble telescope can reasonably be expected to contain tiny moons, and the size of the gap can be used to calculate their expected masses. Several new moons, with the predicted masses, have recently been found by this method. So we see yet again that chaos can be used to make perfectly sensible predictions (which even work). Some people seem to think that because chaos is unpredictable then it cannot make predictions, so that chaos theory cannot be scientific. I do wish they would think about what words *mean* and not just look at how they are spelled.

King Oscar's Answer

When King Oscar asked about the long-term stability of the Solar System, he kicked off the whole subject of chaos. In consequence we have discovered that his question can be interpreted in many ways, and that affects not just the answer, but how sensible the question is. For example, the dynamics might be chaotic: we would now consider this to be stable, in a global sense, whereas King Oscar would probably have considered it to be unstable because of its irregularity. So instead of splitting hairs about concepts of stability, let's consider a really drastic scenario. If we wait long enough, a billion years or more, could the overall arrangement of planets in the Solar System change? Could Mars collide with the Earth? Could Mercury be ejected from the Solar System altogether? Could Venus end up inside the orbit of Mercury?

Five years ago we lacked the computational power to investigate this question on the necessary timescale: even the Digital Orrery was too slow. But in computing nothing stays the same for long, and the Digital Orrery has now been superseded by faster hardware and more sophisticated software. As a result, our understanding of the chaotic nature of our Solar System has increased substantially. For example, we now know that it isn't just Pluto that follows a chaotic orbit. Jacques Laskar of the Bureau des Longitudes in Paris (where Laplace worked, oddly enough) has shown that the entire Solar System is chaotic. To a mathematician, once one body in the Solar System is chaotic then the whole system is chaotic too, because if the position of Pluto varies wildly then the global state of the whole Solar System also varies wildly. An astronomer, however, can reasonably ask whether the wild motions of Pluto affect anything else. 'If we ignore Pluto, is the rest of the Solar System stable?' Prior to Laskar's work the computations indicated that on the whole the answer seemed to be 'yes'.

In the mathematical equations of celestial mechanics, it is not only planets and other bodies that can move. 'Orbital elements' – parameters that specify the shape, size, or position of orbits – can also move. Those motions are just as important as the motions of the planets themselves for making sense of the mathematics. On astronomical timescales, planets move quickly round the Sun, where 'quickly' means that human beings can observe the motion over the course of a few days or months. But the position of the orbit also moves, much more slowly – so slowly that Tycho Brahe did not observe any change,

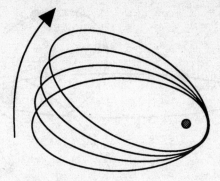

Figure 113   Precession of an elliptical orbit.

which was fortunate for the development of science since otherwise Kepler would never have found his ellipses. Figure 113 illustrates the most important such motion, known as *precession* of the orbit. In effect what is happening is that a periodic motion round an ellipse is being perturbed into a quasiperiodic one in which the ellipse itself rotates.

Laskar found a way to eliminate the fast orbital motions, so that he needed only to study the slow precession of the orbits. He took Newton's equations and the theory of relativity – about a page of mathematical formulas – and performed a huge computerized algebraic operation, expanding everything in series. The final equation had 150,000 polynomial terms and covered 800 pages. That might seem a backward step, but the motion is only implicit in the mathematical laws, and Laskar's series expansion makes it one stage more explicit. An 800-fold expansion is a small price to pay. The last step was to study the 800-page equations numerically. Because the fast motion had been eliminated, this could be done much more efficiently: instead of having to update the planets' positions at intervals of half a day, Laskar could get away with a 500-year interval, speeding up the calculation by a factor of 300,000. Calculating the motions of the planets over 200 million years required only a few hours on a fast computer.

Laskar discovered that the motion of all the planets, especially the inner planets: Mercury, Venus, Earth and Mars, is chaotic. For the Earth, an initial uncertainty about its position of 15 metres grows to only 150 metres after 10 million years – but over 100 million years the error grows to 150 million kilometres.

This is the distance between the Earth and the Sun.

So there's a practical limit to the predictability of planetary motions. Over a 10-million-year span, predictions are perfectly possible, but over a

Figure 114   Changes in eccentricity make the shape and size of a planet's orbit unpredictable.

100-million-year span they're not. Even if we could take into account every tiny asteroid, we would suffer a 150-million kilometre error after 160 million years instead of 100 million years.

None of this means that Earth's *orbit* becomes irregular. It is the Earth's position in its orbit that is chaotic: the orbit itself doesn't greatly change. Far into the future, the timing of summer and winter will become unpredictable, but the Earth will continue to orbit between Venus and Mars. So this kind of chaos has a rather weak effect on the general appearance of the Solar System. If, on the other hand, the *distance* variable is subject to chaos, then the planet's orbit changes radically; for example, the planet might wander out of the Solar System altogether. In fact, as we shall see shortly, it is not so much the distance of the planet from the Sun that matters as the *eccentricity* of its orbit – how fat or thin the ellipse is. It is changes in eccentricity that lead to changes in distance (Figure 114).

Today we know how to solve the dynamical equations for the Solar System over billions of years. During that time, nothing much changes for the big planets – Jupiter Saturn, Uranus, Neptune. Their motions stay regular. But it's different for the inner planets. The orbits of the Earth and Venus change their eccentricity by a maximum of about 8 per cent – but this takes at least 5 billion years, maybe longer. Mars is dramatically different: in 5 billion years' time its eccentricity could become 20 per cent or more, bringing it close to the orbit of the Earth with the danger of a collision or a near collision.

Mercury's eccentricity could change enough to cause the planet to cross the orbit of Venus, and after a close encounter Mercury could even be ejected from the Solar System.

So finally we have an answer to King Oscar's problem, one that he probably wouldn't have liked – although the 5-billion-year timescale does give us a bit of breathing space.

Saved by the Moon?

Another discovery made by Laskar's group suggests that, without the Moon, the Earth would have been a far more uncomfortable place for life to get started. The line of argument is only semi-serious – it attempts to reconstruct an alternative history for the long-term movement of the Earth by discarding the influence of the Moon in a particular simplified model, and there's no reason to place a great deal of credence on the details. But it does suggest that our unusual 'double planet' might possess life-sustaining advantages that we have not hitherto appreciated.

The central character in the story is the tilt of planetary axes. All planets revolve about an axis, which is usually tilted relative to the plane in which the Solar System roughly lies. The angle of tilt is called the *obliquity*. The present-day obliquities of the planets are a strange, patternless list:

Mercury	0°
Venus	178°
Earth	23.44°
Mars	23.98°
Jupiter	3.12°
Saturn	26.73°
Uranus	97.86°
Neptune	29.56°
Pluto	⩾50°

When a celestial body spins about an axis, that axis itself is subject to a slow rotational movement known as *precession* (the same word is used for the slow rotation of elliptical orbits because the same underlying mathematical phenomenon – quasiperiodicity – is involved). The Earth's axis precesses through one rotation in 26,000 years. One consequence of precession is that the 'pole star' changes. Today the star Polaris appears to be fixed in the sky as the Earth rotates, because the Earth's axis points towards it; but in AD 14,000 the fixed star will be Vega. Another consequence is the precession of the equinoxes. Equinoxes are the times of year at which day and night have equal length, and these too change slowly, as first discovered by Hipparchus

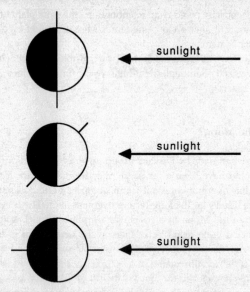

Figure 115   The amount of sunlight that hits a given part of a planet depends upon the angle of inclination of the axis.

around 200 BC. Precession also affects the position of the Earth in orbit, relative to the constellations that make up the zodiac.

Figure 115 shows that the amount of sunlight that hits a given part of a planet depends on the planet's obliquity. Combining this effect with the planet's annual rotation around its Sun we see that the *total* sunlight per year falling on a given latitude also depends on the obliquity. So changes in obliquity have a big effect on climate. Climatologists are fairly convinced that very tiny changes in the Earth's obliquity are one of the main causes of ice ages – so even though the Earth's obliquity is remarkably constant, its effects on climate can be large.

Why is the Earth's obliquity remarkably constant? The answer is that we have a nice big Moon which keeps us steady. Precession is caused by the combined tidal effects of the Sun and the Moon. About two thirds of the tidal effect is due to the Moon and one third to the Sun. If we eliminated the Moon we would reduce the tidal effect by two thirds. Then the Earth would still precess, but its period of precession would be 75,000 years instead of 26,000. Now 75,000 years comes close to the period of other planetary motions, and the resulting jumble of resonances would produce a huge chaotic zone, big enough to destabilize the Earth's orientation completely. Its obliquity

Figure 116   Plots of obliquity (measured in degrees along the horizontal axis) against the rate of precession α of the orbit (measured in seconds of arc per year along the vertical axis) for four cases: Mercury, Venus, Earth, and Mars. The figures along the right-hand vertical scales are the planet's rotation period in hours.

would no longer be constant but it would vary between 0° and 90°. The chaotic zone would cover the entire phase space.

Laskar has calculated the chaotic zones and associated resonances for the obliquities of various planets. Some of his conclusions are shown in Figure 116. Here we see four plots of obliquity, measured in degrees along the horizontal axis, against the rate of precession of the orbit, measured in seconds of arc per year along the vertical axis: one plot for each of the inner planets Mercury, Venus, Earth, and Mars. The thin lines indicate non-chaotic behaviour, the dark regions of dots are the chaotic zones. The chaos takes the form of a slow irregular drift in the horizontal direction, meaning that the obliquity varies erratically while the precession rate remains virtually unchanged.

The Earth – thanks in part to the influence of the Moon – has a precession rate of 55″/yr (seconds of arc per year) and obliquity 23°, lying well inside a large non-chaotic zone. Laskar argues, on reasonable grounds, that if the Moon were not present then the Earth would probably be within the chaotic zone. For Mars, on the other hand, the precession rate is currently 8.26″/yr and the obliquity is 24° – well inside Mars's chaotic zone. In fact the calculations indicate that the obliquity of Mars must be varying chaotically between 0° and 60°. Mercury is no longer in the chaotic zone, but Laskar's analysis shows that at some time in the past its obliquity probably varied chaotically between 0° and 100°. Tidal friction has slowly driven it into its present, far more stable, state.

Venus is a very strange place, and it poses many puzzles for astronomers, one being its obliquity of 178°. Venus is upside down: its rotational motion is *retrograde*, in the opposite direction to most of the other planets, although its motion in orbit is in the same direction as the rest. Laskar shows that chaotic changes to Venus's obliquity occurring in the distant past may have driven it into this curious state.

Laskar has argued that chaotic changes in obliquity are inimical to life – possibly sufficiently so to prevent it evolving in the first place, even if conditions were otherwise suitable. Imagine what the climate on Earth would be if the obliquity was 90°. The northern hemisphere would have sun all the time and it would be extremely hot. Then six months later there would be no sun at all, and it would be very cold. Now consider whether life and its subsequent evolution could ever have got started on a planet whose obliquity was perpetually changing. Chaotic variations of obliquity might well have rendered planet Earth uninhabitable, were it not for the stabilizing presence of its sister planet, the body that we call the Moon. 'You are a child of the universe. No less than the trees and the stars, you have a right to be here.' Laskar's work suggests that while the stars may have a right to be here, our own presence, and that of the trees, may owe an awful lot to cosmic coincidence.

13

The Imbalance of Nature

There is no bound to the prolific nature of plants or animals but what is made by their crowding and interfering with each other's means of subsistence. Were the face of the earth vacant of other plants, it might be gradually sowed and overspread with one kind only, as for instance with fennel; and were it empty of other inhabitants, it might in a few ages be replenished from one nation only, as for instance with Englishmen.

Thomas Malthus, *An Essay on the Principle of Population*

There was once a man who kept a jar full of flies.

Yes, the world is full of bizarre obsessions, but this wasn't one of them. He wasn't some eccentric keeping unusual pets. He was a scientist, studying how a population of blowflies, limited by space and food, would change with time. His name was A. J. Nicholson; his subject was ecology. We hear the name 'ecology' a lot these days, usually in association with 'green' politics: *the* ecology, the environment in which we – and the rest of creation – spend our existence. Ecology, as a subject, is the study of this environment, especially the interactions between animals and vegetable species within it.

Some days, there would be close on 10,000 blowflies in Nicholson's container. At other times, the population would drop to a few hundred (Figure 117). The fly population would outgrow the space in the container, and then the number would crash steeply; but then, with plenty of space available, the flies would breed anew. After thirty-eight days or so the cycle would repeat; never quite the same, but fluctuating around a periodic rhythm.

The rhythms and non-rhythms of animal populations have always been vitally important to humanity. Unexpected plagues of locusts cause famine and death. Other pests, be they rabbits, kangaroos, or opossum, can devastate farmlands and orchards. Populations of bacteria and viruses – disease epidemics – also fluctuate from year to year. One of the longest available time-series is that of lynxes and hares in Canada, compiled from the records of the Hudson Bay Trading Company.

Figure 117   Fluctuations in a population of blowflies. Horizontal timescale is measured in days.

The cicada belongs to the order Homoptera or sucking insects. Most of the Homoptera are very short-lived, but not three species of cicada. The adult females drill holes in trees and lay eggs. They hatch after a few weeks. The nymphs drop to the ground, dig into the soil, and start to feed on the tree's roots. They stay underground for seventeen years; thirteen years in some species. Then they emerge from the soil and metamorphose into adults.

The adults live only a few weeks. Adults seem to be the nymphs' way of reproducing.

What are they up to? It's a real puzzle. One speculation is that a prime number like 13 or 17 avoids resonating with other, shorter cycles of potential predators. But that's guesswork.

Some of these fluctuations are regular, some are not. Is the dynamical image just a metaphor? Or should the phrase 'population dynamics' be taken more literally? When the only phenomenon is periodicity, it's almost impossible to answer that question. But with the advent of chaos, much more stringent tests are available. Do we observe the footprints of chaos in the irregularities of populations?

Very probably.

Sharks and Shrimps

The idea that an ecological system is driven by some kind of dynamic has been around for a long time. Vito Volterra, an Italian mathematician, spent the First World War in the Air Force developing dirigibles as weapons. He was the first to propose using helium rather than inflammable hydrogen in airships. When the war ended, he directed his thoughts into peacetime channels, inventing mathematical models of the interaction between predators and prey. He found a system of differential equations to explain why the fish population of the Adriatic fluctuated periodically.

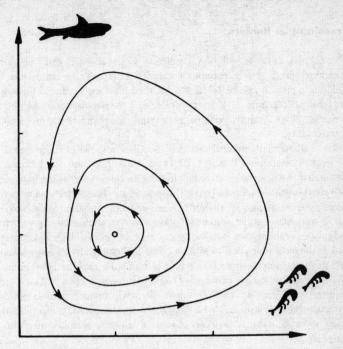

Figure 118   Volterra's predator – prey cycles.

Volterra's cycles (Figure 118) can be made plausible by a purely verbal argument. Suppose that a small number of predators – sharks, say – infest waters containing a large number of prey – shrimps. I use these terms only for vividness. The shrimp population is limited by the available food; but in addition it can be reduced by predation. The shark population, on the other hand, is limited by the number of shrimps. Initially, there's plenty of shrimp, so the shark population grows sharply. The shrimp population begins to drop as sharks deplete it. Soon there are too many sharks and not enough shrimp. Starving sharks die for lack of food and float, bloated by decay, to the surface. Their numbers decline. The relative absence of predators allows the shrimp to reproduce faster, and the shrimp population explodes. The cycle is now ready to repeat.

Transluminal Rabbits

It's a cliché that the growth of a population, in the absence of any restraint, is exponential. If – as is commonly claimed – the average family has 2.3 children, a growth rate of $2.3/2 = 1.15$, then after n generations there will be $(1.15)^n$ people. Since $(1.15)^5$ is very close to 2, the population doubles every five generations. At thirty years per generation, the population grows tenfold every century.

The earliest mathematical model of population growth can be found in the work of Leonardo of Pisa, in 1220. Leonardo is often known as 'Fibonacci', although that name was given to him in the 19th century by the mathematical historian Guillaume Libri and has no historical basis. His model was somewhat tongue-in-cheek, a puzzle rather than a serious piece of mathematical ecology; but it anticipates some important ideas. It was about the reproductive behaviour of rabbits. Not in its biological sense, but numerological. Leonardo took as the basic unit a pair of rabbits – a natural enough hypothesis. Assume that in the beginning there is one pair of immature rabbits. These mature for a season. Every season after, they beget one immature pair, which in turn matures for a season. And of course, all newly mature pairs beget one immature pair per season as well. Suppose that rabbits and their procreative urges never die. How many pairs of rabbits will have been begotten after n seasons?

Suppose there are M_n mature pairs and I_n immature pairs season n. Then we start out in season 1 with $M_1 = 0$, $I_1 = 1$. The growth laws are:

$$I_{n+1} = M_n$$
$$M_{n+1} = M_n + I_n$$

That is, in season $n + 1$ the M_n mature pairs give birth to M_n immature pairs, which produces I_{n+1}; and the I_n immature pairs from the previous season mature, adding themselves to the existing pool of M_n mature pairs, to give the formula for M_{n+1}.

If we tabulate these numbers we get

n	M_n	I_n	Total
1	0	1	1
2	1	1	2
3	2	1	3
4	3	2	5
5	5	3	8
6	8	5	13
7	13	8	21
8	21	13	34

and so on. These are the famous *Fibonacci numbers*, each being the sum of the previous two. What we see here is a discrete dynamical system. The time-interval is a season; the state of the system is the pair of numbers (M_n, I_n). The growth-law is the dynamic.

The equation has an exact solution. If we introduce the *golden number* $\tau = \frac{1}{2}(1 + \sqrt{5}) = 1.618034 \ldots$ then it can be proved that

M_n is the nearest integer to $\tau^n/\sqrt{5}$,
I_n is the nearest integer to $\tau^{n-1}/\sqrt{5}$.

I don't want to go into why this is, but you can check it on your calculator if you don't believe me. The point is that, to a very good approximation, Leonardo's model has the population growing by a factor of 1.618034 each season.

Once more this is exponential growth. If nothing happened to check it, then after 114 generations the total volume of rabbits would exceed that of the known universe. Long before that, the earth would be submerged beneath a sphere of rabbits, expanding faster than light!

The Limits of Growth

This is of course absurd. In practice, some external influence will come into play to limit the population to more sensible numbers. The availability of oxygen, for example. But more likely, lack of space, or lack of food, or both.

Thus Leonardo's discrete dynamics must be modified, to provide a cut-off at high populations. In the ecological jargon this is 'density-dependent population growth' because the birth-rate depends on the density – the ratio of the actual population to the maximum that the environment will support – of the creatures present.

Leonardo's model is discrete not just in time – seasons – but in the number of rabbits. It's a little simpler to analyse equations that are continuous in rabbits (but remain discrete in time). To do this, replace the number of rabbits by its ratio x to the maximum population. Now x ranges between 0 and 1. It does so in very tiny discrete steps; but if the maximum population is, say, a billion, then the steps have size 0.000000001, and you'll hardly notice them. Neither will your computer.

The simplest models of population growth are iterative, just like Leonardo's: the density of population in a given season depends in a predictable fashion on that in the previous season. In other words, we have an iterative model, a discrete dynamical system, of the shape

$$x_{n+1} = F(x_n)$$

where x_n is the density in season n and F is some specific mapping.

An enormous variety of mappings F have been proposed, each attempting to capture some alleged facet of the reproductive process. If you approach them in a classical spirit, you'll begin with the impression that each mapping should lead to highly distinctive dynamics. So you'll try to devise methods to test which fits the data best, in the hope of pinning down the best model and thereby learning something about the underlying biology.

That could be a mistake. Most mappings in the literature have one thing in common: they define a single-humped curve. On a qualitative level, therefore, they all behave just like the logistic mapping. In particular, various striking features – notably the fig-tree with its period-doubling cascade – will occur in all of them. The 'Feigenvalue' 4.669 will likewise appear in all of them. So will periodic cycles with periods differing from 2^n. So will chaos.

This isn't to say that the different models can't be distinguished at a quantitative level; but you have to realize that in experimental ecology it's hard to get really good data. So you've got serious problems. It's probably best to accept that the experimental evidence, such as it is, detects a whole class of models rather than any individual one.

Anyway, the upshot of all this is that even the simplest models of population growth in a restricted environment can generate periodicity and chaos. As we've seen, periodicity is common in real populations. So is random fluctuation, which poses a pretty problem: how much of it is due to external influences, and how much is genuine deterministic chaos?

Combination of Circumstances

The first person to appreciate just what was involved here seems to have been Robert May, whose paper in *Nature* – with its impassioned plea for wider appreciation of the complex behaviour of simple models – has already been mentioned. In *Proceedings of the Royal Society*, vol. 413A (1987) May offered a few thoughts on why it took so long for people to spot what in essence ought to have been obvious to anyone with a desk calculator, or even pencil and paper.

> Given that simple equations, which arise naturally in many contexts, generate such surprising dynamics, it is interesting to ask why it took so long for chaos to move to centre stage the way it has over the past ten years or so. I think the answer is partly that widespread appreciation of the significance of chaos had to wait until it was found by people looking at systems simple enough for generalities to be perceived, in contexts with practical applications in mind, and in a time when computers made numerical studies easy.

This remark echoes my own, earlier, that it requires a combination of circumstances – time, place, person, culture – for a new idea to take root. And, as May goes on to say, *some* of these circumstances held good long ago. But not all of them. As a result, the subject never acquired enough sense of identity to be perceived *as* a subject at all. The same is true of fractals: although various pieces of the puzzle had lain around for generations, it took the special talents of Benoît Mandelbrot to put them together and convince people that the resulting picture was worth having.

In fact, several population biologists were in some sense aware of chaos in the 1950s. For example P. A. P. Moran studied insects in 1950 and W. E. Ricker studied fish populations in 1954. They found stable solutions, periodicity, and even chaos. But at that time the interest was in the stable solutions; and the chaos – observed only through laborious work on desk calculators – was neither understood nor trusted.

But by 1970 the necessary combination of factors had come together. From then on, it was impossible *not* to notice chaos occurring in numerical simulations. Anyone who has played around iterating mappings on a computer – a very easy problem to program – will find that the biggest difficulty is often avoiding chaos, rather than finding it.

Except, of course, when you're deliberately looking for it.

Bacteria are everywhere, but without a microscope, you'll never see them. Galaxies are everywhere, but without a telescope, they look like slightly blurred stars. Subatomic particles are not only everywhere, but every*thing*: despite which, it takes multi-million dollar accelerators to show they exist. In the history of science, the invention of new instruments has always led to immediate progress. Here the crucial piece of equipment was the computer. But instruments alone are not enough. It takes the wit of a scientist to recognize that what his new instrument has revealed is important. And it takes even more wit to work out *why* the instrument is revealing what it does.

Blow-by-blowfly Account

Let's take a closer look at Nicholson's blowfly data.

Nicholson fed his flies a uniform but restricted protein diet. When their population was high, there wasn't enough food for the flies to breed properly. Not many eggs were laid, and the fly population crashed. The resulting smaller generation of blowflies had plenty of food, so the population bounced back.

Above, I argued that predator–prey interaction can produce cyclic behaviour. The same kind of argument would lead us to expect a periodic oscillation of Nicholson's blowfly population. And indeed, across a period of two years, the main feature of the experimental data is a fairly regular oscillation with a period of around thirty-eight days.

But it doesn't *just* do that.

Many of the peaks are double, M-shaped rather than Λ-shaped. This suggests that an additional high-frequency motion is superimposed on the basic period.

The height of the peak modulates, in a fairly regular way, in a pattern that repeats every three peaks. A small peak is followed by a medium one is followed by a large one, and then the cycle repeats.

Moreover, after the first 450 days or so, the oscillations become ever more irregular.

If you think – as used to be the common view – that regular cycles are the most complicated things a natural population should do, left to its own devices, then you have to find extra factors to explain Nicholson's data. Was the food supply really constant? Were there disease organisms present? How accurate was the count?

But we now know that *all* of the effects observed in the blowfly data are common in discrete nonlinear dynamics. Periodicity, quasiperiodicity, chaos.

Many biological phenomena involve time delays. A disease organism, for example, undergoes a period of incubation. So between the time a person becomes infected, and the time he starts to show symptoms, there may be a lengthy delay. (For chickenpox it's fourteen to fifteen days, for AIDS, between five and ten years.) Breeding cycles include periods of gestation. An animal, deprived of its food supply, first works through its built-in store of surplus fat; only then does serious starvation set in.

May showed that a very simple model, incorporating time-delay effects, could be made to mimic the blowfly population's thirty-eight-day explosion–crash cycle (Figure 119). The smooth theoretical curves, and the jagged experimental data, fit reasonably closely.

George Oster took the analysis further. In his model, there are two main

Figure 119   A time-delay model of the basic oscillation in the blowfly data.

factors influencing the population size. The first incorporates a delay: it's the 'gestation period' – the delay while an egg matures into a reproducing adult. The other is a nonlinear dependence of the adult reproduction rate on food supply. The results produced by the model (Figure 120) include stable states, periodic states of various periods such as 3 and 6, and well-developed chaos.

One way to mimic time delays dynamically is to use a model with two age-classes. In fact Leonardo's rabbit model is just like this: the classes are

Figure 120   Multiple periods and chaos in George Oster's model of population cycles.

the immature and mature pairs. The delay arises because immature pairs don't breed in their first season. Instead, they spend the time transmuting themselves into mature pairs. But the growth rates in Leonardo's model are linear, with no cut-off at high populations, so his rabbits breed exponentially. Oster's model contains a nonlinear cut-off.

The model can generate periodic bursts of egg-laying, superimposed on the basic cycle, and leading to an M-shaped double peak in the population. It can also make the height of the peak modulate; and it can produce chaos. Oster went on to obtain quantitative agreement with data, not just qualitative. So the entire range of dynamics that Nicholson observed may be just a consequence of a single deterministic law. No additional unaccounted-for effects are needed.

Beetle Dynamics

If you want to pin down the occurrence of chaos in real ecosystems, the obvious method is to camp out beside some chosen ecosystem, observe the population numbers, carry out a Ruelle – Takens reconstruction, and – behold! – find a strange attractor. Unfortunately this is not a sensible procedure. It would be hopeless for 'field data' – measured in a meadow, or a forest, anywhere in the wild – because such systems are constantly disturbed by outside influences. There you are watching the foxes and rabbits, and along comes a bear, a pack of foxhounds, or a school picnic. In fact the mere presence of an observer can affect the ecosystem's behaviour. You can no more expect chaos theory to apply to such a system than you can expect the laws of geophysics to apply to the tea-lady's trolley when it bashes into your seismometer. In the physical sciences you can try to avoid such outside effects by performing carefully controlled laboratory experiments, designed to keep the tea-lady safely away from sensitive apparatus. But that trick isn't so effective in ecology. Laboratory ecosystems are much less precise than physical apparatus, if only because of the Harvard Law of Animal Behaviour, which states that 'Under carefully controlled laboratory conditions, experimental animals will do what they damned well please.'

In 1995 J. M. Cushing of the University of Arizona decided to deal with this problem by making sure that they *don't* do what they damned well please. His methods are sometimes a little drastic, so they are carried out on a creature that has not yet attracted much sympathy from animal rights organizations: the flour beetle *Tribolium*. These tiny beetles are a pest for the flour trade, because they get into the stuff and render it inedible – or at least unfit for human consumption, which is not quite the same thing. Cushing's team formulated a theoretical model, whose variables are the populations of larvae, pupae, and adults, and whose parameters are mortality rates, the rate of egg cannibalism (you know, I'm not sure flour beetles *deserve* rights), and suchlike. But instead of simply turning the whole lot loose inside a container in a laboratory, the experimentalists took steps to ensure that the *parameters* remained constant. If, for example, the death rate became higher, they would top up the supply of beetles; if it became too high, the appropriate number of beetles would be eliminated. (This can be done humanely, if you're worried: just take them out and pension them off in a nice flour-bag in the country.) Ditto for egg cannibalism.

As dynamical systems theory would lead you to expect, it takes a while for the system to settle on to its attractor – or, as Cushing put it at a conference in Utah, 'There's always a beginning stage in which beetles do goofy things.' However, once the goofiness has abated, the laboratory system and the predictions of the model equations resemble each other very closely indeed. There are realistic models in which the population can be chaotic without repeatedly crashing to near-zero levels, and the real experiment seems to behave in a similar way for the appropriate parameter ranges. Steady states and period-3 cycles have been observed. Final results are not yet in, but it looks as if beetle dynamics can be chaotic too.

The mere possibility of chaos has had a big impact on how ecologists go about their work.

Until recently ecologists had assumed, at least implicitly, that the natural state of a population is steady, but in practice this desirable state of affairs – the 'balance of nature' – is upset by density-dependent effects and environmental noise. The problem for the experimentalist was to extract, from noisy data, the underlying steady or periodic states. But if the same simple dynamics that gives rise to steady states and periodicity can also give rise to chaos, then that underlying state may itself be chaotic, and the problem of extracting the underlying structure becomes far more subtle.

In the past, ecologists have tended to look at averaged quantities, asking how the averages relate to each other. This is a bit like the thermodynamic approach to a gas: emphasize averages such as temperature and pressure. It works pretty well for gases, and rather badly for populations. That may be because populations contain fewer creatures than gases do molecules. Environmental noise (predators, climate, availability or not of suitable food) really acts on individuals. Changes in population also occur on the level of the individual. Moreover, the population dynamic itself can vary dramatically according to very local effects.

M. P. Hassell and May have studied how a garden pest, whitefly, is distributed on viburnum bushes. Their data led them to conclude that a three-tier mechanism is operating. First, the distribution of the insect is very patchy. Second, within each patch the density can vary, so density-dependent dynamical effects vary from patch to patch. Third, environmental noise can affect each patch differently.

To analyse such a system it is important to do the dynamics first, and then average the results, rather than take averages first and then do the dynamics. For example, if you choose a dozen patches, each with a different population density, and see how the average population size varies from generation to generation, you will *not* expect to see the same pattern that would occur for a uniform population of average density. This is because the dynamic is nonlinear, and nonlinearities don't respect averages.

As an analogy, consider a car travelling a distance of 30 km at a speed of 20 km/h, and returning at 60 km/h. What is its average speed? If you just add the speeds and halve you get 40 km/h. But that's wrong. It takes the car one and a half hours to go out, half an hour to come back, a total of two hours. So the average speed is $60/2 = 30$ km/h. The reason that adding and halving doesn't give the right average is that speed is proportional to the reciprocal of time, and this is a nonlinear relation. In other words, you have to average in the right place.

So chaotic dynamics raises entirely new, and difficult, problems for the interpretation and analysis of data. But it's better to have a clear problem, however difficult, than to live forever in a fool's paradise.

The Web of Life

Theoretical ecologists have finally started to take the ideas of chaos and related nonlinear phenomena on board, but many environmentalists, nature lovers, politicians, and managers still seem to be trapped in a way of thought that was outmoded forty years ago. In 1995 the environmental critic Stephen Budiansky put it like this:

> Open any popular ecology book or a fund-raising letter from an environmental organization and you will not have to read very far before you encounter some reference to 'the balance of nature'. If these precise words do not appear, their equivalents are sure to: the notion that all species are connected in a delicate 'web of life' easily broken by man; that predator and prey, if left undisturbed, will perfectly regulate one another, and that forests, once saved from the rapacity of loggers, will resume their eternal march of succession towards a stable 'climax' community of towering old growth . . .
>
> Many nature lovers . . . express genuine surprise when told that practising ecologists have not taken such ideas seriously for decades. The idea prevalent at the turn of the century, that plants form 'com-

munities' to which they are bound by tight associations of mutual dependence, and that these communities develop and age in a rigid succession of stages, has been especially slow to die. This is despite the fact that, since the 1950s, ecologists have regarded it as little more than a fairy tale.

One reason for the attractiveness of such fairy tales is a misunderstanding of the nature of stability. It is clear that a viable ecosystem must be stable in some sense, otherwise it would not continue to exist (which is what 'viable' means). Until recently, the paradigm for stability was the steady state. In consequence, we find many people arguing that because an ecosystem is a complex web of interactions, the loss of any part of that web will destroy its stability – because ('obviously') it will affect that steady state.

This argument is wrong on many counts (which does not mean that its conclusion is always wrong). Even a steady state may move rather than collapse. When chaos is present, *nothing* is terribly obvious. Chaos theory does not tell us that losing part of such a web will destroy its stability, and it doesn't tell us that losing part of such a web will *not* destroy its stability. What it does tell us is that this whole line of reasoning is an oversimplification and we should find a better way to think about the stability of ecosystems. The concept of an attractor provides a natural and useful image; it alerts us to the vast range of possible types of 'stable' behaviour that might occur and provides techniques for thinking about them and investigating their occurrence. For example, removing part of an ecosystem may just change its attractor a little, and if so, it will remain stable (in the sense that any attractor is stable – maintaining the same 'texture' of behaviour but not necessarily in the same temporal sequence) and hence remain viable. It may also cause the attractor to disappear, or change in a dramatic way. You can't tell just by looking at the size of the change that you make; it depends on whether the dynamics is near a 'bifurcation point', that is, whether it is highly sensitive to changes in parameters. And if the attractor does change slightly then it may indeed become impoverished; on the other hand, it may actually end up richer. For example, the rich and diverse ecology of the lakes in the African Rift Valley has been seriously damaged by the introduction of the Nile perch, a generalized predator. If you removed that particular creature from the ecological web, you would start to restore the diversity that is currently disappearing down the Nile perch's throat. 'Ah, but the Nile perch was introduced by humans.' True, but the general point remains valid: if a particular creature's presence makes an ecosystem less diverse than it would be in that creature's absence, then removal of the beast will increase the diversity, not reduce it.

Most people want easy answers, most politicians and pressure groups want easy slogans. Ecosystems are too complex for those. Chaos theory does not solve the problems posed by such complexity, but it does make us aware that they are present. For example, most of the world's fisheries have been so badly managed that they are over-fished and their stocks are seriously depleted. The entire industry is hovering on the brink of disaster: in 1992 the Canadian government banned, probably for a decade, all cod fishing on the Grand Banks, because there were hardly any adult cod left. Among the causes of disappearing fish stocks are fishermen exceeding quotas and quotas being set too high for political reasons, but another is government officials believing (a) the simple ecological models in the textbooks, and (b) the figures for fish populations in their files. Unfortunately many of the models are nonsense, and most of the figures are out of date – often by several years. We know that merely introducing a delay into a feedback loop can drive a system away from a stable steady state and into chaos, and something like this seems to have happened in fisheries management. Better models and more up-to-date data would be a good start.

The management and regulation of ecosysems is a truly difficult issue. (Would you want a government snoop telling you what to plant in your garden? How many pets you can have?) The stability of chaotic attractors could, for example, be cited as justification for the idea that human interference cannot actually wreck an ecosystem at all. Budiansky has his own argument in this direction:

> Nearly 90 per cent of the Atlantic coastal forest of Brazil has been cleared, but rather than the predicted loss of half of all species the actual recorded loss of species is zero. Despite extensive surveys by zoologists, not a single known species could properly be declared extinct. Indeed several birds and butterflies believed 20 years ago to be extinct were recently rediscovered.

I think I'm following his line of thought, and if so I disagree. There's nothing wrong with the statement, as such, but I'm not happy about the implications he is clearly expecting us to draw. The survival – somewhere – of threatened species is all well and good, but it doesn't mean that we should continue to clear forests willy-nilly. The coelacanth was thought to have become extinct millions of years ago, was rediscovered safe and well, and is now back on the verge of extinction because of indiscriminate fishing methods. If I were a coelacanth I wouldn't be too impressed by the fact that my species had previously been declared extinct erroneously: I'd be more worried that this time around the verdict would prove correct.

Obviously many business concerns would be delighted if everybody accepted that massive destruction of habitat is safe because chaos theory –

or any other ecological theory – says so. But all that chaos theory tells us is that the response of an ecosystem to disturbances is a much more difficult problem than we used to imagine. We need to know an awful lot more about ecosystem dynamics before we can decide whether any particular activity is 'safe'. In the meantime, we are saddled with an ethical dilemma rather than a scientific one. Is it sensible to risk irreversible damage to the environment, just because it makes money for a few businessmen? And we need to recognize the dangers of slogans. 'Diversity' is not just a matter of the number of species present; it is also important which species are present, how many members of those species there are, and over what territory they are free to range. In fact, if you were to dig up any random square metre of land almost anywhere, and catalogue every single species in it, you would find that about half were unknown to science. The new ones would be tiny microorganisms, new kinds of mites, and so on, rather than a new big cat or pygmy rhinoceros, but the diversity would be a lot bigger than you had thought based on the species you knew about. This is all very well, but of course it is almost entirely irrelevant to the question of environmental damage, which is about losing known species, or reducing their numbers, not just totalling up the numbers that you know about. And geographical spread matters too: you could have a wonderfully diverse ecosystem inside a tiny enclosure in the middle of a desert, but it wouldn't be the same as a thriving rainforest.

Thanks in part to the discovery of chaos, and in part to other equally important influences, theoretical and field ecologists are beginning to grapple effectively with these problems. In the meantime, it would be better if the rest of us stopped leaping to unjustified conclusions on the basis of an over-simple mental picture. And you can be just as simple-mindedly fixated on the incredible multidimensional complexity of life as you can be on the crass one-dimensional simplicity of money.

Chaos in Epidemics

Bacteria and viruses are living organisms, and the way their populations fluctuate can be very important indeed. In a measles epidemic, it is ultimately the population of measles virus that determines the extent and severity of the infection. So population dynamics has direct applications to epidemiology. The remarks in the previous section but one, for example, apply essentially unchanged to the epidemiology of AIDS, which you can hardly fail to be aware is a nasty and fatal disease syndrome thought to be caused by the Human Immunodeficiency Virus (HIV). The spread of HIV is also very patchy, being related to factors such as sexual behaviour, and studies of AIDS based upon average incubation periods and average sexual behaviour may prove misleading. This is a question that deserves looking into, because control – and perhaps even cure – of the disease depends heavily on having good models of how it is transmitted.

I'll return to AIDS at the end of this section. The first evidence for chaos in disease epidemics arose in connection with much more common diseases. The problem of extracting chaotic dynamics from experimental data came up earlier, in the context of turbulence. I mentioned the method of Ruelle and Takens, of concocting sufficiently many 'fake' time-series to reconstruct the topology of the attractor. But the method in principle works on any time-series, not just one obtained in a physicist's laboratory. Extensive time-series of epidemics are available in medical records.

W. M. Schaffer and M. Kot applied the Ruelle–Takens method of reconstructing attractors to diseases. They used data on mumps, measles, and chickenpox in New York and Baltimore, obtained in the days before mass vaccination existed (Figure 121). For each disease there's a time-series, recording the number of cases per month. Their results show that in each case there appears to be a two-dimensional attractor (Figure 122). It has a one-dimensional Poincaré section which strongly suggests the presence of chaos. In fact, the dynamic appears to be controlled by a one-humped mapping, qualitatively similar to the logistic mapping. An independent analysis of

New York

Baltimore

Figure 121   Measles in New York and Baltimore. Raw data on the left, power spectrum on the right.

New York

Baltimore

Figure 122   Reconstructed strange attractors (left) *and Poincaré maps* (right) *for the measles data of Figure 121.*

measles data for Copenhagen, by L. F. Olsen and H. Degn, leads to an almost identical one-humped map, suggesting that the results are not just coincidence. Reverting to populations of larger creatures, Schaffer has also argued that the notorious lynx–hare data from Hudson's Bay, given by the records of transactions with fur trappers, exhibit chaos in much the same manner.

Conventional approaches to the spread of epidemics are based on constructing specific models of the physiological and transport processes involved. The approach by way of chaos complements this, by concentrating on empirical observations and trying to extract the underlying dynamic directly. Its main disadvantage is that rather long time-series are needed, and these are seldom available. Both methods together may do better than either alone.

Subsequent studies using sophisticated new methods of data analysis have not been able to detect chaos *conclusively* in the New York measles data: the indications hover tantalizingly on the border between low-dimensional chaos and randomness. However, according to May, 'the tests . . . suggest that the measles data are indeed best described by a low-dimensional chaotic attractor'. The difficulty in detecting chaos is partly due to the nature of the data, which were not collected with the sensitive phenomena of chaos in mind. The difficulties are related to those mentioned above in connection with averaging in nonlinear systems. The measles data are just like averages, in the sense that they are aggregated from records in different parts of New York City. Totals suffer from the same problems as averages, because an average is just the total divided by the number of instances. Now, we can't tell from the data whether the spatial distribution was clumpy or smooth; we're forced to try to detect the dynamics after the data have been aggregated, rather than before. Unfortunately – for the study of chaos in epidemics, not for the victims of the disease – vaccination is now so widespread that there's no point in collecting more sophisticated data: there aren't enough cases of measles anyway.

The effect of spatial structure shows up clearly if we consider not New York measles data but data from England and Wales (Figure 123). When these data are analysed by exactly the same methods as the New York data it turns out that they can best be modelled not by low-dimensional chaos, but by a simple 2-year periodic cycle plus (rather a lot of) statistical noise. This might, of course, indicate that the dynamics of measles in the UK is different from that in the USA, so that differences in environmental parameters have caused different behaviour in what is basically the same dynamical system. We know that changes of this kind – bifurcations – are commonplace in nonlinear systems. However, David Holton and May noticed that a very different explanation presents itself if the data are separated, city by city. The data cover the period 1948–66 and are pooled from the records of seven cities. The separate data for the five largest cities – London, Birmingham, Liverpool, Manchester, and Sheffield – resemble the New York data closely, and can best be modelled as chaos. The data for the two

Figure 123   Time-series of measles cases in England and Wales, 1948–66. The vertical coordinate gives the difference in the number of measles cases compared to the previous observation.

smaller cities, Bristol and Newcastle, are different: if chaotic, they require a higher-dimensional attractor. The analysis of the separate data thus leads to a different conclusion from that of the pooled data – a very important point in its own right. It is also intriguing, though possibly coincidental, that the populations of Bristol and Newcastle are below the threshold at which measles can remain endemic – present at a reasonable level all the time. The other five cities lie above that threshold. So the dynamics might well involve different parameter values in different cities, and thus be subject to all of the dangers associated with taking averages at the wrong stage of the analysis. This provides epidemiologists with a useful practical insight: they will need to rethink their methods of data collection, presentation, and analysis if they are going to make progress in understanding the nonlinear dynamics of the spread of disease.

HIV Dynamics and AIDS

Those remarks are especially relevant to the investigation of AIDS, a disease with highly unusual characteristics that does not fit conventional epidemiological models at all well. In particular it has a remarkably long incubation period coupled with a relatively high infection rate, which is unusual. Early modelling attempts made the standard assumption that infected individuals suffer a constant level of infection, starting from the time when they first acquire HIV and continuing unchanged until they start to exhibit symptoms of full-blown AIDS (after which they shortly die). However, over the average incubation period of seven to eight years, the incidence of HIV in patients actually fluctuates considerably, as can be deduced from fluctuations in their level of HIV-specific antibodies. Not surprisingly, conventional epidemiological models can't cope with the peculiarities of AIDS.

Holton and May have developed a model that reflects current medical opinion on various features of the disease, but does not assume a constant level of infection. The following discussion is highly simplified and deliberately avoids some of the more sophisticated biological terminology: I've included it to give you a feel for how such a model is built. The HIV virus infects cells known as lymphocytes, part of the body's immune system. Lymphocytes may be either infected, activated, or uninfected. (Activated cells are susceptible to HIV infection but not yet infected.) The body produces new uninfected lymphocytes at a constant rate; these may die or become activated. The population of activated cells increases in two ways: by the activation of uninfected lymphocytes and by a process called 'clonal expansion'. It decreases when activated lymphocytes become infected with HIV. The infected lymphocytes either stay the same or die. The HIV virus replicates within infected lymphocytes, subverting the cell's own genetic systems, and the HIV virus population increases when viral particles are released from infected lymphocytes. HIV viruses may die, be absorbed by activated lymphocytes, or be destroyed by uninfected lymphocytes still doing their normal job in the immune system.

Got that? The point is that you can encode my verbal description into a system of four differential equations that describe how the populations of the three types of lymphocyte and the virus will change. Those equations contain various numerical parameters, representing infection rates, growth rates, and death rates. You can study the equations by mathematical analysis and computer simulation, and see what behaviour they predict, depending on the values of those parameters.

It turns out that when HIV is introduced into an otherwise 'clean' immune system, three different things may happen, depending on the parameter values:

- The virus will fail to establish itself within its host.
- The virus will establish itself, but in such a manner that the system settles into a state where the number of activated lymphocytes remains constant.
- The virus will not only establish itself: it will start to regulate the population of activated lymphocytes.

Figure 124　Variation of HIV virus concentration in the blood as a function of time, according to the Holton–May model.

In the third, and most interesting, case the result is a low population of HIV for long periods of time, which gives way to sudden bursts of chaos in which the population climbs rapidly and then dies back (Figure 124).

The HIV virus does not cause death directly. It damages the immune system, so that it cannot cope with an 'opportunistic infection' by an agent of some other disease. When the HIV population is high, it is easy for an opportunistic infection to become established, so what the model predicts is a lengthy incubation period followed by a dramatic rise in HIV population and the incidence of opportunistic infection. This is what usually happens in reality.

Holton and May emphasize that their model is a simplified caricature of the real disease. However, it possesses the right qualitative features, and it suggests that the exotic epidemiological properties of HIV/AIDS are caused by exotic dynamics – namely, a particular type of chaos. However, the exotic dynamics stems from an entirely mundane and reasonable source – a bit of nonlinearity in the governing equations. Since a simple and natural nonlinear model can generate some of the more puzzling features of the disease, then it seems only sensible to try to fine-tune it. As Holton and May say: 'It is hoped that these caricatures can be systematically refined to reveal behaviour closer to reality.'

Cardiac Arrest!

Epidemiology is not the only potentially important medical application of chaos. Chaotic dynamics has been advanced to model the uncontrolled behaviour of cells that become cancerous, to analyse brain-waves, and to study genetics. There is also a well-developed study of irregularities in heart-beats (Figure 125), and I'll concentrate on that. The work has been done by Leon Glass and his colleagues at McGill University, Montreal.

A normal human heart beats between fifty and a hundred times per minute, every day, year in, year out, without stopping. However, a number of different irregularities can occur in the heartbeat. Some can kill – for example fibrillation, where different heart muscles contract out of rhythm with one

Figure 125   The Wenckebach phenomenon: irregular fluctuations of the heartbeat. Note the lack of regular pattern in the spacing between the wide and narrow spikes.

another. Obviously, it's important to understand the dynamical nature of the heartbeat.

Mathematical models of heartbeats go back to the 1920s, in work of W. Mobitz, Balthasar van der Pol, and J. van der Mark. Van der Pol's model is closely related to his equations for the oscillations of an electronic valve, mentioned earlier as an example of a limit cycle. Van der Pol and van der Mark even encountered chaos; but at the time nobody thought it was significant. So – although it's not widely realized – nonlinear dynamics has had a connection with physiological processes since its earliest days. It's hardly surprising that advances in nonlinear dynamics can suggest new approaches to the heartbeat.

It's highly controversial whether or not chaotic dynamics is responsible for irregularities in the human heart. In a sense, the dynamics doesn't have to get that complicated to kill you: quasiperiodicity, or even periodic oscillations with too large an amplitude, will do the job perfectly well. Or indeed a steady state – which, to be morbid, is where we all end up. It's also not easy to get observational data on fatal heartbeat irregularities: medical staff, naturally enough, prefer to try to save the patient's life rather than measure the details of how he or she dies.

Kicked Rotator

One important type of heartbeat arrhythmia involves the interaction of two regular periodic effects, known as *parasystolic rhythms*. A simple mathematical model, capturing the broad dynamics, is just a forced oscillator. A natural oscillator is stimulated by an outside disturbance that varies periodically: the interesting question is the interaction of the two modes of oscillation. We've already seen, via Smale's horseshoe, that a forced van der Pol oscillator can go chaotic. So it's not unlikely that parasystolic rhythms might do likewise.

The physicists and mathematicians involved in chaos have their own favourite forced oscillator. Like Smale's horseshoe, it's the most stripped-down version of the dynamics that still retains the key features. It's known as the *kicked rotator*. It's more like a stroboscopic snapshot, a discrete Poincaré section, of a forced oscillator. The state of the system is defined by a point on a circle. At each discrete time-step, the angle at which this point sits changes, according to a fixed rule; but in addition a periodically varying disturbance is added. For example, if the angle at time t is x then the angle at time $t + 1$ might be $x + 1 + \sin t$. Here $x \rightarrow x + 1$ is the natural motion of the oscillator, and $\sin t$ represents the effect of the forcing. More generally we might consider $x + k + A \sin t$ where the constant k lets us adjust the frequency of the natural oscillator, relative to the the forcing frequency; and A lets us adjust the amplitude of the forcing.

Something very interesting happens in systems like this, even before chaos sets in. They *phase-lock*. What happens is that the forcing frequency and the natural frequency of oscillation get 'in step' in some simple numerical ratio. For example, three periods of the forcing oscillation may be the same as four periods of the natural one, a 3:4 phase lock. An astronomer would say that they resonate: it's basically the same thing.

When A is zero, that is, forcing is absent, the dynamics is easy to work out. If each time-step just adds k to the value of x then after n time-steps x changes to $x + n$k. If k is a rational multiple of 360° then the dynamics becomes periodic; if it's an irrational multiple then the dynamics is not periodic.

When A is nonzero, the nonlinearity caused by the forcing has the effect of making the periodic solutions persist even when k moves a little away from a given rational value. This leads to regions of phase-locked behaviour known as *Arnold tongues* after the Russian mathematician Vladimir Arnold. These can be seen as distorted triangular regions in Figure 126 below.

Arnold recently told an amusing tale which reveals the attitudes mathematicians used to hold towards physiology. Arnold was a student of Andrei Kolmogorov, a leading figure in Russian mathematics who died in 1987. Arnold says, of Kolmogorov, 'He stood out from the other professors I met by his complete respect for the personality of the student. I remember only one case when he interfered with my work: in 1959, he asked me to omit from my paper on self-maps of the circle the section on applications to heartbeats, adding: "That is not one of the classical problems one ought to work on." The application to the theory of heartbeats was published by L. Glass 25 years later, while I had to concentrate my efforts on the celestial-mechanical applications of the same theory.'

What gives this tale an ironic twist is that Kolmogorov took a very broad attitude towards mathematics, and himself worked on applications to biology.

The Queen Stoops

I must now digress, or at least appear to, because phase-locking requires new mathematical techniques. Well, new in the sense that they haven't been used for this purpose before. In truth they haven't been used for *any* very practical purpose before, although they're among the most beautiful ideas in mathematics. I refer to the Theory of Numbers.

'Mathematics,' said Carl Friedrich Gauss, 'is the queen of the sciences, and arithmetic is the queen of mathematics.' By arithmetic he meant the theory of numbers, not $2 + 2 = 4$, and the tendency of queens not to dirty their lily-white hands was not entirely absent from his mind. The overt subject matter of number theory – the patterns and perplexities of ordinary whole numbers – does not evoke immediate applications to science. 'That subject is in itself one of peculiar interest and elegance, but its conclusions have little practical importance,' wrote W. W. Rouse Ball in 1896. In terms of the common division of mathematics into 'pure' and 'applied', number theory is about as pure as you can get: poles apart from traditional applied topics, such as dynamics.

Not any more.

Number theory explains the beautiful and complex patterns of phase-locking in considerable detail. For example, the order in which phase-locking regions occur can be found by using gadgets known as *Farey sequences*. A Farey sequence consists of all rational numbers p/q between 0 and 1 for which q is at most some given size, arranged in order of size. For instance, when q is at most 5 we have the Farey sequence

$$0/1 \quad 1/5 \quad 1/4 \quad 1/3 \quad 2/5 \quad 1/2 \quad 3/5 \quad 2/3 \quad 3/4 \quad 4/5 \quad 1/1.$$

This is not the only place where number theory occurs in chaotic dynamics. What not long ago was generally held to be the most useless branch of mathematics – as regards practical applications – has suddenly acquired new importance in dynamical systems theory. Ian Percival and Franco Vivaldi have published a beautiful application of classical number theory to chaotic mappings of a torus. And I have heard Predrag Cvitanovic, a mathematical physicist active in chaotic dynamics, say that 'my main reference is Hardy and Wright', the Bible of classical number theory.

Chickenheart

So much for phase-locking. Now the chaos.

Chaos, in a forced oscillator, is the culmination of a series of changes in these phase-locked frequencies. So to study quasiperiodicity and chaos in the heart, Glass and his colleagues devised a kicked rotator model which they thought especially appropriate to the heartbeat, and analysed how it phase-locked.

Not only that: they tested their model experimentally (Figure 126). Not on a human heart, of course. Instead they used a mass of cells from the heart of an embryo chicken. Such cells can pulsate spontaneously, and they correspond to the natural oscillator. In practice the cells of the chicken heart's ventricle are separated and then allowed to reassemble in a culture medium. The resulting aggregates of cells are small – about 200 micrometres across – and pulsate between 60 and 120 times per minute.

Figure 126   Theory and experiment for the kicked rotator model of the heartbeat.

A glass microelectrode is then inserted into the pulsating mass, so that tiny periodic electric shocks can be produced, corresponding to the forcing. In effect, the miniature chicken heart acquires an equally miniature pacemaker. By varying the frequency and amplitude of the electrical pulse, various types of phase-locking, and chaos, can be produced.

The intricate pattern of phase-locking can be recognized in experiments because it's highly structured. Chaos, in contrast, is – chaotic. If an experiment detects, in great detail, the prechaotic phase-locking, and also shows irregular behaviour in the places where the same model predicts chaos, then this is strong – though indirect – evidence that the chaos is there in the real world. You can recognize chaos by the company it keeps.

Glass's results compare extremely well with his theoretical model of a kicked rotator, showing that aggregates of chicken heart cells can be made to beat chaotically.

Medical Mathematics

Obviously, a 200-micrometre aggregate of chick heart cells is not the same as a real heart, nor is an artificial electrical pacemaker the same as the heart's natural one. Having said this, it's remarkable how well the dynamical theory and the physiological experiment agree. It would be hard to argue that chaotic dynamics has no relevance at all to the real heartbeat.

Living organisms display an enormous range of behaviour. Some of it is so complex that it's hard to imagine that mathematics could shed any light on it. I find it difficult to envisage a mathematical theory of mother-love, and I doubt the world would be a better place if some misguided genius were to find one. But others are relatively simple. The dynamics of the heart is surely more approachable than the psychodynamics of emotional response.

Many organs operate like specialized pieces of machinery. Sophisticated machinery, to be sure, far beyond our ability to fabricate or mimic in all respects. But already we can build artificial hearts good enough to keep people alive when their own natural hearts fail. Speaking of the 'machinery' image: it's about time we discarded our Victorian prejudices that a machine is pretty simple and predictable. If chaotic dynamics has any lessons to teach us, one is surely that a simple system can do some very sophisticated things.

Scientists the world over are beginning to realize that the mathematics of dynamical systems has leaped across the yawning gulf between theory and applications. Mathematicians are working out the concepts and techniques to face up to the reality of nonlinear dynamics. This opens up prospects of penetrating to the essence of many dynamical effects in the real world. The physiological workings of the body – heart, lungs, liver, kidneys, thyroid gland, knee-joints, and less obvious pieces of the human machine – are beginning to make mathematical sense.

To understand a malfunction is not the same as to cure it; but as every garage mechanic knows, it's hard to put a fault right if you don't understand what it is. Dynamical systems theory has now established a serious role in the advancement of medical knowledge. There is a rapidly growing research area of 'dynamic diseases', conditions caused not by infection or genetics but by the 'incorrect' dynamics of bodily organs. As Glass says, referring to the workings of the heart: 'A full understanding will only be achieved from the integration of nonlinear mathematics with experimental physiology and clinical cardiology.'

14

Beyond the Butterfly

Prediction is very difficult, especially about the future.

Niels Bohr

The concept of chance has been known to the human race for thousands of years, and it has ingrained itself into our culture. It has given rise to multibillion-dollar industries, ranging from gambling to religion: gambling exploits our fascination with chance (and money), and religion is a protection mechanism that we have devised to make our way in a hostile world where the effects of chance may sometimes be brutally personal. An effective calculus of chance – probability theory – has been in development for at least three hundred years. It has taught us that chance has its own type of pattern, which applies not to individual instances but to 'ensembles' of repeated experiments: statistical regularities that make themselves felt only on average. And we have developed an applied arm of probability theory, which we call statistics, which lets us exploit those patterns in areas such as life insurance and quality control.

Chaos is a relative newcomer, with a prehistory that goes back no more than a century, and it is only thirty years since it began to take off as a coherent subject with its own identity. It bears a complex and often rather confusing relation to chance. It has its own deep patterns too, but unlike the patterns of chance they are not primarily statistical, and they often apply to individuals. 'Chaotic' is *not* just a fashionable new name for 'random'. Not at all. Chaos is a different kind of beast. It may sometimes masquerade as chance, but in essence chaos and chance are poles apart.

Because the underlying regularities of chaos do not lie on the surface, scientists have been forced to devise new mathematical techniques to detect chaotic effects and distinguish them from random ones. This has led to a thriving area of research with applications right across the scientific board, including the vibrations of stars, population dynamics in ecosystems, the fluctuations of epidemics, and the variability of material properties of wire. We'll take a look at these, and other, applications of the new techniques; and along the way we'll gain a deeper understanding of the similarities between chance and chaos and those all-important differences. In addition, we'll discover that chaos need not always be unpredictable. As Niels Bohr was implying when he distinguished 'prediction' from 'prediction about the future', it all depends upon what kind of prediction you're after.

What is Randomness?

First, we'd better take a closer look at what we mean by 'random'. Chaos has taught us that we must be careful to distinguish between what happens in mathematical systems where we assume perfect and infinitely precise knowledge, and what happens in practice when our knowledge is imperfect and imprecise. The meaning of the word 'random' depends heavily upon this distinction.

A dynamical system is one whose state changes over time, according to some rule or procedure which I'll call a *dynamic*. A dynamic is a rule for getting to the 'next' state from the current one. The key to thinking about randomness is to imagine such a system to be in some particular state, and to let it do whatever that particular system does. Then imagine putting it back into *exactly* that initial state and running the whole experiment again. If you always get *exactly* the same result, then the system is deterministic. If not, then it's random. Notice that in order to show that a system is deterministic, we don't actually have to predict what it will do: we just have to be assured that on both occasions it will do the same thing.

For example, suppose the system is a cannonball, being dropped off the edge of a cliff under controlled, repeatable conditions. Suppose that the dynamic is the action of gravity according to Newton's laws. You drop the cannonball and it falls, accelerating as it does so. Obviously if you repeat the same experiment under identical conditions, the ball will do exactly the same thing as before, because Newton's laws prescribe the future motion uniquely. So this system is deterministic.

In contrast, the system may be a pack of cards, and the dynamic may be to shuffle the pack and then take the top card. Imagine that the current top card is the ace of spades, and that after shuffling the pack the top card becomes the seven of diamonds. Does that imply that whenever the top card is the ace of spades then the next top card will always be the seven of diamonds? Of course not. So this system is random.

Even this distinction is not so clear cut when we think about the real world. In fact, it's difficult to imagine circumstances in which you can be absolutely sure that the real world 'is' random rather than deterministic, or vice versa. The distinction is about appearances, not deep realities – an apparently random universe could be obeying every whim of a deterministic deity who *chooses* how the dice roll; a universe that has obeyed perfect mathematical laws for the last ten billion years could suddenly start to play truly random dice. So the distinction is about how we model the system, and what point of view seems most useful, rather than about any inherent feature of the system itself.

In modelling terms, the difference between randomness and determinacy is clear enough. The randomness in the pack of cards arises from our failure to prescribe unique rules for getting from the current state to the next one. There are lots of different ways to shuffle a pack. The determinism of the cannonball is a combination of two things: fully prescribed rules of behaviour, and fully defined initial conditions. Notice that in both systems we are thinking on a very short timescale: it is the *next* state that matters – or, if time is flowing continuously, it is the state a tiny instant into the future. We don't need to consider long-term behaviour to distinguish randomness from determinacy.

Scientists have devised many models of the real world: some deterministic, some not. In a clockwork Newtonian model nothing is *truly* random. If you run a deterministic model twice from the same initial state, it will do the same thing both times. However, we currently have a different model of the universe – a quantum mechanical one. In quantum mechanics – at least, as currently formulated – there is genuine randomness. Just as the rule 'shuffle the cards' permits many different outcomes, so the rules of quantum mechanics permit a particle to be in many different states. When we observe its state, we pin it down to a particular value – in the same way that turning over the top card of the pack reveals a particular card. But while a quantum system is following its own nose, unobserved, it has a random choice of possible futures. So whether we think that our universe as a whole is random depends on what brand of physics we currently espouse, and since we can't actually run the entire universe twice from the same initial conditions the whole discussion becomes a trifle moot.

However, instead of asking 'is the entire universe *really* random?' we can ask a less ambitious question – but a more useful one. Given some particular subsystem of the real world, is it best modelled by a deterministic mathematical system or a random one? And now we can make a genuine distinction. It is clear from the start that any real world system might be suddenly influenced by factors outside our knowledge or control. If a bird smashes into the falling cannonball then its path will deviate from what we expect. We could build the bird into the mathematics as well, but then what of the cat that may or may not capture the bird before it can crash into the cannonball? The best we can do is choose a subsystem that we think we understand, and agree that unexpected outside influences don't count. Because our knowledge of the system is necessarily limited by errors of measurement, we can't guarantee to return it to *exactly* the same initial state. The best we can do is return it to a state that is experimentally indistinguishable from the previous initial state. We can repeat the cannonball experiment with something that looks like the same cannonball in the same place moving at the same speed; but we can't control every individual atom inside it to produce the identical initial state with infinite precision. In fact, whenever we touch the cannonball a few atoms rub off and a few others transfer themselves to its surface, so it is definitely different every time.

So now – provided we remember that only short timescales are important – we can formulate a practical version of the distinction between deterministic chaos and true randomness. A real subsystem of the universe *looks* deterministic if, ignoring unexpected outside effects, whenever you return it to what looks like the same initial state it then does much the same thing for some non-zero period of time. It is random if indistinguishable initial states can *immediately* lead to very different outcomes.

In these terms the cannonball system, using a real cannonball, a real cliff, and real gravity, still looks pretty deterministic. The experiment is 'repeatable' – which is what makes Newton's laws of motion so effective in their proper sphere of application. In contrast, a real card-shuffling experiment looks random. So does the decay of a radioactive atom. The randomness of the card-shuffle is of course caused by our lack of knowledge of the precise procedure used to shuffle the cards. But that is *outside* the chosen system, so in our practical sense it is not admissible. If we were to change the system to include information about the shuffling rule – for example, that it is given by some particular computer code for pseudo-random numbers, starting with a given 'seed value' – then the system would look deterministic. Two computers of the same make running the same 'random shuffle' program would actually produce the identical sequence of top cards.

We can also look at the card system in a different way. Suppose the choice of card is determined by just the first few digits of the pseudo-random number, which is fairly typical of how people write that kind of program. Then we don't know the 'complete' state of the system at any time – only the few digits that tell us the current top card. Now, even with a fixed pseudo-random number generator, the next card after an ace of spades will be unpredictable, so our model has become random again. The randomness results from lack of information about some wider system that includes the one we think we are looking at. If we knew what those 'hidden variables' were doing, then we would stop imagining that the system was random.

Suppose we are observing a real system, and we think it looks random. There are two distinct reasons why this might happen: either we are not observing enough about it, or it truly is irreducibly random. It's very hard to decide between these possibilities. Would the decay of a radioactive atom become deterministic if only we knew the external rules for making it decay (the shuffling rule) or some extra 'internal' dynamic on 'the entire pack of cards'? Fine – but right now we don't, and maybe we never will because maybe there is no such internal dynamic. (See Chapter 16 for some speculations on this topic.)

I repeat, we are in the business of comparing observations of the real world with some particular model, and it is only the *model* that can safelybe said to be random or deterministic. And if it is one or the other, then so is the real world, as far as those aspects of it that our model captures are concerned.

Chance and Chaos

Having sorted out what we mean – or at any rate, what *I* mean – by 'random' and 'deterministic', we can turn to the relation between chance and chaos. This is not a simple story with a single punchline. The main source of potential confusion is the multifaceted nature of chaos: it takes on different guises when viewed in different lights.

On the surface, a chaotic system behaves much like a random one. Think about computer models of the Earth's weather system, which are chaotic and so suffer from the butterfly effect. Run the computer model starting from some chosen state, and you get a pleasant, sunny day a month later. Run the same computer model starting from some chosen state *plus* one flap of a butterfly's wing – surely an indistinguishable state in any conceivable practical experiment – and now you get a blizzard. Isn't that what a random system does? Yes – but the timescale is wrong. The 'randomness' arises on large timescales – here months. The distinction between determinacy or randomness takes place on short timescales; indeed it should be immediate. After a day that flapping wing may just alter the local pressure by a tenth of millibar. After a second, it may just alter the local pressure by a ten billionth of a millibar. And indeed in the computer models that's just what happens. It takes time for the errors to grow – and we can quantify that time using the Liapunov exponent. So we can safely say that on short timescales the computer model of the weather is not random: it is deterministic (but chaotic).

To add to the scope for confusion, in certain respects a chaotic system may behave *exactly* like a random one. Remember the 'wrapping mapping' of Chapter 6, which pulls out successive decimal places of its initial condition using the dynamical rule 'multiply by ten and drop anything in front of the decimal point'? There is nothing random about the *rule* – when presented with any particular number, it always leads to the same result. But even though the rule is deterministic, the behaviour that it produces need not be.

The reason is that the behaviour does not depend solely on the rule: it depends on the initial condition as well. If the initial condition has a regular pattern to its digits, such as 0.3333333 . . . , then the behaviour (as measured by the first digit after the decimal point) is regular too: 3, 3, 3, 3, 3, 3, 3. However, if the initial condition was determined by randomly throwing a die, say 0.1162541 . . . then the behaviour will appear equally random: 1, 1, 6, 2, 5, 4, 1.

In the sense described, the 'multiply by ten' dynamical system displays absolutely genuine random behaviour – exactly as random as the die that produced 1, 1, 6, 2, 5, 4, 1 in the first place. However, it would be a gross distortion to say that the system 'is' random, for at least two reasons. The first is that the measurement we are considering, the first digit after the decimal point, is not a complete description of the state of the system. A more accurate representation is 0.3333333, 0.333333, 0.33333, 0.3333, 0.333, 0.33, 0.3; or in the random case 0.1162541, 0.162541, 0.62541, 0.2541, 0.541, 0.41, 0.1. That second sequence doesn't look totally random if you see the whole thing. The second reason is that it is the *initial condition* that provides the source of randomness; the system merely makes this randomness visible. You might say that chaos is a mechanism for extracting and displaying the randomness inherent in initial conditions, an idea that the physicist Joseph Ford has advocated for many years as part of a general theory of the information-processing capabilities of chaos.

However, a dynamical system is not just a response to a single initial condition: it is a response to *all* initial conditions. We just tend to observe that response one initial condition at a time. When we start thinking like that, we can soon distinguish regular patterns lurking among the chaos.

The most basic is that – for a time – systems whose initial conditions differ by a small amount follow approximately similar paths. Thanks to the butterfly effect this similarity eventually breaks down, but not straight away. If the initial condition had been 0.3333334 then the behaviour would have been 3, 3, 3, 3, 3, 3 – so far so good – and then 4, well, it couldn't last for ever. In exactly the same way, if the initial condition had been 0.1162542 instead of 0.1162541, the two behaviours would also have looked very similar for the first six steps, with the difference becoming apparent only on the seventh. In fact, if we compare the exact values (rather than just our 'observations' of the first decimal place) then we can see how the divergence goes. The first initial condition goes

> 0.1162541, 0.162541, 0.62541, 0.2541, 0.541, 0.41, 0.1

and the second goes

> 0.1162542, 0.162542, 0.62542, 0.2542, 0.542, 0.42, 0.2.

The differences between corresponding values go

      0.0000001, 0.000001, 0.00001, 0.0001, 0.001, 0.01, 0.1,

and each is ten times bigger than the previous difference. So we can actually see how the error is growing; we can watch how the butterfly's flapping wing cascades into ever-bigger discrepancies.

This kind of regular growth of tiny errors – I'll use the word 'error' for any small difference in initial conditions, whether or not it's a mistake – is one of the simplest tests for chaos. The technicians call it the system's Liapunov exponent, named after A. M. Liapunov, a famous Russian mathematician who invented many of the basic concepts of dynamical systems theory in the early 1900s. Here the Liapunov exponent is 10, meaning that the error grows by a factor of ten at each step. (Well, strictly speaking the Liapunov *exponent* is log 10, which is about 2.3026, because the rate of growth is *e* raised to the power of the Liapunov exponent, not the exponent itself. But that's a technical nicety. To avoid confusion I'll talk about the 'growth rate', which here really is 10.)

Of course the growth rate of a tiny error is not always constant: the only reason that the growth here is exactly tenfold at every step is that the dynamics multiplies everything by ten. If the dynamics were more variable, multiplying some numbers by 9 and others by 11, say, then you'd get a more complicated pattern of error growth; but on average and for very small initial errors it would still grow by some rate between 9 and 11. In fact Liapunov proved that every deterministic dynamical system has a well-defined rate of growth of errors, provided the errors are taken to be sufficiently small.

The Liapunov growth rate provides a quantitative test for chaos. If the Liapunov growth rate is bigger than 1, then initial errors, however small, increase exponentially. This is the butterfly effect in action, so such a system is chaotic. However, if the Liapunov growth rate is less than 1, the errors die away, and the system is not chaotic. That's wonderful if you know you have a deterministic system to begin with, and if you can make the extremely accurate observations required to calculate the growth rate from experiments. It's much less useful if you don't, or can't. Nonetheless, we see that deterministic systems behave differently from random ones, and that certain features of that difference lead to quantitative measures of the degree of chaos that is present. The Liapunov exponent is just one diagnostic of chaos. Another is the fractal dimension of the attractor (see Chapter 11). A steady state attractor has fractal dimension 0, a periodic cycle has fractal dimension 1, a torus attractor formed by superposing *n* independent periodic motions has fractal dimension *n*. These are all whole numbers. So if you can measure the fractal dimension of a system's attractor, and you get numbers like 1.356 or 2.952, then that's an extra piece of evidence for chaos. How can we measure such a fractal dimension? There are two main steps. One is to reconstruct the qualitative form of the attractor using the Ruelle–Takens method of Chapter 9 or one of the many variants that have appeared since. The other is to perform a computer analysis on the reconstructed attractor to calculate its fractal dimension. There are many methods for doing this, the simplest being a 'box-counting' technique that works out what proportion of different-sized boxes is occupied by the attractor. As the size of the box decreases, this proportion varies in a manner that is determined by the fractal dimension. The mathematics of phase space reconstruction then guarantees that the fractal dimension of the reconstructed attractor is the same as that of the original one – provided there *is* an original one, which means that your system must be describable by a deterministic dynamical system to begin with.

It is absolutely crucial not to be naïve about this process. You can take any series of measurements whatsoever – say the prices in your last twelve months' shopping lists – push them through the Ruelle–Takens procedure, and count boxes. You will get *some* number – maybe 5.277, say. This does not entitle you to assume that your shopping list is chaotic and lives on a 5.277-dimensional strange attractor. Why not? Firstly, because there is no good reason to assume that your shopping list comes from a deterministic dynamical system. Secondly, because even if it did, your shopping-list data contains too little information to give any confidence in a dimension that big. The bigger the dimension of an attractor, the more data points you need to pin the structure of the attractor down. In fact any fractal dimension over about 4 should be viewed with great suspicion.

Pulsating Stars

By using your head and assembling enough supporting evidence it may become possible to claim, with confidence, the existence of a chaotic attractor in real data. A typical case study is an investigation of the light output of the star R Scuti, carried out in 1995 by Robert Buchler, Thierry Serre, and Zoltán Kolláth, physicists at the University of Florida in Gainesville.

The phrase 'the fixed stars' goes back many centuries, and its connotation is not just that true stars remain in fixed positions – relative to each other – unlike the 'wandering stars' or planets. It also encapsulates the view of stars as unchanging and eternal.

This is of course false: stars are dynamic entities.

The first changing star to be studied scientifically was a nova ('new star', actually a star that has exploded and thereby become bright enough to be seen by the naked eye) observed by Tycho Brahe in 1572. In 1596 Fabricius discovered that the star omicron Ceti seemed to appear and disappear; in

1638 it was realized that it does so periodically. In 1669 the Italian astronomer Geminiano Montanari noticed that the star Algol is subject to sudden decreases in brightness; ancient Arab astronomers may well have known this since the name comes from *al-ghul*, the ghoul or demon. In 1782 the English astronomer John Goodricke discovered that Algol's light ouput changes in a periodic cycle, and also came up with the right explanation: Algol is what we nowadays call an eclipsing binary. It is not one star, but two, rotating about their common centre of mass. Whenever one star gets in the way of the other, as seen from Earth, the total light output diminishes. You can tell from the light output, which is mostly constant, punctuated by two brief periods when it declines (star A goes in front of star B as seen from Earth; star A goes behind star B).

This is one kind of stellar dynamic, but there is another, driven by the star's own nuclear reactions. Goodricke discovered the first such star, delta Cephei, in 1874, while looking for more eclipsing binaries. The character of the changes in its light output is very different from that of a typical eclipsing binary. The light curve of delta Cephei has the shape of a rounded sawtooth: it climbs rapidly, then drops slowly but steadily, then picks up and climbs rapidly again. Variable stars of this type, known collectively as cepheids, are single stars whose light output really does cycle. Astronomers are extremely interested in variable stars, which are used, for instance, to work out distances. The brightness of stars falls off with distance according to the inverse square law, and the period of most cepheids depends on the true brightness in a known manner. So we can observe the period and the apparent brightness, calculate the true brightness, and by comparing these deduce how far away the star must be. In some cepheids, the variation in brightness is not truly periodic, but irregular. In 1987 Robert Buchler and G. Kovács ran computer models of the dynamics of stars of 'W Virginis' type, and showed that their light ouput fluctuates chaotically, for a large range of masses, temperatures, and luminosities. This is true mathematical chaos: when a suitable parameter is varied it arises as the culmination of a period-doubling cascade, and is triggered by a 5:2 resonance of two distinct modes of vibration of the star. This raises the question whether chaos can be observed in the light curves of actual stars.

Critics of 'chaos theory' always seem to think that if chaos really exists then it should be obvious: just look at the data, and the chaos ought to stare you in the face. Unfortunately this is not so. It is never possible to record *all* relevant data on some physical event: instead the observer makes choices, based upon what is convenient to measure and what they deem to be useful. These choices are affected by whatever model the observer may have at the back of their mind, if only tacitly: a statistical point of view leads to the collection of lots of statistical data like averages and standard deviations, a

Fourier model leads to the collection of power spectra, whatever. Data that have not been recorded with a chaotic model in mind – which means virtually all data prior to about 1975 – are seldom in a suitable form for chaos-theoretic analysis. There may be gaps, or the data may be too few in number (most methods of detecting chaos require a *lot* of data), or the observational errors may be too large, obscuring the fine detail that is characteristic of chaos.

These problems are especially difficult in astronomy, where the scheduling of large expensive telescopes precludes long or unbroken runs of data acquisition, and many observations are intrinsically inaccurate anyway. Until 1995 nobody had found solid evidence for chaos in the dynamics of a real star. In that year, however, Buchler, Serre, and Kolláth published an analysis of the cepheid variable R Scuti, obtaining strong evidence that its light output is chaotic.

You will see that this conclusion required a lot more than just looking at 'the data'. Observations of the light ouput of numerous stars, including R Scuti, are made at irregular intervals by amateur astronomers and collected by an international organization known as AAVSO. (Professional astronomers are too busy on more sophisticated things to spend their time collecting routine data like this: the amateurs do a wonderful and useful job for the fun of it.) These observations show that R Scuti displays irregular oscillations with an approximate 'period' of around 140 days. The data are scattered (indicating that errors are common) and there are gaps. To deal with these problems, the data must first be 'preprocessed' to average out the errors and interpolate to fill the gaps. This is done by taking a 'moving average' of values over a period of 2.5 days, and then fitting the resulting 'filtered' data to a type of smooth mathematical curve known as a cubic spline. See Figure 127.

Buchler's team chose a 15-year segment of filtered and smoothed AAVSO data that seemed to them to be representative, having phases of both low and high brightness. These data are shown in *one* of the three pictures in the left-hand column of Figure 128. I will deliberately conceal, for a few paragraphs, which. The other two pictures show 'synthetic' data derived from a

Figure 127   Light output of R Scuti. Dots show individual AAVSO observations, solid line shows result of smoothing and interpolation.

Figure 128 Three light curves and power spectra for R Scuti. Two pairs are synthetic data from the low-dimensional chaotic model, one pair shows real data. Which is the odd one out?

low-dimensional chaotic attractor that Buchler's team reconstructed from the real data. The three pictures in the right-hand column are power spectra, an idea that I described briefly in Chapter 9. To find the power spectrum of a time-series you try to represent it as a sum of periodic sine curves of different frequencies (by Fourier analysis) and then plot the amplitude of the components against their frequencies. You can see that the three power spectra are virtually identical, and that the three time-series all have the same 'texture', indicating that they correspond to very similar attractors.

Right, now: two pairs of pictures come from a chaotic model derived from the real data: the other one *is* the real data. Can you spot the odd man out?

The chaotic model is obtained by a version of Ruelle–Takens phase space reconstruction, but instead of drawing the shape of the reconstructed attractor, this version lets you reconstruct a formula approximating the dynamics that gives rise to that attractor. The method is called 'global polynomial expansion' and was worked out by Buchler and co-workers. They find that the important features of the star's dynamics can be captured by a discrete dynamical system, defined by a mapping which is a fourth degree polynomial. That is, a function containing only products of the variables in which at most four variables are multiplied together in any term. The key point is: it's not terribly complicated. They then subjected this model system to a series of stringent tests. They showed that for most initial conditions it led to dynamics that were stable, in the sense that the values of the variables remained within definite limits, and were not periodic. The time-series generated by the model closely resemble the real time-series in 'texture', and the power spectrum of the model is very similar to that of the real data. (Have you worked out yet which pair of pictures in Figure 128 is the real one? It's the top pair.). In contrast, if 'lower order' approximations by quadratic or cubic equations are used instead, none of these features holds good – the synthetic data look totally different from the real data. But there is more. If the fourth degree model is changed slightly, it *continues* to produce time-series and power spectra that resemble the real data. That is, the model is robust, and its close resemblance to reality does not depend upon choosing parameters very, very carefully. There is also some additional quantitative evidence. The fractal dimension of the attractor for this model is about 3.1, which (a) indicates chaos since it is not an integer, (b) is small enough to be useful, and (c) is consistent with the need to use a fourth order polynomial rather than a cubic one. The Liapunov growth rate is 1.0019, and since this is greater than 1 the synthetic model is subject to the butterfly effect – that is, it's chaotic.

Buchler's team provides a lot more evidence for the consistency of the hypothesis that the dynamics is chaotic, but you get the idea. What you see is a team of informed and intelligent scientists doing their level best to *shoot down* the suggestion that chaos does *not* fit the data, failing, and only then concluding that chaos is present. Their work goes far beyond merely looking at an irregular time-series and declaring it chaotic. Such techniques are now well understood and there is quite a battery of them.

R Scuti's light curve may not have the immediate human relevance of building a better mousetrap, and it may not make you a billionaire overnight, but it's important to astronomers. If R Scuti is varying randomly, then we don't understand it – or other stars like it – and we don't even know where to start. If R Scuti is varying chaotically, with a fractal dimension as low as 3.1, then there are dynamic models simple enough for mere humans to understand that match the star's real dynamics pretty well. Moreover, the same may well be true of other variable stars too. Let me leave the last word to Buchler's team, who write:

> More standard techniques, for example a Fourier sum of 30 optimally chosen frequencies, barely generate a synthetic signal of similar properties. While such a fit can *serve as an interpolation*, when a synthetic signal is generated as a continuation, the appearance is quite different from the R Scuti light curve; the Fourier amplitude spectrum is the same, but does not have the proper phase relationships. Standard maps such as the deterministic linear ARMA schemes fall short of producing synthetic signals that are both stable and have the same appearance as the data. On the other hand, it is always possible to cook up stochastic schemes that successfully model the data. However, when our approach was tested on such an apparently similar signal, the polynomial map was unable to produce a decent synthetic signal. [That is, if you start with data that you know are stochastic, because you made them that way, then Buchler's methods do *not* indicate chaos.] Furthermore it would take a strong *deus ex machina* to cause disturbances of that size in a star. In our opinion such stochastic models appear *ad hoc* compared to a fourth order nonlinear map.
>
> We conclude that the R Scuti data are chaotic and that they are well reproduced by a four-dimensional map. In view of the high noise level in the observational data the important question arises as to whether our results carry over to the actual star. We think they do for the following reasons: First, the main result, namely the dimension, exhibits a certain robustness to the smoothing process. Second, all other explanations appear contrived in comparison to the simplicity of a 4D map. In support we also note that a cursory analysis of another RV Tau-type star, namely AC Her[culis], similarly indicates the presence of a chaotic attractor of the same low dimension, this despite the substantially different appearance of the pulsations.

Short-term Prediction

If you can take a chaotic time-series, process it mathematically, and extract a model that fits it reasonably well – as Buchler's team did with their pulsating star – then you are well on the way towards predicting what it will do in the future. The butterfly effect precludes long-term prediction, but the determinism of chaos implies that it is predictable in the short term. In order to make such predictions you need either a good model for the dynamics or a time-series of actual observations that explores the entire attractor – in the sense that near any point in the attractor you can find some observed values.

In the first case you just plug your initial value into the model and see what it does. Liapunov's theory shows that the actual system will do something similar, to begin with, and the size of the Liapunov exponent tells you roughly how far away the 'prediction horizon' is, beyond which your predictions will cease to be valid. If the Liapunov exponent is small then errors grow more slowly than they do if it is big, so the prediction horizon varies inversely with the Liapunov exponent. This is how we know, for example, that predicting the weather much more than four days ahead will be very, very hard indeed.

In the second case you play a game that goes back to Lorenz, the one that he called 'analogues', which I mentioned in Chapter 7. Suppose your initial state is a point A in phase space (or equivalently a set of observations A). Then you look back through your records to find a previous point (or set of observations) B that is very close to A. This is always possible if the previous records 'explore the whole attractor', as I'm assuming for simplicity; but even if they don't, you might get lucky. At any rate, you then see what happened last time, starting from B, and you take it as your prediction of what will happen now starting from A (Figure 129a). This is prediction based on previous experience.

It may have occurred to you that there is a more intelligent way to do this, which is to take *several* points from the records – not just B but C, D, and E, say. If these points are spread around A, forming the corners of some small region that encloses it, then you can follow what happened to those corners last time around, and – provided they are close enough to A and provided you don't try to forecast the future beyond the prediction horizon – you can assume that the future of A lies somewhere within the region bounded by the previous futures of B, C, D, and E (Figure 129b).

Indeed you can do better still: you know how A sits relative to B, C, D, and E; and Liapunov's theory of the short-term dynamics near a given point tells you that the points in the forward trajectory starting at A – the future behaviour that should occur if we start at A – will bear the same relationship to the points in the forward trajectories starting at B, C, D, and E

(a) (b) (c)

Figure 129   Three methods for predicting a chaotic trajectory: (a) find a previous point nearby and see where that went; (b) find several previous points nearby and use where they went to determine a region in which the prediction must fall; (c) like (b) but interpolate the trajectory to give the best-fitting estimate.

(Figure 129c). Indeed Liapunov's theory is based on the fact that in any deterministic system the local dynamics is approximately linear, even though the large-scale dynamic may be nonlinear – and usually is. For example, if A is one third of the way between B and C, then it's not a bad approximation to assume that its trajectory follows a path one third of the way between those of B and C, because that's what happens in a truly linear system. It's a kind of interpolation scheme that gives you an optimal guess for what a point not in your records will do, by comparing it with nearby points that are in your records. This method is known as *tessellation*, because you 'tile' phase space with tiny regions whose corners occur in your records, and use those records to predict the future of each tile. You can come up with multidimensional versions of this idea, and if you want to be more sophisticated you can use nonlinear interpolation. This is how Buchler's method of global polynomial expansion works.

In the last few years many different short-term forecasting schemes for chaotic systems have been devised, each one exploiting particular features that distinguish chaos from randomness. I'll show you a few examples of how the tessellation method performs in practice, based on work of Alister Mees at the University of Western Australia. Figure 130a shows data from an electroencephalogram (EEG), which is a device for recording 'brain waves', traces of the electrical activity of the brain. The solid line shows actual data, the dotted line shows the result of repeatedly making short-term predictions based on the actual data using the tessellation method. You can see that the errors are generally small. Figure 130b shows actual and predicted data for the monthly changes in the number of cases of measles. Figure 130c shows actual and predicted values for one part of the Hudson's Bay data, mentioned in Chapter 13 – here the annual numbers of lynx pelts brought to the trading post by fur trappers.

Figure 130 Chaotic time-series (solid lines) and their short-term predictions (dashed lines): (a) EEG of a resting subject; (b) measles epidemic differences; (c) lynx pelt numbers. In (c) the horizontal scale indicates the year (0 = 1800).

All of these methods assume that the system being predicted is deterministic; they can't cope with sudden disturbances imposed from outside. But no method can cope with such disturbances – except crystal-gazing, and I don't believe anybody has ever made that work. Other applications include sunspot numbers and financial data. However, don't expect the tessellation method to warn you about earthquakes or to predict stock-market crashes. Both of those involve a degree of outside interference, and understanding them will probably require different data from the obvious historical records, if it is possible at all.

The Real World

Some of you may feel that pulsating stars and lynx pelts are all very well, but what about chaos in the real world?

I'm always amazed by the number of people who assume that *they* live in the real world and everybody else lives in fantasy land. There are businessmen who tell you to live in the real world of business; bankers who tell you to live in the real world of banking; advertising executives who tell you to live in the real world of advertising; journalists who tell you to live in the real world of newsprint; doctors who tell you to live in the real world of medicine priests who tell you to live in the real world of their own brand of god . . .

Me, I spend a lot of my time in the real world of research mathematics. I could say, 'Come on, guys, stop playing silly schoolboy games with stock-market derivatives, those are meaningless fictions. How about proving some *real* theorems?' But I've (just about) got more sense. Let's have a little respect the other way, hey? Many people think that scientific research is an artificial activity, and even more seem to think that abstract mathematics is just intellectual game-playing. 'Get out of your ivory towers and live in the real world!' they scream. Usually over the television networks that wouldn't exist without generations of research from scientists, engineers, and mathematicians.

Ho hum.

The thing is, if you have even a cursory understanding of the history of science you know that rather a lot of what constitutes today's real world began its life in the ivory towers. And I'll tell you something else: my world of mathematics *is* just as real as your world of barkeeping, bookkeeping, or beekeeping. You spend five years trying to prove that there are spiral solutions to reaction–diffusion equations, in between giving lectures to 150 noisy students and helping to raise enough grant money to keep yourself in a job and your university on the black side of the balance-sheet, and you'll find my world pretty real too. In fact you'll find it resembles your own rather closely.

Let's retain the 'ivory tower' terminology – not because it's accurate, but because it stands for a common point of view. How long should it take for a new mathematical concept to make its way out of the ivory tower of pure research to earn its keep in the real world of industry and commerce?

A hundred years is not unusual.

I'm not sure how you're going to react to that. You may conclude that mathematicians must be a pretty disorganized lot, in serious need of some simple managerial skills. But before you march in with your three-dimensional technicolor histograms, business-plans, and flip-charts, I'd like to point out that the only effect they are likely to have is to extend the process even further. There are a lot of people who think that they can manage creativity – but I know no creative people who think that. In fact one reason *why* it takes a hundred years for mathematics to turn into money is that mediocre bureaucrats who lack the necessary imagination get in the way and slow the process down.

In any case, what you should be surprised at is how *fast* the process of assimilation of new mathematics into human culture is. *Only* a century.

I'm serious. In engineering circles it is not at all unusual for a really new concept to take eighty years or more to go from the drawing-board to a saleable piece of kit. For instance, the fax machine was invented around 1890 but didn't become a worldwide method of communication until 1990. It took eighty years to get the principle of xerography, which lies behind photocopiers, working effectively in a commercial product. Lasers fared better – only forty years from the initial gleam in an experimentalist's eye to ubiquitous existence in CD players. When it first appeared in the ivory towers the laser was derided as 'a solution looking for a problem'. It was. And its creators found an enormous range of important problems that it solved. Among them were holograms, whose operating principle was discovered by Dennis Gabor in 1947 but which had to wait for the invention of the laser before they could end up as a security device on credit cards.

Leonardo da Vinci designed a helicopter, but he couldn't make it fly.

And that's just the engineering step. Engineering builds on physics, which in turn takes its inspiration from mathematics. At each step it takes time for new ideas to sink in, for potential to be recognized, for potentialities to be realized. Often it requires a new generation that grew up with the ideas and feels familiar with them. Interest in binary arithmetic predates its application in computer hardware by at least five centuries. From the wave equation to television took two hundred years. When Arthur Cayley invented matrices in 1855 he said (I paraphrase) 'well, here's a mathematical idea that will never be useful for anything'. Today Cayley's matrices are indispensable in economics, statistics, mechanical and electronic engineering, geology, astrophysics, and almost certainly in the design of running shoes. (Don't believe it? Computer Aided Design makes heavy use of matrix algebra to rotate three-dimensional objects. Case closed.)

Chaos is doing far better: only twenty-five years from the realization that there was a new *area* of mathematics to serious applications in the commercial sector. Fractals did even better, only fifteen years or so.

Current commercial applications of chaos and fractals include a chaos-theoretic method for extracting meaningful conversations from recordings made in a noisy room – guess who the main users are likely to be – and Michael Barnsley's fractal method for compressing the data defining visual images mentioned in Chapter 11. There are commercial companies that employ chaos-theoretic data analysis to advise investment banks on market movements. There are chaos-theoretic studies of wear on train wheels. One Japanese company has discovered that dishwashers are more effective if their rotor arms move chaotically. As I'll discuss in the next chapter, NASA engineers have exploited chaos in the three-body problem to send a 'dead' satellite on an encounter with a comet using only the tiny amount of attitude-control propellant that was left – far too small for any conventional rendezvous trajectory to be feasible. And there are dozens, maybe hundreds, of practical chaos-theoretic ideas under development in the world's laboratories – ranging from ways to transmit unbreakable code messages to novel Earth–Moon trajectories that require only half the fuel needed for what had previously been thought to be the most economical orbit possible.

You want dollars-and-cents payoff? There are glimmerings now. Give chaos a few more decades, let the next generation of scientists get really used to it: you won't be disappointed. And it will still be producing the goods faster than many industrial R. & D. projects.

The FRACMAT Project

I do have inside knowledge of one commercial exploitation of chaos, and I'm going to tell you about it because it sheds a lot of light on what goes on when you try to transfer a piece of mathematical technology out of the learned journals and into something that can be sold. If you think that this is a straightforward process requiring nothing more than managerial skills – well, welcome to the real world.

The application that I want to tell you about is a machine called FRACMAT. As I write, there is only one such machine in the world, but a commercial order has just been placed for the second and the number of enquiries is in the nineties. FRACMAT applies chaos-theoretic techniques of phase space reconstruction to solve a problem that has plagued Britain's spring-making industry, and the wire industry that serves it, for at least twenty-five years.

That problem is: how can you tell, quickly and cheaply, whether a consignment of wire can be coiled successfully into springs?

Yes, springs. Bed-springs, car valve springs, and the things that make ballpoint pens explode when you stupidly unscrew them.

Before I got involved in FRACMAT I had no idea how springs were made – and even less of an idea how wire was made. Basically, you make a spring by feeding wire, at speed, into a coiling machine. This machine is about the size and shape of two filing cabinets. The wire begins as a loose coil, known as a wap, about a metre across, which lies horizontally on a rotating turntable called a swift. The coiling machine draws it along through a series of rollers, and runs it past two tools. One bends the wire through a quarter of a circle, and the other nudges it sideways. A third cutting tool snips off the spring when it has fully formed.

You make wire, by the way, by starting with rather thick wire called 'rod' – about as thick as a pencil – and drawing it through a series of twenty or thirty rotating dies whose holes get successively smaller. You lubricate the dies with soap and there are a lot of other tricks that only wire-makers know.

I don't know how you make rod.

Anyway, back to springs. A normal coiling machine can make three or four springs every second. Some specialist machines that use fine wire can coil 80,000 high-precision springs an hour – that's 22 per second. The best way I can think to describe what a spring looks like when it is being coiled is to imagine what a corkscrew looks like as it begins to poke through the cork, when viewed from *inside* the bottle. (I know a corkscrew shouldn't poke throught the cork at all: like I said, this is the *best* image I can find – I didn't claim it was perfect.) What you see (or would see in slow motion) is a helical length of wire that turns and lengthens. Unlike the corkscrew, however, this wire enters the coiling machine more or less straight, and is then nudged in two separate directions to form a coil: a quarter turn round the eventual helix, and a slight push along its axis.

Springs may look crude, but they are very precise components, and they have to be the right size, shape, and strength. There are springs everywhere. A video-recorder contains hundreds. Car engines contain between eight and thirty-two valve springs, depending on design – no doubt motor aficionados will tell you that some contain more, and which, in enormous detail, probably at a party. The collision-detecting device that sets off a car's airbag, to save the driver's face from intimate contact with the steering-wheel, is basically a ball balanced on a few springs. You wouldn't be too pleased if your airbag went off by mistake, so those springs have to be very precise, allowing the device to distinguish reliably between hitting an obstacle and driving over your local council's traffic-calming measures. Nowadays manufacturers are starting to put airbags in off-the-road vehicles, whose normal mode of operation is only marginally distinguishable from colliding with an avalanche. The airbag must be triggered only when the vehicle hits a rock big enough to stop it. So those springs have to be very precise indeed.

All of which poses a quality-control problem.

It takes a skilled operator a lot of time to 'set' a coiling machine – that is, to adjust it so that it produces the right design of spring. Four to six hours is not unusual, and that's using computerized adjusters instead of spanners as in the Old Days. And that's also assuming that the wire has 'good coilability' – which means that if you adjust the machine right, it actually will form into springs. What the operator does is to coil some test springs, run them through the entire manufacturing process – including, if appropriate, heat treatment, hardening, galvanizing, and machining (say to get the ends flat). Then the resulting batch of springs is given a statistical test to see if it's of acceptable quality. If not, the operator tries to guess what went wrong, resets the coiling machine, and has another go . . . If the wire won't coil, this is never going to work, but the operator may take 12 hours or more before becoming convinced that the wire is at fault.

The spring-maker then has to convince the supplier to take it back. Actually that's not a big problem: the supplier always takes it back, to preserve customer relations. But wouldn't it be nice if the supplier could test its coilability – reliably and quantitatively – in advance? Then there'd be no argument. And then they could sell 'guaranteed coilable' wire for a higher price . . .

The problem is that – prior to FRACMAT – there was no quick and easy way to distinguish good-coilability wire from poor-coilability wire. All the wire sent to the spring-makers by the wire-makers passes all of the standard quality-control tests, things like material composition and tensile strength. Even so, about 10 per cent of wire that passes these tests has poor coilability (Figure 131).

Not so long ago I knew none of this, of course. I was living in the ivory tower, playing intellectual games about reconstructing chaotic attractors and the like. Silly, useless stuff – should have got out and done something sensible like running market surveys. Some time in 1991 I started to get telephone calls from an engineer named Len Reynolds who worked for an outfit in Sheffield called SRAMA – Spring Research and Manufacturers' Association. (It has recently changed its name to Institute of Spring Technology.) World-wide, the manufacture of springs is carried out by relatively small companies, and the same goes for the wire-making companies that supply them with raw materials. These small companies have banded together to create their own joint R. & D. arm, which is SRAMA. There are many other similar trade associations – for instance SRAMA shares a building with CATRA, a research association for the cutlery industry.

Len worked for SRAMA. SRAMA had devised what ought to be an

Figure 131   Samples of springs made from (a) good-coilability and (b) bad-coilability wire: springs coiled from good wire are much the same size, those from bad wire vary considerably. Test springs coiled on a mandrel (above) show greater variability in poor wire.

effective test for coilability: force the wire sample to form a long spring by winding it round a long metal rod, or mandrel. Even poor-coilability wire can't fail to form some kind of coil when it's wound on a mandrel, just as spaghetti cannot fail to coil when you twirl a spoon in the middle of a plateful. You can't *manufacture* springs that way – too slow. But you can manufacture test springs. And having done so, you can try to decide whether the test spring looks like what you'd expect from good-coilability wire.

There are several ways to do this. One is to get an experienced engineer – the traditional 'Old Fred' who has been with the firm since the year dot and knows where all the bodies are buried – and show him the spring: he either nods or shakes his head. It works – if Fred has a 'wise eye' – but you can't build Fred into an international quality standard.

Another is to measure the spacings between successive coils on the test spring: engineers call these spacings *pitches*. Experiments show that on the whole, good-coilability wire makes a test spring with nice, regular coils, whereas poor-coilability wire makes a test spring with erratic spacings. SRAMA had invented a machine that did this with a laser micrometer, and fed the resulting list of numbers into a computer. They had then tried every statistical test in the book, and a few not in the book, trying to separate the good wire from the bad.

Nothing worked. Not as well as they wanted.

Len had figured out why, in general terms. It wasn't just the statistics of the spacings that mattered, but the order in which they came. Let's say that a single coil is 'fat' if it's a bit wider than it ought to be, and 'thin' if it's a bit too narrow. Then, simplifying hugely, wire that produces successive coils something like

fat / thin / thin / fat / fat / thin / fat / thin / fat / thin / thin

will probably have good coilability, whereas wire that goes

fat / fat / fat / fat / fat / thin / thin / thin / thin / fat / thin

won't. The reason is that a real spring consists of several coils. In the first case, the errors tend to cancel out. In the second case, what you get is a complete spring that is too long, followed by one that is too short, and neither are any use.

This is an oversimplification, of course, but the basic idea is right. Figure 132 shows measured sequences of spacings for some typical wire samples, one very good, the other very bad. You don't have to be a genius to tell the difference; but it's in the no man's land in between that the problem lies.

At any rate, Len reasoned that the key to coilability is *sequential* variability of the material properties of the wire, not statistical variability.

But how can you quantify this?

Figure 132   Time-series of spacings of coils: (a) good-coilability wire; (b) bad-coilability wire.

He had picked up a copy of the first edition of *Does God Play Dice?*, because (like you) he's the sort of person who would naturally be interested in such things. One day he was reading the description of phase space reconstruction, in which a time-series of measurements is turned into a geometrical shape by an appropriate mathematical algorithm. It occurred to him that the sequence of coil-spacings produced by SRAMA's laser micrometer is, in effect, just a time-series. Indeed it represents the temporal sequence in which the coils were formed – but the mathematical point is that it's an ordered sequence, and thus can be analysed *as if* it were a time-series. So he tried out Ruelle–Takens phase space reconstruction – and found that the difference between good- and bad-coilability wire seemed to be staring him in the face. The reconstructed attractor generally resembled an elliptical blob. If that blob was nice and compact, the wire was good; if not, then the wire was bad. Figure 133 shows the reconstructed attractors for the two wires of Figure 132.

Now, these 'attractors' don't look pretty and fractal, like the Lorenz attractor or the Hénon attractor. They're just fuzzy clumps. But then, so is the third attractor in Figure 76, which is actually quasiperiodic, or the 'dripping tap' attractor in Figure 80, which definitely is chaotic. As Len pointed out to me over the phone, it doesn't *matter* whether the time-series for a test spring is genuinely chaotic or not, because that's not the question. The question is to find a quantitative method for distinguishing good wire from bad – and, refining the idea, to say *how* good or bad it is. Ruelle–Takens reconstruction doesn't just work for a chaotic time-series. It works for any time-series whatsoever; and what it does is to provide a rigorous mathematical way to

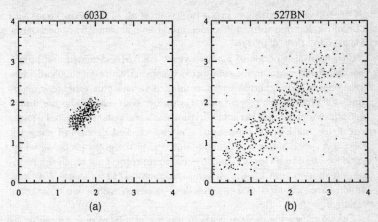

Figure 133   Attractors reconstructed from Figure 132 by the Ruelle–Takens method: (a) good-coilability wire; (b) bad-coilability wire.

characterize the type of sequential variability occurring in that time-series. So Len's idea was not to use chaos theory to prove that the spacings of test springs are chaotic – which might be true and might not, but either way is irrelevant to coilability. It was to take a mathematical technique invented within chaos theory, and use it to provide a quantitative characterization of the sequential variability of the coils on a test spring.

His first few tries, carried out on a home PC, suggested that this might work. I thought this was all very interesting, and encouraged him to keep going, but I never dreamed that I would become seriously involved. Everything changed when Len found out about the Department of Trade and Industry's 'Carrier Technology' programme, which was offering grants for technology transfer projects that might benefit UK industry. To cut a long story short, we applied for a grant, and got it. We've since got a second, bigger one for the next phase, which is in-line computer control of the coiling machine plus an extension of FRACMAT to the strip metal industry – but that's another story. Over a period of two years a team at SRAMA (Len Reynolds, Derek Saynor, Mike Bayliss, and others) joined forces with a team at Warwick University (myself, Mark Muldoon, Matt Nicol). They were supported by several wire manufacturers who supplied samples of wire, told us which they thought were good or bad, tested out prototype equipment in their factories, and attended regular project meetings. A few key people from the DTI kept a close and helpful eye on what we were doing. Over a year or so what started out as a bunch of people with very different backgrounds and prejudices turned into a highly effective team with a common (and idiosyncratic) language. My off-the-cuff characterization, in one project meeting, of springs as 'chaos with a bit of randomness thrown in' ended up on several company noticeboards. 'Dead right,' they all told me.

Over a project period of just two years, SRAMA designed and built a machine that would automatically form a test spring (with about 500 individual coils) on a mandrel, and Warwick embodied various chaos-theoretic algorithms for phase space reconstruction, together with some more traditional algorithms for detecting periodic variations, in a computer program. I wanted to call the machine MANDRELBOT, but we chickened out and named it FRACMAT – the acronym originally attached to the grant proposal, which stood for FRACtal MATerials, an early (and in restrospect not totally appropriate) attempt to sum up the design concept for DTI officials. We really should have called it TAKMAT – Ruelle–TAKens reconstruction of MATerial variability.

FRACMAT (see Figure 134) is about the size of a large desk turned on its side. It incorporates a number of innovations not originally foreseen, such as a friction-measuring device whose sequential measurements can also be phase-space reconstructed. We added this because (a) it was relatively easy

Figure 134   The FRACMAT machine for quality control of wire.

and (b) for some kinds of wire, such as stainless, it is the sequence of friction measurements that conveys the main information on coilability. The entire project timetable and budget had to be rejigged to fit it in, and the software had to be 'doubled up' to handle both kinds of data.

We had teething troubles. On one occasion a memory upgrade stopped the computer communicating with the motors. It turned out that the upgrade had speeded up the computer, and the card that drove the motors was out of synch with the motors themselves. We solved that one by adding a time delay to the computer. What we would have done if the upgrade had slowed the computer down, nobody knows. Some of our software used (legally purchased) commercial packages; one of those had a nasty habit of overwriting a key area of memory if the operator pushed the wrong key. We spent weeks trying to disable that key, and eventually settled for a recovery routine that automatically put back the overwritten file. We had to grapple with the difference between the units favoured by scientists and those traditional in the wire industry; we had to anticipate what would happen to our beloved prototype when a seventeen-year-old apprentice got his grubby paws on it, and stop him doing whatever piece of cleverness he might have in mind . . .

Anyway, we did all that, and then we built the machine, and then we tested it and changed it, and sent it out to participating companies for trials, and . . . finally, within budget and on time, we had a working machine.

FRACMAT's computer both controls the operation of the machine and analyses the results. It has two motors: one turns the mandrel about which the test spring will be coiled, and the other, via a worm drive, moves the wire along the mandrel so that it is always being coiled in the right place. The computer counts the number of turns and stops when a required number – usually 500 – is reached. For thick wire the machine can coil only 300 turns or so, in which case a second test spring is made to provide the missing 200. The completed spring is transferred (by hand, and with a few jiggles to make sure it settles freely) to a shelf above which the laser micrometer can track rapidly along, measuring all the spacings. Virtually instantly, the machine reconstructs the attractor, quantifies how long and how broad it is, and plots out the results on a 'classification diagram' (Figure 135) that determines how good or bad the wire's coilability is. The entire test takes about three minutes – not so long ago it took SRAMA two days to make the same measurements.

Once we had a working machine, we were able to run large numbers of tests quickly and easily. In particular we could do 'blind' testing where the manufacturer knew – from their own attempts to set a coiling machine – how good or bad the wire was, but we didn't. In every case FRACMAT performed beautifully. On one occasion we were sent two test wires, one supposedly good, one bad – but the FRACMAT measurements placed them at exactly the same point on the classification diagram. Suspecting a trick, we phoned the

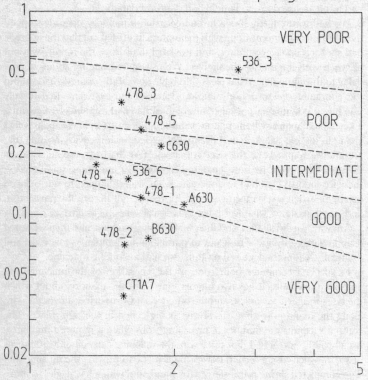

Figure 135   Classification diagram for wire quality. Coilability of wire depends on two descriptors: one statistical, one chaotic. Asterisks mark observations of wire samples of known coilability.

company and told them that we thought the two samples were identical. There was a long silence, followed by 'How the devil do you know that?' – or words to that effect. The company itself had only just discovered that one sample, which they had thought poor, was actually good: their coiling machine had suffered a fault. The samples had come from different suppliers, but on tracking them back it turned out that the two suppliers had each received them from the same wire-making company.

On another occasion a company that used wire to make goods other than springs (I can't say what for commercial reasons, indeed only SRAMA knows and they won't even tell *me*) arrived with four wire samples, which they had put through their manufacturing process and found to be good, fairly good, fairly poor, and very poor. Although they weren't intended for spring-making, the problem was definitely related to sequential variability of material properties, so the FRACMAT test seemed worth trying. Again, this was a blind test: we didn't know which sample was which until after we'd told the company. Would FRACMAT be able to tell which was which? No problem, they came out on the classification diagram in exactly the right places.

These anecdotes dramatize what we have established in numerous tests: the concept of the machine is valid, it really does let you predict coilability.

FRACMAT is now the subject of a UK patent application. It won joint second prize in the 'innovative metrology' section of the Metrology for World Class Manufacturing Awards in Birmingham in 1995. On the evening before the very last day of the project, just as if scripted, the first commercial order for the machine was placed. And if it is adopted as widely as we anticipate, it should save the UK spring-making sector tens of millions of pounds every year.

Welcome to the real world.

Not the 'real world' of small-minded accountants who won't spend a penny unless they can guarantee a 15 per cent return tomorrow: the *real* real world, in which ideas can make their way from the ivory tower to the factory floor, and imagination can turn mathematics into money.

15

Von Neumann's Dream

Control had done it. Control had started the hue and cry. There was no other explanation.

John le Carré, *The Spy Who Came in from the Cold*

'That's not unpredictability – that's control.' Around 1950, when the great mathematician John von Neumann first heard of the sensitivity of the weather to small perturbations, that's what he said – or words to that effect. The precise words seem not to have been recorded, but von Neumann's idea is reported in Freeman Dyson's 1988 book *Infinite in All Directions*:

He said, as soon as we have good computers we shall be able to divide the phenomena of meteorology cleanly into two categories, the stable and the unstable. The unstable phenomena are those which are upset by small disturbances, the stable phenomena are those which are resilient to small disturbances. He said, as soon as we have some large computers working, the problems of meteorology will be solved. All processes that are stable we shall predict. All processes that are unstable we shall control . . . This was John von Neumann's dream.

That is, since we know that you can interfere with the weather in a small way but produce a big effect, then there ought to be an economical way to produce whichever big effect you want. All you have to do is make the correct small effect. Swat the right butterfly: stop a hurricane.

Except that the vortices created by your swatter may start a bigger one.

In more scientific terms: because chaotic systems are inordinately sensitive to small changes, they are also sensitive to errors in those changes.

Nevertheless, von Neumann had the right idea. We don't yet know how to use it to control the weather – which is probably a good job when international diplomacy can't really deal with allocating the blame for acid rain. You can't just pick local weather conditions and string them together to get coherent global weather. If the transatlantic winds dump all their rain on Ireland, they can't also dump it on Norway. The European Union's Common Agricultural Policy is bad enough – imagine what bureaucrats would make of a Common Weather Policy.

We do, however, know how to use von Neumann's idea to control simple turbulence, irregular heartbeats, brain waves, nerve impulses, and artificial satellites. In the future we may well use it to control the flow of turbulent air past an aircraft wing, the population of codfish off the coast of Newfoundland, or the migration patterns of locusts in North Africa. And we may employ it to send supplies to our newly constructed Moonbase using only half the fuel required by today's methods.

The 'it' here is a dramatic new method for controlling chaotic dynamical systems, reasonably known as *chaotic control*. It realizes von Neumann's dream of making the butterfly effect work in your favour. The key idea was published by Edward Ott, Celso Grebogi, and James Yorke in 1990, and it triggered a huge amount of research – for example a review published in June 1993 lists 78 articles. But that was just the beginning, and new results and applications continue to pour out.

Chaotic control opens up a totally new direction in a classical area of engineering mathematics known as control theory. The underlying idea of control theory is that if a dynamical system does not naturally do what you want, then you should modify it so that it does. You do this by monitoring the state of the system, comparing it to what you want, and applying repeated corrections to push it back where you think it ought to be. A simple example of a control system is the thermostat in a heater. The 'natural' dynamic for a heater is either to pour out huge amounts of hot air (when it is switched on) or to do absolutely nothing (off). Neither is terribly useful on its own, but if the heater alternates between the two it can be persuaded to produce a comfortable level of heat. A thermostat uses a cunning device – typically a 'bimetallic strip' made from two different metals which expand at different rates when they get hot – to switch the heater off when the room gets too hot and switch it back on again when it gets too cold.

To witness another example, get a broom and try to balance it upright on your hand. Mathematically speaking, a vertical steady state *exists*, but it's unstable. If you balance the broom upright, let go, and do nothing, then no matter how carefully you poise it, it falls over. But if you waggle your hand about underneath the broom, you can easily keep it upright for a minute or more. Random waggles won't work, however: instead you have to keep an eye on the direction in which the broom starts to fall, and move your hand a suitable amount in that direction to restore verticality. In fact it is probably more effective to overshoot a little so that the broom rocks backwards slightly, so that you keep it wobbling to and fro about the desired vertical state.

Otherwise you'll end up chasing the broom across the kitchen and crashing into the wall.

Control theory systematizes such techniques. Control engineering builds the hardware to make them work, which nowadays usually involves a lot of electronics – mainly sensors and computers. It has been extremely successful: the space shuttle would fly like a brick were it not for computer-assisted control. Until recently, control theory mainly concentrated on stabilizing simple dynamics, mainly steady states. Its mathematical viewpoint was relentlessly linear, and it relied upon approximating the dynamics near an unstable equilibrium by some linear system. Then you changed the linear system by tinkering with the hardware, to render that equilibrium stable.

Chaotic control also attempts to stabilize simple dynamics, but it does so within the context of a nonlinear, chaotic system. It also uses linear approximations, but in a very different way.

You don't invent a new method of control just by fiddling around and hoping something will happen: you need to start with a new *idea*. Ott, Grebogi, and Yorke started with an observation well known to the pure mathematicians: chaotic attractors usually contain vast numbers of (unstable) periodic points. By exploiting the mathematicians' theoretical understanding of chaotic attractors, they came up with a method for stabilizing those normally unstable periodic points.

Pinball Wizard

One informative image of a chaotic attractor is a pinball machine. The pins push the ball away if it tries to approach them, so it has to move in a highly complex manner. The unstable periodic points in a chaotic attractor have much the same effect. So a point on a chaotic attractor is bouncing around in phase space like the ball in a pinball machine, perpetually coming close to unstable periodic points and being pushed away. And if there are *no* stable periodic states nearby, then the motion is forced to be a good deal more complex than anything recognized in classical dynamics. Whence chaos.

It's not especially obvious that chaotic attractors contain periodic points – but it *is* clear that such points must be unstable. If not, they would be attractors in their own right, and – at least according to the usual conventions of the subject – you can't have one attractor contained in another. In fact, not all chaotic attractors contain periodic points, but it's astonishing how often they turn up. Because they are unstable, they don't make their presence evident unless you look for them in the right way; but when you do, you usually find a lot of them. For example, I'm going to convince you that the logistic map $f(x) = 4x(1 - x)$, which is the map of Chapter 8 with k = 4, has *infinitely many*

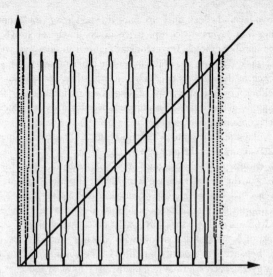

Figure 136   The fifth iterate of the logistic map $4x(1 - x)$.

periodic points inside its attractor – which is the unit interval, comprising all numbers between 0 and 1. Indeed, the periodic points are dense in that interval, meaning that as close as you wish to any point in the interval there are periodic points – but that's a bit harder to demonstrate.

Suppose you want to find the period-5 points of the logistic map. The easy way to do this is to look at the fifth iterate $f^5(x) = f(f(f(f(f(x)))))$ and find its fixed points. These are the places where the graph of f^5 crosses the diagonal line. Figure 136 shows what you find. The graph of f^5 is a line that zigzags from 0 to 1 and back again many times. Clearly such a curve must cross the diagonal many times – and whenever it does, you have found a period-5 point.

Well, not quite. The period might *divide* 5, but the only possible divisor is 1, and period-1 points are fixed points, steady states. You can easily find those by looking at the graph of f, which crosses the diagonal twice: once at 0, on the way up, and once at $x = 3/4$. (Or you can solve the equation $4x(1 - x) = x$: either $x = 0$, or you can cancel the x from both sides to get $4(1 - x) = 1$, so $1 - x = 1/4$, so $x = 3/4$. Easy!) Apart from these, every other crossing point is a period-5 point.

How many crossings are there? By counting, you will find that the graph of f^5 wiggles up and down 32 times. Now $32 = 2^5$, and this is no coincidence. In general the graph of the nth iterate f^n wiggles up and down 2^{n-1} times. This happens because f folds the unit interval over on top of itself, to produce a kind of two-layered effect: once up, once down. Then f^2 folds that over on itself, giving four layers: twice up, twice down. Each successive iteration doubles the number of folds. If you look at Figure 136 sideways, you can see all 32 folds stacked one on top of the other like a pile of towels. The diagonal line cuts each of these layers precisely once. So there are 32 points whose period is either 5 or 1. Exactly two of them have period one ($x = 0$ and $x = 3/4$), leaving 30 genuine period-5 points. These form six separate period-5 cycles, each containing five points.

You can play the same game with *any* iterate f^n of f, so altogether you can find infinitely many periodic points.

(A cute digression follows: ignore it if you wish and skip to the next paragraph. For the fifth iterate, we found $2^{32} - 2$ period-5 points, splitting into six cycles each containing five points. Luckily, 5 does divide $2^{32} - 2$, or we'd be in trouble. But it's not luck: it's the logical consistency of mathematics. Either the whole of mathematics falls apart, or 5 *has* to divide $2^{32} - 2$. Suppose that instead of 5 we look for points with a prime period p. By the same reasoning, there are $2^p - 2$ points of period p. These must split into a number of period-p cycles, each containing p points, so p must divide $2^p - 2$. That's not at all obvious, but we can test it: when $p = 7$ we have $2^p - 2 = 128 - 2 = 126 = 7 \times 15$; when $p = 11$ we have $2^p - 2 = 2046 = 11 \times 186$. Wow! In fact, by following our mathematical noses we've stumbled upon a piece of number theory that was known in ancient China, and generalized by the famous Pierre de Fermat in his 'little theorem' of 1640. All of this is completely irrelevant to this chapter, but it's fascinating to see that dynamics can be used to prove theorems in number theory. This is not an isolated case: there is a whole new area of mathematical research centred around this idea, uniting some of the deepest problems in both areas.)

Back to the dynamics. If you think of a chaotic attractor as a kind of pinball machine whose pins are the unstable periodic points, then you begin to see that chaos is much more flexible than regular behaviour. The balls cannot escape from the pinball machine – the attractor as a whole is stable – but they can pursue all kinds of interesting trajectories through it; and as von Neumann realized, tiny disturbances can switch them from one to another. So with very little effort you can rapidly redirect the dynamics and produce big changes. You can't do this with, say, a periodic *attractor*: tiny perturbations just take you back where you started. Some instability is crucial. Think of Andre Agassi getting ready to receive a serve. He doesn't know whether the ball is going to go to his left or to his right, so instead of standing still, he dances irregularly from side to side. Because his bobbing and weaving lacks pattern, his opponent can't anticipate his intentions, but because he is already in motion, and trying as hard as he can to be moving in all directions at the same time, he can respond rapidly, no matter where the server hits the ball.

Ott, Grebogi, and Yorke were among the first to investigate the extraordinary flexibility of chaotic dynamics. And they realized that the pinball analogy suggests a method for controlling chaotic systems with equal flexibility and efficiency of effort.

Back in the Saddle Again

To see how their idea works, we must take a closer look at the structure of chaotic attractors. We now know that – usually – they contain a vast set of periodic points, all of them necessarily unstable. However, there are more ways than one to be unstable. Think for a moment of a steady state of a continuous-time system in a two-dimensional phase space. We saw in Chapter 6 that there are three types of steady state: sink, source, and saddle. Of these, both sources and saddles are unstable – but there is a sense in which saddles are *less* unstable than sources.

If you perturb a point sitting at a source, then whatever you do to it, it will start to move away. The vertical position of a broom is like this – once it starts to fall sideways, then it doesn't matter what the direction is, it will keep going. This means that a periodic point inside a chaotic attractor can't be a source: if it were, then it would create a hole in the attractor, consisting of those points that get pushed away from the source and never return. But if a point lies inside such a hole, then it can't be on the attractor. So it has to be a saddle.

Moreover, saddles are very different from sources. The phase portrait near a saddle has two special pairs of lines, its separatrices: see Figure 39 in Chapter 6. Along one pair of lines, the arrows run away from the steady state – instability. But along the other pair, the arrows point back towards the steady state – stability. So there are two distinct (pairs of) directions. Perturb north–south, and the system moves further away; but perturb east–west, and it will come back. Saddles don't create surrounding holes; they just redirect the flow, like traffic-lights. A – rather contrived – example of a saddle is a broom with a dustpan tied to it by a short bit of string. Balance the broom upright (an unstable position) so that the dustpan hangs downwards from it (a position that is stable provided the broom doesn't move). If you perturb the broom, the whole thing falls over, but if you perturb *only* the dustpan, it returns to its original state. Every saddle point in the plane has these two special curves. Every saddle point in a higher dimensional phase space also has two such sets, only now they are multidimensional surfaces rather than curves, known as the *unstable* and *stable manifolds*. (Simpler terminology, unfortunately less popular, is *outset* and *inset*.)

Now, it's true that at a saddle the perturbed point returns to the steady

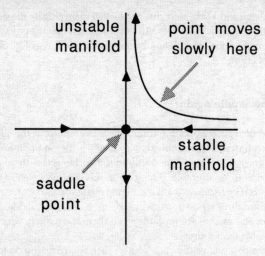

Figure 137   Motion in the neighbourhood of a saddle point. The phase point approaches the steady state along a path that remains close to the stable manifold, stays near the saddle point for a while, and finally moves away along a path that is close to the unstable manifold.

state *only* when the perturbation moves it along the stable manifold. But if it is *close* to the stable manifold, it begins by moving back towards the steady state and only after a rather long time does it get caught up by the unstable manifold and fall away (Figure 137). Imagine the broom with its attached dustpan. Balance the whole thing very carefully so that the broom is toppling only very slowly, and pull the dustpan to one side. *First* the dustpan moves rapidly back towards the vertically downwards position, *then* the whole shebang collapses. This means that perturbations around saddles come in two flavours. There are 'initially stable' ones, which – at first – start to move back towards the steady state. And there are the rest, which are 'completely unstable' and simply move away.

Ott, Grebogi, and Yorke realized that you could use the initially stable perturbations to control a system near a saddle point. Suppose something happens to perturb it. If it's an initially stable perturbation, you're laughing; for the moment at least, you can safely do nothing. If it's a completely unstable perturbation, however, you'd better react. The easy way to do this is to impose a new perturbation of your own, so that the perturbed point falls inside the 'initially stable' zone. You do the same thing to an initially stable perturbation that starts to wander away as it gets trapped by the unstable manifold. And to keep everything simple, what you actually do is try to move the perturbed point smack bang on top of the stable manifold. That won't

Figure 138 Principle of OGY chaotic control. Objective: make the current point remain steady at the saddle point. Grey lines show original saddle point and separatrices. The current point is controlled by displacing the original system (solid lines) so that the displaced dynamic makes it move on to the stable manifold of the original system.

actually work, but you can expect to come close; and then – before the error causes problems – you try again. Repeatedly you adjust the position of the perturbed state, always aiming to keep it on – or close to – the stable manifold.

The final step is to realize that you can play the same game with the unstable periodic points inside a chaotic attractor. They too have a saddle-like structure, in the sense that some perturbations are initially stable whereas others are not. So you monitor the state of the system, repeatedly trying to perturb it back on to the stable manifold. What you *don't* do is try to perturb it back to the steady state itself, which is what conventional control theory tries to achieve. Why bother? The stable manifold is easier to hit – it's a whole curve instead of a single point – and as long as you come close, the natural dynamics is guaranteed to improve things even more. Whereas, if you try to push the system back to the steady state in one go, your error may place it back in the region of completely unstable perturbations, and you're in trouble.

You don't need to know the equations for a model of the system to carry out this procedure. You can estimate where the stable manifold goes by analysing observational data. Of course if you do have a good model it helps.

The actual details are not quite this simple. The method described above is called 'Proportional Perturbation Feedback', or PPF, and was introduced more recently in the control of heartbeats, as described below. The original Ott–Grebogi–Yorke (or OGY) method is a bit different. Instead of perturbing the state point itself, you temporarily perturb the entire dynamical system that is driving it. And you arrange for the point that it will move to *next*, rather than where it is now, to be on the stable manifold. Figure 138 illustrates the principle for a saddle point in the plane.

It really is a terribly natural idea – provided you think of the geometry in phase space and exploit the existence of a stable manifold. In fact, it's such a simple idea that it eluded everybody until Ott and colleagues had the genius to think of it. The simple ideas are always the hardest to find.

Off on a Comet

The discovery of chaotic control by Ott, Grebogi, and Yorke was followed by a whole series of laboratory implementations – things like controlling a magnetic ribbon that would normally oscillate chaotically. These experiments showed that the method works on a real, mathematically untidy system subject to random 'noise', but they happened in the laboratory.

The first important application of chaotic control in the outside world – *very* outside – happened before the method was invented. (Don't be surprised: nearly every important idea has a prehistory in which some variant of it is used without it being clear to anybody that there is a general principle involved.) In 1985 some of NASA's engineers got a bright idea about how to make a 'dead' satellite rendezvous with a comet. It was a case of chaotic control, and it illustrates the dramatic efficiency of the method compared to more classical ones. Spacecraft are not simply fired off and left to follow the desired orbit – for the obvious reason that they won't. Initial errors will build up, and things like the solar wind will perturb the craft away from where you want it to be. Virtually all spacecraft have a degree of manoeuvrability, and it is provided by tanks of hydrazine fuel. This can be bled off through valves to pass over a catalyst that turns it into gas, which is expelled to act like a tiny rocket, pushing the craft gently in the desired direction.

The satellite ISEE-3/ICE began its career as the third International Sun–Earth Explorer and later was renamed the International Cometary Explorer, by which time its stock of hydrazine had run so low that it had pretty much been written off. It was effectively dead. This was a pity because comet

Giacobini-Zinner was approaching and the instruments on ISEE-3, as it then was, could have been used to study it. Sadly, the dead satellite was about fifty million miles away from the right place.

So the engineers decided to move it.

With the usual approach, this would have been impossible – the satellite was dead, its fuel reserves far too low for such a manoeuvre. But NASA's engineers realized that there was still enough hydrazine left to make a few really tiny adjustments to the orbit. The trick was to arrange for their effect to be out of all proportion to the amount of fuel consumed. This meant putting the satellite into an orbit whose stability was very delicate, and making course corrections when it was in especially critical positions. They discovered, using computer simulations, that if they repeatedly flew the satellite past the Moon, they could nudge it into an orbit that would meet that of the comet. It took five separate lunar fly-bys to pull the trick off. And although the engineers weren't explicitly using that language, it worked because of the chaotic nature of the three-body problem – the bodies in this case being the Earth, the Moon, and the satellite. An orbit that passes close to the 'neutral point' between Earth and Moon, where their respective gravitational fields cancel out, will be unusually sensitive to small perturbations. Not caused by a random flap of a butterfly, but by a carefully chosen squirt of hydrazine.

It was a form of chaotic control, the first such exploitation of the butterfly effect. The mission was a great success, achieving the first ever cometary encounter, and paving the way for more elaborate missions such as those that made Halley's comet resemble a hive surrounded by buzzing bees. No fewer than five separate spacecraft made rendezvous with comet Halley – two Russian probes, *Vega 1* and *2*, two Japanese probes, *Suisei* and *Sakigake*, and the European *Giotto*.

A similar trick was used to get the *Galileo* probe to Jupiter when the money for a special propulsion unit, intended to boost it out of Earth orbit, was cut. It was swung by Venus, the Earth, Venus again, and Mars before finally heading towards Jupiter, gaining speed with each encounter. But on that occasion the butterfly effect was less crucial than the 'slingshot effect', by which the probe gained kinetic energy at the expense of the planets that it passed by. (Planets are big – they can afford to donate a little energy to a tiny probe.)

The latest such idea to emerge – and it has emerged from the Ott– Grebogi–Yorke stable – again makes use of chaos associated with the neutral point between Earth and Moon. Now the task is to send payloads from low Earth orbit to the Moon's surface. The conventional solution, long assumed to be the most efficient, is an orbit called a Hohmann ellipse (Figure 139a), which touches the low Earth orbit at one end and the Moon's orbit at the

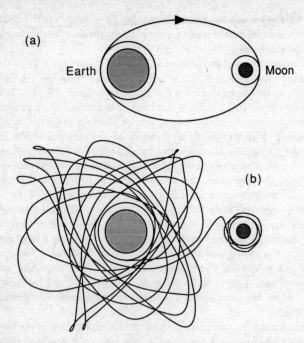

Figure 139   Orbits from Earth to Moon: (a) the Hohmann ellipse, widely thought to be the most efficient possible; (b) a more efficient chaotic orbit (schematic).

other. However, it's not really necessary to get the payload all the way to lunar orbit. If you can get it to the neutral point, then it will – if it's pointing the right way – *fall* the rest of the way to the lunar surface. Jim Meiss and Erik Bolt carried out computer experiments that revealed the existence of a complicated chaotic trajectory requiring much less fuel than a Hohmann orbit.

The only snag: it would take 10,000 years to reach the Moon.

But that's where chaotic control comes in. Meiss and Bolt found a systematic way to apply tiny, carefully timed nudges, so that the orbit could be persuaded to reach the Moon in two years, with a 50 per cent saving in fuel compared to a Hohmann ellipse (see Figure 139b). Their orbit starts from a parking orbit round the Earth, circles the Earth 48 times in a highly irregular manner, is captured by the Moon, which it circles a further 10 times before settling into a stable lunar parking orbit. This convoluted orbit is not much use for flying passengers to the Moon, but it can convey 83 per cent more payload for the same amount of fuel as a Hohmann orbit, so it would be entirely reasonable for sending durable supplies to a lunar base. Of course, we don't have such a base yet, but if we don't blow ourselves to kingdom come we surely will eventually. But the real message is more general. We now know that the motion of celestial bodies is chaotic. We can't change that, and we can't ignore it.

But we can use it.

The Intelligent Pacemaker

In 1992 a team composed of Alan Garfinkel, Mark Spano, William Ditto, and James Weiss showed that chaotic control can be used to suppress irregularities in a beating heart.

Medically, irregularities of the human heart are controlled using artificial pacemakers. These are boxes of electronics that deliver electrical pulses to the heart's own pacemaker system, triggering regular contractions. Now, it would seem reasonable to expect that the way to make a heart beat regularly is to supply regular pacemaker impulses, and that's what the designers of the most artificial pacemakers assume. However, it's not entirely true, because the heart is a *nonlinear* oscillator, and when you stimulate a nonlinear oscillator with a periodic signal it may decide to do something else – including, possibly, chaos. So, in principle, a regular stimulus might actually create an irregular heartbeat instead of a regular one. (By the way, if *you* have an artificial pacemaker, please don't start worrying: your pacemaker is working fine, otherwise you'd have other things on your mind than reading this.) It would obviously be nice to devise an 'intelligent' pacemaker that could respond to what the heart is doing, in any circumstances, rather than just trying to impose a fixed rhythm. That means that you have to understand how to control the nonlinear dynamics of a real heart.

The work of Garfinkel and colleagues is the first serious step in this direction. It's not just theoretical: they tried it out on part of a rabbit heart. This was kept in an active state by passing a solution of oxygen over it, and stimulated with brief electrical pulses of about 10–30 volts. Left to itself, the piece of heart tissue would beat regularly. To induce irregular beating they administered a drug known as ouabain, which affects the rhythms of the heart, sometimes combined with epinephidrine, another drug. The normal rhythm of a heartbeat is a pulse that rises rapidly, levels off flat, and drops rapidly back to near zero. The effect of ouabain is to add an oscillatory 'tail' of electrical activity, so that instead of dying back to zero the electrical current wobbles to and fro. This tail lasts long enough to affect the growth phase of the next beat. This is very similar in general terms to a dripping tap, which we know becomes chaotic when the flow of water is suitably rapid. In the dripping tap, chaos

Figure 140   Reconstructed attractor for chaotically beating heart tissue, and time-series for the same tissue with and without chaotic control.

sets in when drops detach fast enough to leave behind a wobbling droplet, so that when the next drop grows it is affected by the wobble. It is the same with the heart: the oscillating tail from the previous beat interferes with a delicate phase in the growth of the next one, and the result is – among other things – chaos.

Garfinkel's team observed the timings of the heartbeats, concentrating on the 'interbeat intervals' – the times between successive beats. For various reasons the original OGY method of chaotic control is not appropriate for this particular system, so they devised a variant, which – as I said earlier – they called Proportional Perturbation Feedback. They used PPF to alter the timings of the electrical pulses that were stimulating the beating heart tissue, and they carried out experiments to see whether the method could bring the heartbeat back from chaos to regularity.

Figure 140 shows, on the left, a reconstruction of the attractor for chaotically beating heart tissue obtained by plotting successive interbeat intervals against each other. On the right, it shows what happens to the sequence of interbeat intervals when a 'chaotic control' system is switched on, and then off again. When the control system is off, the heartbeat is irregular. When it is on, the heartbeat is (approximately) periodic. The difference is very clear.

Babloyantz's Brainwave

The heart is not the only bodily organ to be controlled by electrical pulses. Muscles are another example, but the most obvious is the brain itself.

A number of people, among them Agnessa Babloyantz from the University of Brussels, have studied brain dynamics, looking for chaotic features.

The brain is an intricate network of nerve cells, or neurons; for example, Figure 141 shows a tiny segment of the cerebral cortex, one of the brain's information-processing regions. For many years neurologists have known how to monitor the large-scale activity of the brain using electroencephalogram (EEG) recordings. Electrodes are attached to the scalp and they register the electrical activity of the brain. Figure 142 shows an EEG recording from a normal seventeen-year-old girl while asleep. The ticks along the top line indicate time intervals of one second, and the numbers 1–8 show the choice of electrode positions for the eight channels of the instrument, corresponding to the eight time-series shown. This technique is quite old; it was initiated by Hans Berger in 1929. Berger thought that one day by analysing EEGs you might be able to read people's thoughts. He was so scared that future people would be able to read *his* thoughts that he destroyed all of his own EEGs. However, EEG telepathy is not very likely, because an EEG trace is an average over relatively large regions of the brain. You might as well try to read a James Bond novel after it has been torn into chunks and each has been put through a food-processor. But EEGs do contain clues to the overall dynamics of the brain.

An EEG record is a time-series, and is therefore amenable to phase space reconstruction techniques such as the Ruelle–Takens method so, we can work out the topology of the dynamical attractor that corresponds to an EEG – thereby extracting some of the clues to brain behaviour contained within it. In 1985 Babloyantz and co-workers showed that different types of brain behaviour yield visibly different attractors. Figure 143 shows several different kinds of EEG trace, together with the corresponding reconstructed attractors. Observe that in all cases the attractor looks chaotic, but considerable differences in the geometry of the attractors are apparent, depending on the mental state of the subject. When the eyes are open the EEG signal has low amplitude and high frequency, and the corresponding attractor is very complex and multidimensional. In contrast, if the eyes are closed, then the brainwaves change to higher amplitude and lower frequency – called alpha waves. The corresponding attractor has a well-defined structure; it is low-dimensional and chaotic. As we drop off to sleep the brain waves become high amplitude and lower frequency. This deep-sleep stage is followed by REM (Rapid Eye Movement) sleep, during which we dream, and the waves start to resemble those that occur when the eyes are open.

The brain is a highly complex system, and it is not surprising that the attractors obtained by this procedure are rather irregular. To obtain more 'tidy' attractors it is necessary – in the current state of technology – to study simpler systems. One example of such work is a study of the alga *Nitella flexilis*

Figure 141   Nerve cells and their connections in a tiny piece of cerebral cortex. (Ignore the numbers and letters, which are irrelevant here.)

Figure 142   EEG from a normal seventeen-year-old girl during natural sleep.

carried out by H. Hayashi, M. Nakao and K. Hirakawa in 1982. They stimulated the alga's 'giant internodal cell' with a periodic electrical signal, and measured the pattern of electrical activity across the membrane. Figure 144 shows typical results. The top picture (a) shows periodic activity, but (b) shows chaotic activity.

EEG records can distinguish various types of brain pathology. One is the form of epilepsy called 'petit mal', in which there are a few seconds of high amplitude, very regular, activity. In that state brain activity seems to be locked together as a single whole. The corresponding attractor looks like a limit cycle around which there is some noise. Figure 145 shows an EEC recording of a human epileptic seizure of 'petit mal' type. There is a very clear sudden transition from a relatively random time-series to one with a strong element of periodicity. The reconstructed phase portrait Figure 146 shows that the time-series is not precisely periodic; instead it to some extent resembles the Rössler attractor (see Chapter 9, Figure 79), which has a generally cyclic form but spreads out into a band.

Epilepsy lasts only a few seconds, but there is another kind of brain pathology called Creutzfeldt-Jacob disease (CJD). In the UK it seems mandatory nowadays to describe CJD as 'the human form of mad cow disease', although its links to bovine spongiform encephalopathy (BSE) are somewhat conjectural. (This is not to say that the handling of BSE by several successive Tory governments has been anything other than grotesquely irresponsible.) It is thought that in both CJD and BSE there are virus-like particles, called prions, which enter the brain and destroy it. The EEG of a CJD patient shows a very regular wave with a slightly chaotic attractor – rather more so than in epilepsy – and the brain has neither motor nor cognitive power.

Figure 143   EEG traces when the brain is in various states, and the corresponding reconstructed attractors.

Brains Need Chaos

When one looks from eyes open, to REM sleep, to epilepsy and CJD, it is tempting to conclude that the brain's cognitive power is highest when its attractor is most complex, and diminishes as the attractor becomes less chaotic. So what is the attraction (pun intended) of chaos for the brain? Babloyantz thinks that chaos is necessary to brain function, because the brain processes

Figure 144   Pattern of electrical activity across the membrane of the alga Nitella flexilis *(above) when the alga's giant internodal cell is stimulated with a sinusoidal signal (below): (a) low-frequency stimulus produces periodic oscillations; (b) high-frequency stimulus produces chaotic oscillations.*

information, so it has to switch rapidly between one state and another. We have seen that this kind of flexibility is characteristic of chaotic systems, because more regular dynamics cannot change state anything like as quickly. So it looks as if the brain has to be chaotic in order to function properly.

In 1994 a team of six scientists – Steven Schiff, Kristin Jerger, Duc Duong, Taeun Chang, Mark Spano and William Ditto – examined the use of chaotic

Figure 14.5   EEG recording of a human epileptic seizure of 'petit mal' type. Inset shows a magnified segment of the time-series. Numbered 'channels' refer to particular electrodes.

Figure 146   Reconstructed attractor for a human epileptic seizure.

control on tissue from the hippocampus (part of the brain) of a rat. They found that it could be used to render chaotic dynamics periodic, and that it could also be used 'backwards' to suppress periodic behaviour – occurring in pathologies such as epilepsy and CJD – and replace it by more normal-looking behaviour. All this is a long way from medical application, and the ethical questions surrounding control of brainwaves are even more tangled than those surrounding weather control; but it is clear that the technique of chaotic control has promise in a surprisingly wide range of areas.

It is also becoming clear that the 'chaotic control' techniques often work – in the sense of producing stable, regular behaviour – on systems that may not be chaotic in any technical sense. This is probably true of the work on brain tissue, and it has very recently been observed by David Christini and Jim Collins of Boston University in the chaotic control of nerve impulses. They find that apparently random sequences of nerve impulses can be rendered periodic by pretending they are chaotic and applying the usual methods of chaotic control.

Ha. If chaotic control works when there's no chaos, doesn't that imply that chaos is useless?

This sort of question is usually asked by people who dislike the concept of chaos and would prefer it to go away, and the underlying assumption seems to be that if a method inspired by chaos theory works when there isn't any actual chaos, then we can forget about chaos as an important concept.

I see it the other way round. The concept of chaos proves its worth by leading researchers to create new ideas, in this case chaotic control. If those methods happen to work on non-chaotic systems *as well*, all the better. After all, statistics assumes events are random, but nobody thinks statistics is somehow diminished as a concept if it also works on some events that prove not to have been random at all. Death is largely non-random: overwhelmingly it strikes at those who are terminally ill. This doesn't stop actuaries using a random model to predict life expectancy, and it doesn't stop statisticians proclaiming the result as a triumph of applied probability theory.

I have in my cupboard a bottle of stuff that was invented to clean tar off people's clothes after they had foolishly sat on beaches adjacent to the English Channel, where marine traffic is among the densest in the world and a lot of it is oil tankers heading for the Isle of Sheppey. I've discovered that it also removes the little bits of supermarket sticky label that adhere to paperback books when you try to peel the labels off. I don't think that this makes the concept of tar irrelevant.

In a 1994 'news and views' article about this work in *Nature*, Frank Moss ended on a cautious but optimistic note:

> ... Both control and anticontrol, and maybe new diagnostic and therapeutic technologies, [may] become feasible. Because of the great potential benefit, the search for chaos in biological and medical settings has been energetically pursued; it has not yet been convincingly demonstrated, although some results are encouraging ... The detection of recurrent, unstable cycles, and thus of chaos, in a statistically convincing way might just become routinely possible with biological preparations or even with patients. These new experiments ... open the door to this grand possibility.

16

Chaos and the Quantum

> The mathematician plays a game in which he himself invents the rules while the physicist plays a game in which the rules are provided by Nature, but as time goes on it becomes increasingly evident that the rules which the mathematician finds interesting are the same as those which Nature has chosen.
>
> Paul Adrien Maurice Dirac

> 'I used to be uncertain, but now I'm not so sure.'
>
> T-shirt slogan

When Einstein made his celebrated remark about God not playing dice, he was referring to quantum mechanics. This differs in fundamental ways from the 'classical' mechanics of Newton, Laplace, and Poincaré, upon which nearly all of the discussion until now has focused. Einstein made his famous statement in a letter to the physicist Max Born, and at greater length it went like this:

> 'You believe in the God who plays dice, and I in complete law and order in a world which objectively exists, and which I, in a wildly speculative way, am trying to capture. I firmly *believe*, but I hope that someone will discover a more realistic way, or rather a more tangible basis than it has been my lot to do. Even the great initial success of the quantum theory does not make me believe in the fundamental dice game, although I am well aware that your younger colleagues interpret this as a consequence of senility.

Chaos was unknown in Einstein's day, but it was the kind of concept he was seeking. Ironically, the very image of chance as a rolling cube is deterministic and classical, not quantum. And chaos is primarily a concept of classical mechanics. How does the discovery of chaos affect quantum mechanics, and what support – or otherwise – does it offer for Einstein's philosophy? Answers to these questions are, for the moment at least, highly speculative. There is some interest among physicists in what they call 'quantum chaos', but quantum chaos is about the relation between non-chaotic quantum systems and chaotic classical approximations – not chaos as a mechanism for quantum indeterminacy. Quantum chaos is not what this chapter is about: the central thrust of the chapter is the possibility of changing the theoretical framework of quantum mechanics altogether, replacing quantum uncertainty by deterministic chaos, as Einstein would have liked.

It must be admitted at the outset that the vast majority of physicists see no reason to make changes to the current framework of quantum mechanics, in which quantum events have an irreducibly probabilistic character. Their view is: 'If it ain't broke, don't fix it.' However, hardly any philosophers of science are at ease with the conventional interpretation of quantum mechanics, on the grounds that it is philosophically incoherent, especially regarding the key concept of an observation. Moreover, some of the world's foremost physicists agree with the philosophers. They think that something *is* broke, and therefore needs fixing. It may not be necessary to tinker with quantum mechanics itself: it may be that all we need is a deeper kind of background mathematics that explains why the probabilistic point of view works, much as Einstein's concept of curved space explained Newtonian gravitation. Of course Einstein's general relativity actually went beyond Newtonian mechanics and changed the mathematics of gravitational theory as well as the interpretation, but along the way it explained the philosophically incoherent Newtonian appeal to forces acting at a distance, replacing it by the inherent curvature of space acting locally. And Newton's theory can be recovered as a very good approximation to general relativity, valid when the curvature of space is small. So maybe a new framework for quantum mechanics will accommodate the existing, highly successful probabilistic viewpoint *exactly*; maybe it will even reveal it as an *approximation* to something deeper but essentially different.

Maybe.

The philosophers – being philosophers – mainly think that it is the interpretation of quantum mechanics that needs to be fixed. They aren't mathematicians or physicists, so they are happy with the mathematics and the physics. The physicists aren't terribly interested in a reinterpretation, however philosophically superior it might be, unless it yields radically new physics; but several major figures are convinced that quantum mechanics itself is in need of a fundamental reformulation that goes well beyond mere tinkering. They believe that despite its immense success in predicting the outcome of experiments, quantum mechanics needs to be rebuilt from the ground up. And some mathematicians, perhaps excited by the prospect of interesting new kinds of mathematics, agree.

Before we can understand these issues, we must take a stab at understanding quantum mechanics as it is currently taught to all budding physicists – which is an ambitious task. You should therefore be warned that everything I say should be considered as an informal paraphrase of something much more specific and technical, and that the full story involves very sophisticated mathematics. For a more extended discussion of quantum physics that steers clear of mathematical technicalities I recommend John Gribbin's *In Search of Schrödinger's Cat*.

Quantum mechanics was not invented merely to shock the conservatives. It was forced upon physicists by the results of a large number of careful experiments that demonstrated the inadequacy of Newtonian mechanics. The very existence of electrons inside atoms, for example, provides such evidence. In a classical model of the atom, electrons are electrical point charges orbiting a central nucleus made from protons and neutrons. But in classical physics, a moving electric charge must radiate some of its energy as electromagnetic waves, so that electrons could not continue to orbit the nucleus for very long. Instead, they would spiral into the nucleus and disappear, losing their electric charge in a collision with an oppositely charged proton. Atoms would fall apart and disappear.

Since this doesn't happen, something is wrong. It *could* be the image of orbiting point charges, but nobody has ever managed to fix that up in a way that fits experiments. So maybe it's classical physics that is wrong. Maybe moving electrically charged particles do *not* radiate their charge away.

And a lot of experiments showed that they don't.

It took a long time, and a lot of experimentation, to convince physicists that quantum mechanics was necessary, and that it worked. Indeed it is arguably the most successful scientific paradigm ever, and it should not be dismissed lightly, however strange it may appear to be.

On the other hand, nothing is sacred in science, and nobody need feel inhibited about questioning the prevailing wisdom. Neither should they be outraged if somebody else does it.

I wish the bulk of scientists would remember that.

Waves, Particles, and Quanta

Before physicists ran up against awkward experimental evidence to the contrary, their view of the physical universe was straightforward. There was matter, which was composed of particles, and radiation, which was composed of waves. Matter possessed mass, position, and velocity. Mass could be any positive real number, and position and velocity existed in a space–time continuum, meaning that position and velocity coordinates (relative to some choice of axes) could be arbitrary real numbers, positive or negative. Or, to put it another way, space and time were infinitely divisible. So were associated quantities such as energy; and in principle you could measure all of these quantities as accurately as you wished.

Waves were different. A wave possessed a frequency (number of waves per second) and an amplitude (height of wave) – or, in the case of electromagnetic waves such as light, a more complicated system of amplitudes of both the electric and magnetic fields observed in various directions. Don't worry if you find that hard to visualize: the famous American physicist Richard Feynman said he never really did grasp it: all he could do was play with the mathematical equations and come up with simplified and incomplete analogies.

In particular, an electron was a particle, and light was a wave. You could tell because electrons bounced off other bits of matter like little rubber balls, whereas when two light rays met they formed 'interference fringes' in which the wave patterns became superposed (added together). The easiest way to get a feel for this phenomenon is to drop two pebbles into a still pond, and watch the circular ripples cross each other. What you get is Figure 30 of Chapter 5, a characteristic interference pattern. If you see this kind of pattern, it's a sure-fire bet that waves are responsible.

It soon became apparent, however, that there are circumstances in which light behaves like a stream of particles rather than a wave. One is the photoelectric effect, in which light impinging on a suitable substance produces an electric current. Then it transpired that electrons sometimes behave like waves: if you pass pairs of electrons through fine parallel slits you get interference fringes. So the distinction between waves and particles started to blur.

Another piece of evidence for the particle-like nature of light was contained in a theory announced by Max Planck in 1900, to the effect that the energy of an electromagnetic wave is *not* infinitely divisible. For light of a fixed frequency there is a definite minimum energy; moreover, the only possible values for the energy are whole number multiples of that minimum value. It is as if energy can only come in tiny packets of fixed size, and every light 'wave' is made up from an integral number of packets. Planck called these packets 'quanta'. A new fundamental physical constant, now known as *Planck's constant* and denoted by the letter h, captured the relationship between the frequency of the light and the energy of one quantum. In fact, the energy of a quantum of light is equal to its frequency multiplied by Planck's constant.

Planck's constant is *tiny*: 6.626×10^{-34}. (This in units is called Joule-seconds – for our purposes it doesn't matter what those are, it's still tiny.) So although energy is not infinitely divisible, you can divide it up quite a lot before you get down to the irreducible lumpiness of individual quanta. Only when you

particle-like wave packet   **wave**

Figure 147   Schematic picture of a wave packet and a wave. For actual quantum mechanical wave functions, the vertical 'direction' in the graph is the complex plane or a multidimensional complex vector space. The horizontal direction represents space.

get down to the thirty-fourth decimal place, do you discover that the universe will not subdivide forever. This explains why the universe had previously looked infinitely divisible.

The distinction may seem a fine point, but it had a crucial consequence. If you assume that energy is infinitely divisible, you get the wrong value for the energy radiated by a 'black body' – perfect radiator – at high frequencies. Indeed, you get infinity. Planck discovered that if you assume that energy comes as integer multiples of h times frequency, then you get a finite value – and one that agrees with experiment.

Wave Functions

How do we reconcile the two contrary attributes of matter, wave and particle? Erwin Schrödinger took the view that a particle was a sort of concentrated wave, which in most circumstances acted as if it was localized in a small region of space, and travelled through time as a coherent blob or 'wave packet'. However, in certain circumstances the wave packet could become more extensive, giving the appearance of a conventional wave (Figure 147). But what was doing the waving? Ocean waves are waves of water, electro-magnetic waves are waves in the electrical and magnetic fields. According to Schrödinger, quantum-mechanical waves are waves in a – possibly multi-dimensional – mathematical space of complex numbers. (A complex number is what you get when you extend the usual real number system by including a new number i with the property that $i^2 = -1$.) The properties of this quantum wave were therefore defined by a *wave function*, usually denoted by the Greek letter ψ (psi). At each point (x, y, z) of space and each instant t of time the value $\psi(x, y, z, t)$ is some complex number – or complex vector in the multidimensional case.

What corresponds to the motion of particles or waves through space?

Schrödinger wrote down a simple differential equation that the wave function must satisfy. Schrödinger's equation, as it is now called, determines the propagation of the quantum-mechanical wave function through space and time, by specifying how it changes from its current value as it moves into the future.

Schrödinger's equation is linear, meaning that solutions can be superposed to give more solutions. This fits the wave-like behaviour of matter quite nicely, but at first sight it doesn't look so good for particles. However, a collection of several separate particles can reasonably be viewed as a superposition of the states that you would have got if each individual particle had existed on its own. What about interactions between particles when they come very close together? Those are precisely the circumstances in which microscopic matter stops behaving like a conventional wave or particle, and it is here that Schrödinger's equation proved its worth, predicting things like the energy levels of the hydrogen atom with exquisite accuracy.

The linearity of quantum mechanics had some curious consequences, which sound bizarre but fit experiments perfectly. For example, an electron possesses a feature known – rather misleadingly – as 'spin', because it is in a sense analogous to a spinning ball. When you measure electron spin relative to some choice of 'axis' you always get the value $+\frac{1}{2}$ or $-\frac{1}{2}$ – nothing else. Don't worry about what units spin is measured in: basically it's just a number. You can interpret the difference as analogous to that between clockwise spin and anticlockwise spin, if you wish; but this analogy rapidly breaks down when you superpose spin states. For definiteness, let's use a coordinate system whose axes point north, east, and up, which sounds friendlier than x, y, and z. An electron can simultaneously have spin $+\frac{1}{2}$ about the north axis and spin $-\frac{1}{2}$ about the east axis, say. If you try to do this with a tennis ball, spinning it anticlockwise about the north axis and clockwise about the east axis, you'll find that the combined motion is just a spin about a *different* axis. This is because, on macroscopic scales, a rotating ball is always rotating about a single axis. The electron, however, really does seem to behave as though it is rotating about both axes simultaneously. It is almost as if there are two ghost electrons, one spinning about the north axis and one about the east, and the real electron is a combination of the ghosts (Figure 148). Indeed you really need three ghosts, because we haven't yet added in spin about the 'up' axis. In many different kinds of experiment, this ghostly picture of an electron has been shown to give an accurate picture of reality.

I use the word 'ghost' because you can't actually measure all three spins simultaneously. The very notion of measurement poses nasty problems for quantum mechanics, even though it is central to any experimental test. Physicists know what *actually* happens when you measure a quantum object. You get a number. One number. You can make another measurement to

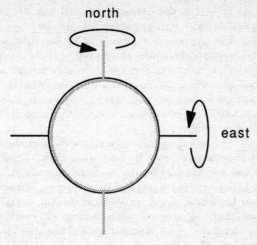

Figure 148   Superposed spin state of an electron. The component eigenstates are shown in black and grey respectively.

get a second number, but you can't necessarily assume that the first number remains the same as it was. For example, if you measure the 'north' spin of an electron you may get $+\frac{1}{2}$. Then you measure its 'east' spin to get $-\frac{1}{2}$. Now go back and remeasure its 'north spin'. You don't necessarily get $+\frac{1}{2}$ again. In fact, you get $+\frac{1}{2}$ half the time, and $-\frac{1}{2}$ the other half – and plus and minus seem to occur completely randomly. It is as if measuring the 'east' spin somehow interferes with the 'north' spin, randomly switching it to plus or minus.

Some quantum variables are independent: if you measure one it does not affect the other. But some aren't, and among them are spins about different 'axes'.

What quantum physics lacks, however, is a clear theoretical description of what happens when a measurement is made on a quantum system. For example – speaking very roughly and ignoring some technical restrictions – you can superpose 50 per cent of the state 'east spin = $+\frac{1}{2}$' and 50 per cent of the state 'east spin = $-\frac{1}{2}$', to get a valid state of an electron. (Note that these spins are about the *same* axis.) But it's not an electron with spin zero; *no* electron has spin zero. Electrons always have spin $\pm\frac{1}{2}$ in any direction. Neither is it an electron spinning about some compromise axis half-way between east and north. It's like two ghosts, one rotating clockwise about the east axis, the other anticlockwise. Or rather, two half-ghosts. You can't detect both ghosts at once, however. If you make a measurement, you just pick up one of the half-ghosts and promote it to full ghosthood.

How do we know that this strange superposed state actually occurs? Certainly there must be something wrong with the standard view of quantum mechanics if it does not, because we know that the superposition principle applies to the equations of quantum theory. But what about experiments? Well, the best we can do is to prepare lots of electrons that are supposed to be in this combined state, and measure their east spins. What we find is that – apparently randomly – we get either $+\frac{1}{2}$ or $-\frac{1}{2}$. So on average we get 50 per cent of spin $+\frac{1}{2}$ and 50 per cent of spin $-\frac{1}{2}$. We interpret this as evidence for the 50/50 combined state.

It's a bit like having a theory about coins that move in space, but only being able to measure their state by interrupting them with a table. We hypothesize that the coin may be able to revolve in space, a state that is neither 'heads' nor 'tails' but a kind of mixture. Our experimental proof is that when you stick a table in, you get heads half the time and tails the other half – randomly. This is by no means a perfect analogy with standard quantum theory – a revolving coin is not exactly in a *superposition* of heads and tails – but it captures some of the flavour.

Cat in a Box

It is all very curious. Why does a measurement of the (east) spin of an electron always produce either $+\frac{1}{2}$ or $-\frac{1}{2}$, even if the electron is actually in a superposition of those states? It becomes even more mysterious if you try to build the measuring apparatus into the quantum equation. A 'spinometer' – spin-measuring device of some kind – is made out of the same sub-atomic particles as the electron whose spin it is to measure. Suppose that when the electron has spin + the apparatus is in quantum state P (for plus), whereas when the electron has spin − the apparatus is in quantum state M (for minus). Then, by the linearity of the quantum equations, when the electron is in quantum state 50 per cent of $+\frac{1}{2}$ plus 50 per cent of $-\frac{1}{2}$, the apparatus should be in quantum state $\frac{1}{2}P + \frac{1}{2}M$.

But it's not. It's a spinometer, so it must always be in state P – and state P alone – or state M. There is still a trace of the superposition of quantum states, however, because the apparatus seems to choose the states P and M at random – and it is in each state for about half of the measurements. The *average* state, over many experiments, is $\frac{1}{2}P + \frac{1}{2}M$. Somehow, when you work with actual measurements using macroscopic apparatus, the superposition principle ceases to function. There is a conceptual mismatch between the microscopic world of the quantum, and the macroscopic world of the spinometer. But quantum mechanics is supposed to apply to all objects, micro or macro, isn't it?

The favoured way to get round this difficulty is to introduce an interpretation of the measurement process that does not attempt to model the apparatus as a quantum system at all. Instead, it simply *accepts* the (mysterious) fact that a spinometer either yields $+\frac{1}{2}$ or $-\frac{1}{2}$, but never a mixture. And it argues that what the apparatus does is to *collapse* the wave function down into one or other of its component parts. Those parts are known as *eigenfunctions*, and the corresponding states are *eigenstates*. Those words mean something rather special in the mathematical formalism, but what they will signal to us is that there are certain 'special' wave functions (eigenfunctions) out of which all the others can be constructed by superposition, and it is those special states (eigenstates), and only those, that you can observe. The measurement process starts with a superposition of eigenstates, such as $\frac{1}{2}P + \frac{1}{2}M$, and 'collapses' it to either P or M. Which one occurs is irreducibly probabilistic. To find out about the coefficients $\frac{1}{2}$, you have to repeat the experiment and calculate probabilities.

Don't confuse those $\frac{1}{2}$'s with the spin values. Spin is always $\pm\frac{1}{2}$, but these $\frac{1}{2}$s arise because I chose to use 50 per cent of each state. If instead I'd used 25 per cent of P and 75 per cent, of M, then I'd be talking about $\frac{1}{4}P + \frac{3}{4}M$. Again any measurement would collapse this down to either P or M, but now – on repeating the experiment many times – we'd find that P turns up one quarter of the time, and M three quarters.

This point of view on the measurement process for quantum systems is called the Copenhagen interpretation, and it was advanced by Niels Bohr in 1927. Although intended as a pragmatic solution to a conceptual difficulty facing practising physicists, it has led to considerable philosophical mysticism about the role of the human mind as an observer of quantum reality, and suggestions that the universe doesn't really exist unless a human being is looking at it. Personally, I think this is silly: what I want to understand is how the *spinometer* manages to avoid being in state $\frac{1}{2}P + \frac{1}{2}M$, which is the heart of the mystery. A human mind observing the spinometer is a secondary stage: the conceptual problem arises even if there are no humans in the loop at all.

Erwin Schrödinger seems to have had the same feeling, and in 1935 he tried to demonstrate what he considered to be the absurdity of the Copenhagen interpretation with his famous 'cat in a box' thought experiment. Einstein called it 'the prettiest demonstration' that the Copenhagen interpretation is an incomplete representation of the real universe. Imagine a box that contains a source of radioactivity, a Geiger counter to detect the presence of radioactive particles, a bottle of (gaseous) poison, and a (live) cat. These are arranged so that if a radioactive atom decays and releases a particle, then the Geiger counter will detect it, set off some kind of machinery that crushes the bottle, and kill the unfortunate cat. From outside the box, an observer cannot determine the quantum state of the radioactive atom: it may either have decayed or not. So according to the Copenhagen interpretation the quantum state of the atom is a superposition of 'not decayed' and 'decayed' – and so is that of the cat, which is part alive and part dead at the same time. Until, that is, we open the box. At this instant the wave function of the atom instantly collapses, say to 'decayed', and that of the cat also correspondingly collapses instantly to 'dead'.

It sounds ridiculous, and that's how Schrödinger intended it. It's my $\frac{1}{2}P +$ $\frac{1}{2}M$ neatly parcelled up in a box with a cat in the starring role. But other quantum physicists didn't think it was ridiculous at all. Quantum theory, they argued, really is strange. Maybe it's so strange that you really can have a cat that's half alive and half dead provided nobody looks at it to determine which. And for every objection they had an answer. Why not put a movie camera in the box to film the cat? Afterwards, you can develop it and see whether the cat died or not. But no: until you open the box, the film is itself in a superposed state, part a film of a living cat, part of a dead one, and only when you open the box – well, you get the gist.

In recent years, experimentalists have devised some very cunning experiments to try to find out just when the wave function of a quantum system inside an impenetrable box collapses. However, it is utterly impossible even to write down the quantum wave function for something as complicated as a cat. Indeed, the helium atom, with two electrons, two protons, and two neutrons, is already too complicated. It is utterly impossible to write down the quantum wave function for a spinometer, too. Or for a Geiger counter. If you want your experiment to match up with theory, you have to replace Schrödinger's cat by a microscopic quantum system – such an electron. So this is what the experimentalists did, and then they found cunning ways to deduce what went on inside the box before it was opened, and whether opening the box changed anything. And some of them said: 'Yes, it is a superposition that collapses only when you open the box.' And others said: 'No, the electron-cat was dead all along, you just didn't know that until you looked.' Because, even when you replace the cat by a microscopic quantum system, ultimately the observations must be interpreted for macroscopic creatures to comprehend, and the same experiment can be interpreted in different ways. The Copenhagen interpretation tells you that only observations possess physical reality, but at the same time it tells you that a key feature of the mathematical formalism, the wave function, cannot be observed completely, and therefore is unreal.

No wonder it leaves room for differences of interpretation.

I strongly suspect that Schrödinger thought that the wave function – all of it, not just its collapsed eigenfunctions – was real. After all, he invented it and wrote down the equation for how it evolves: I'd be surprised if he thought it was merely a mathematical fiction. Definitely he introduced his cat quite deliberately, to dramatize the gulf – not understood then *or now* – between quantum microdynamics and classical macrodynamics.

Today (perhaps because generations of undergraduate physicists have been dragged through the cultural myth that Schrödinger was wrong about the cat) most people use Schrödinger's thought experiment for a purpose very different from what its originator intended. They use it to show us how very weird the quantum world is. 'That poor cat *really* is in a superposition of alive and dead until you open the box.'

What the experiments tell us – as strongly as any experiments about quantum systems tell us anything when their very interpretation is in doubt – is that this is true of small-scale quantum systems like electrons. *We do not know that it is true of cats, and it is almost certainly false.* Not because of anything to do with consciousness, feline or human, but because a cat is a macroscopic system and 'alive' and 'dead' are macroscopic properties. Macroscopic properties do not superpose. The humorous fantasy writer Terry Pratchett, in *Witches Abroad*, locks the witch's cat Greebo in a box. When the box is opened, it transpires that there are *three* states for a cat in a box: alive, dead, and absolutely bloody furious. There is a serious point to this observation. Do you know what the macroscopic state of a cat will look like when its quantum state is $\frac{1}{2}$alive + $\frac{1}{2}$dead? I don't. Maybe it is absolutely bloody furious.

All of these quantum-level experiments are very interesting, but they do not test Schrödinger's contention. He was trying to tell the physics community that the problem of measurement cannot be resolved by grafting a wholly artificial collapse of wave functions on to an otherwise elegant but linear mathematical structure. Instead, it is about the nature of *macroscopic* objects built up from quantum particles. This is *why* he introduced a cat instead of, say, an electron. We inhabit a world of macroscopic objects, which obey classical mechanics much better than they do quantum mechanics.

Why do they do that?

It is believed that a phenomenon called decoherence, related to the fact that the quantum wave function has complex values but our observations must be real, which causes large collections of quantum particles to behave in a classical manner when they are observed in a classical manner. If so, then what happens to the cat is straightforward. It is not in a superposition of states. It is in just one of them – but you don't know which until you open the box. The reason has nothing to do with the cat. It is actually the Geiger counter, one of the other macroscopic systems inside the box, that relies upon decoherence to decide which state the radioactive atom is in. After that, it's classical dynamics all the way. The bottle breaks because of classical interactions triggered by the classical machinery attached to the classical Geiger counter, the classical cat dies because of classical interactions with the classical poison. *We* don't know what's happened, not until we open the box; but the classical systems inside the box *do* 'know' what's happened.

In fact, even at the quantum level, the crucial step happens at the detector, not at the opening of the box. Leonard Mandel has carried out experiments showing that a photon can be switched from wave-like behaviour to particle-like behaviour – which a Copenhagenist would consider to be a collapse of its wave function – without a human observer being aware of this collapse *at the time it happens*. In other words, before anyone opens the box, the cat (photon) is already dead (particle-like). The measurement is completed as soon as the apparatus produces a classical yes/no answer, not when a scientist looks at that apparatus.

The EPR Paradox

The problem of measurement is intimately bound up with another celebrated quantum-mechanical difficulty, which also goes back to Einstein. I'm going to explain it now, because – as we'll shortly see – it provides an excellent test-bed for chaotic replacements for quantum indeterminacy.

In 1935 Einstein, Boris Podolsky, and Nathan Rosen – like Schrödinger – asked whether the quantum description of reality might be missing an essential ingredient. And, again like Schrödinger, they convinced themselves that the answer was 'yes'. The scenario that they came up with requires two particles to interact with each other and then fly apart, interacting with nothing else. Now, at any instant each particle is in a definite position and has a definite momentum, though we cannot measure both simultaneously. When they are close together we can simultaneously measure the distance between them (to be sure they really are close together) and their total momentum: the rules of quantum-mechanical measurement permit this because those two quantities are independent. Later, when they are far apart, we suddenly measure the momentum of one of them, thereby collapsing its momentum wave function to a definite value. However, the equations of quantum mechanics imply that the total momentum of the two particles is conserved. Therefore the momentum of the second particle *also* takes on a definite value at the instant we measure the momentum of the first. Measuring one particle collapses the wave function of the other, because we know what the total has to be.

It is as if there is some kind of instantaneous communication between the particles. But this kind of action at a distance would violate the principle that no information can travel faster than light, a principle known as *locality*.

Einstein, Podolsky, and Rosen felt that 'no reasonable definition of reality could be expected to permit this'. Bohr, on the other hand, saw no difficulty. Until you actually measure what the second particle is doing, you have no right to consider it as being in any particular state at all. It is therefore meaningless to ask whether its wave function has collapsed, and an alleged collapse that you cannot actually observe cannot be considered as the passage of information.

Bohr's point of view prevailed. However, even as I write this section I can't help getting a very strong impression that all of these arguments are full of logical loopholes. For example, can we really be *sure* that, having measured the total momentum of the two particles, it really must stay constant? How do we know they haven't interacted with any other particles if we're not allowed to observe them? For such objections there is usually a reasonable answer within conventional quantum mechanics, but I also get another strong impression, that somewhere inside all the arguments is what philosophers would call a category error – an inconsistency of interpretation in which properties valid on one level of discussion are applied at a different one. For example, we know that the speed of light is a limiting factor for classical particles in special relativity, but how can we be sure that it applies to information (whatever that means in a physical context) carried by quantum particles? And I'm not at all happy with the idea that the collapse of the wave function of the second particle must happen simultaneously with that of the first. Maybe the total momentum is only conserved when it is measured. That is, it might fluctuate to different values when nobody is looking, only to collapse back to its apparently conserved value when you do look. I don't see why this is any harder to swallow than the collapse of the wave function itself; but if it's true, then we are not entitled to deduce anything about the second particle unless we observe it, thereby changing its wave function anyway. (The one thing we can be sure of is that a measurement changes the wave function, because to measure something you have to make it interact with something else, namely the measuring apparatus.)

So there is something very fishy about the whole story. Yes, I know, maybe it only seems fishy if you're hidebound by classical thought-processes. I accept that.

I still think there's something fishy.

Bohm's Interpretation

Don't take my word for it: look at the work of David Bohm. In 1952 Bohm attempted to resolve the EPR paradox in a novel way. Instead of arguing about interpretations of quantum mechanics, he reformulated the underlying mathematics. Louis de Broglie, one of the founders of quantum mechanics, devised a similar but more limited scheme some years earlier. One pillar of Bohm's scheme was to endow the wave function with physical meaning. To him it was not just a mathematical gadget that operated 'behind the scenes' – it was out there on centre stage along with the particles and waves themselves. I strongly suspect Schrödinger would have agreed, for reasons explained earlier; and at one time Paul Dirac had similar views, for in 1935, in the second edition of his classic *The Principles of Quantum Mechanics*, he wrote: 'One of the most satisfactory features of the present quantum theory is that the differential equations that express the causality of classical mechanics do not get lost, but are retained in symbolic form, and indeterminacy appears only in the application of these equations to the results of observations.'

Unfortunately we can't measure a quantum wave function directly, but when it comes down to brass tacks we can't measure *anything* directly; what we do is infer its properties from coherent theories of how the universe works. For example, when we weigh chemicals on an old-fashioned balance we do so on the assumption that the law of the lever holds good, that the numbers stamped on the little brass weights have a definite physical meaning, and indeed that there is a concept 'weight' to be measured in the first place. Copenhagenists seem happy enough for the state of a particle to be a superposition of eigenstates, but they don't endow the superposed state with the same physical reality as the eigenstates themselves. The Copenhagen interpretation can be rendered in mathematical terms as follows: particles obey Schrödinger's equation for the wave function, *except* when measurements are made.

Bohm's idea is simpler and more elegant: particles obey Schrödinger's equation for the wave function, period. But he assumes that in addition to the wave function, a particle also has a definite, physically meaningful, position. He also throws in a new mathematical equation that determines the relation between the wave function and the particle's motion, and another that instructs observers how to incorporate their own ignorance of the positions into their observations.

In Bohm's theory, the laws of physics are totally deterministic. Quantum indeterminacy is not a sign of anything irreducibly probabilistic about the universe, but a sign of the inescapable ignorance of the observer – human or otherwise. Schrödinger's cat, as I have suggested already, is either alive or dead – but *we* don't know which until we open the box. Bohm proved mathematically that this kind of ignorance is just what you need to reproduce the standard statistical predictions of quantum mechanics. You just average out your ignorance and see what's left. For instance, our ignorance of the east spin of an electron, having just measured its north spin, is total. We have absolutely no idea what the east spin should be. So, on average, it is just as likely to be $+\frac{1}{2}$ as $-\frac{1}{2}$; and that's exactly what you find in experiments.

The wave function 'knows' the state.

We don't.

Another interesting feature of Bohm's formulation is that superposition loses most of its meaning. A combination of wave functions is just another wave function. In the same way, the fact that Ankara in Turkey has latitude 39°N and longitude 33°E does not imply that it is a superposition of Valencia in Spain and Kampala in Uganda – two other cities, one at latitude 39°N, longitude 0° and the other at latitude 0°, longitude 33°E.

Bohm's theory is simple, natural, and free of the strange *ad hoc* assumptions of the Copenhagen interpretation, in which the laws of nature are apparently temporarily suspended when somebody makes a measurement. So how would you expect it to be received by the physics community? In 1994 David Albert, a trained physicist who became professor of philosophy at Columbia University, wrote:

> Despite all the rather spectacular advantages of Bohm's theory, an almost universal refusal to consider it, an almost universal allegiance to the standard [Copenhagen] formulation of quantum mechanics, has persisted in physics . . . Many researchers have perennially dismissed Bohm's theory on the grounds that it granted a privileged mathematical role to the positions of particles. The complaint was that this assignment would ruin the symmetry between position and momentum, which had been implicit in the mathematics of quantum theory up until then – as if ruining that symmetry somehow amounted to a more serious affront to scientific reason than the radical undermining, in the Copenhagen formulation, of the very idea of an objective physical reality. Others dismissed Bohm's theory because it made no empirical predictions (no obvious ones, that is) that differed from those of the standard interpretations – as if the fact that these two formulations had much in common on that score somehow transparently favoured one of them over the other. Still others cited 'proofs' in the literature . . . all of which were wrong – that no deterministic replacement for quantum mechanics of the kind that Bohm had already accomplished was even possible.

Bohm's theory is definitely suffering from what the palaeontologist Stephen Jay Gould calls the Panda Principle, because it is involved in the evolution of the red panda's 'thumb'. The panda, at some stage in its evolution, managed to convert its thumb into a claw just like all the others. When, later on, it needed a thumb, it had to evolve one from a knobbly part of its wrist. The Panda Principle holds that once something has become established, it cannot easily be displaced by something else, even if the alternative offers advantages.

If I wanted to defend the conventional physicists, I would expand on Albert's remarks – pointing out, for example, that symmetry between position and momentum is not merely a pretty feature of the mathematics of the quantum, but a central property that leads to a deep and informative viewpoint known as 'symplectic structure', which provides a useful link between classical and quantum mechanics. However, Bohm's theory does not actually affect the *mathematics*: it just adds a layer of interpretational gloss. The symmetry is lost, not in the formalism, but in the interpretation of the formalism. So you can have your symplectic structure and swallow Bohm too.

What does worry many physicists about Bohm's theory is that – like the EPR paradox – it has an aspect of non-locality. The wave function of a particle is spread out over all of space, and it reacts instantly to any interaction with another particle. This is of course also true in conventional quantum mechanics – whose wave function obeys exactly the same equation as does Bohm's. However, in Bohm's interpretation the wave function is a real physical thing. In the Copenhagen interpretation it is a mathematical fiction; only its component eigenfunctions can be observed, and those only one at a time. We could cover a lot of pages analysing the pros and cons here, but again it seems to me that the discussion is misdirected. What is missing from both the Copenhagen interpretation and Bohm's theory is any understanding of how macroscopic measuring devices (such as Geiger counters and dead cats) produce determinate values. Observations detect eigenstates, not arbitrary (that is, in the Copenhagen interpretation but not Bohm's, superposed) states. Why?

In recent years there have been several attempts to describe, mathematically, how a quantum state evolves (decoheres) during a macroscopic measurement process. The physicists involved include L. Diosi, N. Gisin, G. C. Ghirardi, R. Grassi, P. Pearle, A. Rimini, and I. Percival. In all of these theories the interaction of a quantum system with its environment produces an irreversible change that turns the quantum state into an eigenstate. However, all of these theories are probabilistic: the initial quantum state undergoes a kind of random diffusion which ultimately leads to an eigenstate.

Albert Einstein would be distinctly unhappy about irreducible randomness showing up at any level of quantum theory – even if, as here, it is confined to the measurement process. These theories do at least try to fill in the glaring hole in both the Copenhagen interpretation and Bohm's, but they are not fully deterministic. God may not play dice, but apparently Geiger counters and cats still do.

Dice, here, are quite an appropriate image. When you roll a die (sorry, I point-blank refuse to say 'a dice'), any one of the six possible faces may end up on top. The result of throwing a die is like an eigenstate – it is a special state that is selected by the measurement process. (Glory be, maybe it *is* an eigenstate.) Dice have several faces: quantum systems have several eigenstates. A Copenhagenist would say that the presence of a table mysteriously causes a die to 'collapse' to one of the states 1, 2, 3, 4, 5, 6, and that the rest of the time it is in a superposition of those eigenstates – which is, of course, a mathematical fiction with no intrinsic physical meaning. Bohm would say that it does have a physical meaning but you can't observe it – at least with any conventional apparatus, and perhaps not at all. Diosi-to-Percival would say that as the die rolls along the table its state randomly jiggles, and eventually settles down to one of 1, 2, 3, 4, 5, or 6.

Who is right?

The image of dice suggest that they might all be right –

– and all wrong.

Chaos teaches us that anybody, God or cat, can play dice deterministically, while the naïve onlooker imagines that something random is going on. The Copenhagenists and Bohm do not notice the dynamical twists and turns of the rolling die as it bounces erratically but deterministically across the tabletop. They don't even see the table-top. Bohm does think that what the die is doing is real, even if unobservable; the Copenhagenists don't even think that. Diosi-to-Percival notice the erratic jiggles of the die, and characterize them statistically as a diffusion process, not realizing that underneath they are actually deterministic.

Nobody tries to write down the equations for a *rolling* die.

Why not?

One good reason is that they think it can't be done.

Bell's Inequality

Perhaps we can explain the strange behaviour of fundamental particles without recourse to irreducible randomess. Why not equip each article with its own deterministic 'internal dynamic'? This should not affect how the particles interact with one another, but it *should* affect how the particle itself behaves. Instead of rolling the quantum dice to decide when to decay, a radioactive atom might monitor its internal dynamic, and decay when that dynamic attains some particular state. Before the advent of chaos we could not even contemplate playing that kind of trick, because the only known behaviour for the internal dynamic was too regular – steady, periodic, or quasiperiodic. The statistics of radioactive decay simply wouldn't fit such a model, let alone the more subtle aspects of interfering wave functions and the like. But chaos gets over that particular difficulty very neatly, and it suggests that what matters is not *whether* God plays dice – but *how*.

Physicists refer to such theories as 'hidden variable' theories, because the internal dynamic is not directly observable and the variables that define its phase space are in effect concealed from the observed reality. Bohm's theory

angle A

electron source

Figure 149   An experiment to test Bell's inequality.

is a kind of hidden variable theory, with the unobservable details of the *real* wave function as hidden variables.

There is a celebrated proof that no 'hidden variable' theory can be consistent with quantum mechanics. (How does this affect Bohm's attempt? See below.) Its theoretical aspects were devised by John Bell in 1964. Preliminary experimental confirmation came in 1972, and the last plausible experimental loopholes were closed in 1982.

Bell's argument can be formulated in a number of different ways: I will choose one that is close to the spirit of our discussion so far. This addresses the possibility that the observed state of a particle evolves, during the act of measurement, according to a deterministic dynamical process, and that the observed eigenstate is the result of this deterministic evolution. Another of David Bohm's contributions was a fairly realistic scenario for such a thought experiment, and only later did people start to realize that it was so realistic that actual experiments along similar lines might be possible.

Bohm's scenario requires a source of spin $-\frac{1}{2}$ particles (such as electrons) produced in pairs and moving in opposite directions: one stream travels north, the other south, at the same speed. You can therefore keep track of which particles started off close together: call these 'corresponding' particles in the two streams. A spinometer measures the spin of the northbound particles in the 'up' direction; another one measures the spin of the southbound particles in a direction inclined at an angle A to 'up' (in the up/east plane) (see Figure 149). By combining these measurements it is possible to work out a 'correlation function' $C(A)$ which measures how closely the spins in one stream match the spins of corresponding particles in the other. If $C(A)$ is 0 then the spins (measured in the up direction) of northbound particles are statistically independent of the spins (measured at angle A) of the corresponding southbound particles. If $C(A) = 1$ then both spins are the same for corresponding particles: if the northbound particle has spin $+\frac{1}{2}$ then so does the corresponding southbound one; and similarly for spin $-\frac{1}{2}$. If $C(A) = -1$, the spins are perfectly anticorrelated: the spin of any northbound particle is the exact opposite of that of the corresponding southbound one.

For the sake of argument, Bell assumes that the observed values of the spin are not random, but are determined by 'hidden variables' – some *deterministic* dynamical system whose variables are not observed. (In my analogy with a rolling deterministic die, the dynamical state of the die while it is rolling involves hidden variables such as angular velocities, which you simply do not see if you observe only the final steady state of the die.) Suppose that you run the experiment twice: once with the spinometer in the second stream set to angle A, and once with it set to a different angle B. Bell did a calculation to see how the correlation function – which now depends in a deterministic manner on the dynamics of the hidden variables, a fact that you can exploit in the calculation – changes when angle A is replaced by angle B. And what falls out of the mathematics is an inequality, now called Bell's inequality:

$$|C(A) - C(B)| < C(A - B) + 1$$

(Here $|C(A) - C(B)| = C(A) - C(B)$ if this is positive, and $C(B) - C(A)$ if not.) There is a relation between the correlation functions observed at angles A, B, and $A - B$.

The upshot is that, in this kind of experiment, any system whose apparent randomness is driven by a hidden deterministic dynamic must reveal its status by satisfying Bell's inequality. There is a definite constraint on the kind of correlation function that you can get.

What subsequent experiments showed is that the observed correlation function does *not* satisfy Bell's inequality. This was widely held to be definitive proof that quantum mechanics is *unavoidably* probabilistic. You *can't* plug the measurement gap by making God's dice deterministic.

However, there are loopholes. Bell's proof involves a number of assumptions, most of which he stated rather carefully. (Mathematical proofs *always* involve assumptions, and the wise scientist understands what they are before believing that they confirm or deny some aspect of physical reality.) In particular, the proof assumes the principle of locality – that no information can travel faster than light. So – at least in this particular formulation – Bell's inequalities do not rule out a Bohm-type theory, because Bohm's theory is not local.

Nonetheless, most physicists took the view that since experiments confirm that quantum systems violate Bell's inequality, there is little point in hoping to embed conventional quantum mechanics in a deterministic theory. Non-locality is, after all, a bit hard to swallow. (Forget that Schrödinger's equation is non-local – Schrödinger's wave function is a mathematical fiction. Remember that Bohm's theory is non-local – *his* wave function is supposed to be real.

Non-locality behind the scenes is acceptable; out in the glare of the spotlights it's not.) And, to be fair, a deterministic underpinning for quantum mechanics would be *much* more palatable if it really did render God's dice deterministic and local. You don't want dice thrown on a table in Las Vegas to instantly communicate with dice on a table in the bar at Mos Eisley spaceport (in a galaxy far, far away, you'll recall). It may not affect the gamblers, who can't use that kind of information – but it's philosophically untidy.

And Bell's inequality rules out *all* local, deterministic, hidden variable theories.

Or so everyone thought.

Dice and Determinism

In the next subsection I'm going to describe a new idea which gets round Bell's inequality in a very clever way. As preparation, let me first develop the argument that along with that mainstay of probability theory texts, the fair coin, the 'dice' metaphor is one of the most inappropriate ever invented. At least, unless we revise our idea of randomness.

I'm talking of an ideal die, a perfect inelastic cube, thrown on to a perfectly flat inelastic surface, subject to some precise law of friction, and obeying Newtonian mechanics. I have to do that to introduce the mathematics precisely. It seems to me that whatever makes a real die random ought to show up in this model too. Putting on Laplace's hat, however, it's clear that Vast Intellect could work out the final rest state of the die the moment it's thrown. With a video-camera and a supercomputer we ought, at least in principle, to be able to predict the outcome before the die does.

This isn't entirely a fantasy. J. Doyne Farmer, an American chaologist, developed a theory of the roulette wheel which improves considerably on pure chance. He's having trouble getting the casinos to let him play, though.

Anyway, if you can predict exactly what will happen, where does the randomness come from?

I can't do calculations for a die, but I'll do them for a simplified coin, close enough to show what's involved. The coin is a line segment of unit length, confined to a vertical plane. When it is tossed, starting at ground level, it's given a vertical velocity v and also a rotation rate of r turns per second. When it returns to ground level, it freezes: whichever side is then uppermost is considered to be the result of the toss.

If g is the acceleration due to gravity, then the coin takes $2v/g$ seconds to return to the horizontal, and so makes $2rv/g$ turns. The boundary between heads and tails occurs at exact half-turns, that is, when $2rv/g$ is half an integer. If this integer is N, then the head/tail boundary is given by $vr = gN/4$.

Figure 150   Initial conditions for a spinning coin, striped according to its eventual fate. Black = heads, white = tails.

If I could control the values of r and v exactly, then I'd be able to make the coin land whichever way up I want. However, *in practice I can control these values only within limits*. For example, suppose that I can keep v between 480 and 520 cm/sec, with r between 18 and 22 revolution per second. How does the outcome – heads or tails – depend on v and r?

You can get the answer from the formula above. The rectangle of possible values of v and r divides into stripes: black for heads, white for tails (Figure 150).

Any *known* values of the initial velocity and the rate of spin give a unique answer. Not only is the outcome deterministic – I really can tell you, in advance, what it is.

But if all I know is that v and r lie within the given range, I can't prescribe the outcome. The best I can do is think of the rectangle as a kind of dartboard. Each coin-toss is like throwing a dart: if the dart hits a black stripe, I get a head, if white, I get a tail. If the darts are distributed uniformly over the rectangle, then the probability of a head is the proportion of the total area covered by black stripes.

In others words, the source of the randomness lies in the choice of initial conditions. Unless I can control them *exactly*, I can't make a precise prediction.

Here Laplacian determinism breaks down again – but in a subtly different way. The model coin isn't a chaotic system. It's a perfectly regular one.

What we see here is that, associated with any deterministic dynamical system, there is a probabilistic system that offers a kind of 'coarse-grained' representation. Instead of telling us exactly which point in phase space the system occupies at a given instant, it tells us just the probability that the point lies in a given region at some instant. The study of such probabilities, called invariant measures, goes back to the early days of statistical mechanics, when mathematicians and physicists were trying to understand gases as complex collections of molecules, bouncing madly off each other. Invariant measures explain why gases have well-defined average properties like density and pressure. You might say that before we understood the molecular basis of matter, the only things we knew about the dynamics of gases were probabilistic. Afterwards, we realized that the probabilities are derived from a deterministic – but incredibly complicated – underlying dynamic. So statistical mechanics does have a hidden variable theory, whose variables are the positions and velocities of the gas's component molecules.

Could quantum theory be similar? Our experience to date makes us think that it is irreducibly probabilistic, but where do the probabilities come from? Probabilities are patterns, of a kind, and it is actually rather bizarre to think of probability as a primary physical concept when in every case where we understand the deep structure, probabilities arise from a deterministic dynamic as invariant measures. Indeed the existence of well-defined statistical patterns is evidence for a kind of order that becomes apparent only when we average over long timescales. What makes the system's distant future resemble its past, even on average? If it's *really* random, why aren't they just different? If a radioactive atom decays in a manner that has well-defined statistical regularities, where do those regularities come from? To say they are fundamentally probabilistic, and leave it at that, is simply to postulate a pattern that ought to be explained.

For deterministic chaos, in contrast, there is a clear mathematical explanation of the associated probabilities and their statistical regularities. *We know where they come from*: they arise as invariant measures. Although we don't understand everything here that we'd like to – for example the existence of technically 'nice' invariant measures is widely conjectured but narrowly proved – we see very clearly that it is the determinism of the dynamics that makes the future look similar to the past. The reason is recurrence: deterministic systems keep returning close to their previous states. So, paradoxically, it is the underlying determinism of the system that makes probabilities applicable. In order for a coarse-grained model to fit, there has to be something fine-grained to coarsen.

It seems to me that a truly random system ought not to exhibit patterns at all, not even on average. I'm aware that this view is a minority one, and I'm also aware that in some ways it conflicts with Gregory Chaitin's beautiful 'algorithmic information theory', which defines randomness in a manner that necessarily implies statistical regularities. I'm aware that complexity theory (see Chapter 17) suggests that patterns need not 'come from' anywhere, and

I'm inclined to agree. But I still think that there's a grain of truth in the idea. I think that the existence of statistical regularities in quantum-level matter needs to be explained, not simply assumed; and some kind of chaotic hidden variable theory would fit the bill –

– if only it weren't for Bell's inequality.

However, there are more ways to Bell a cat than choking it with correlations. In principle, we can get round Bell's inequality. I don't know if this can be done while remaining consistent with every known experimental result about quantum mechanics – that needs further research. However, despite Bell's inequality, we don't have to give up at the outset. Bell's inequality tells us that *certain kinds* of 'hidden variable' extension of conventional quantum theory can't possibly work, but it doesn't rule out every conceivable extension or alternative. It is a constraint that tells us a little bit about what kind of hidden variable model we can introduce.

I have a feeling that physicists may have been too easily impressed by a mathematical theorem. Mathematicians know that theorems have hypotheses, assumptions that you have to make before the theorem works. Indeed mathematicians spend a lot of time writing down their hypotheses very carefully, thereby infuriating the physicists, who prefer to be sloppy (they call it 'appeal to physical intuition') and leave the hypotheses tacit. This is fine for everyday nuts-and-bolts physics, but I think it can be horribly misleading when we get down to fundamental philosophical issues, where some careful logic-chopping is necessary.

A careful study of the requisite hypotheses reveals potential loopholes in the proof of Bell's inequality. There are several hidden assumptions, technical things like the uniqueness of the probability measure used to compute the correlation function and the convergence of various infinite series and integrals. These loopholes are probably pluggable, but very recently it has been shown that at least one is not.

Let's take a peep inside Pandora's quantum box.

Riddled Basins

In 1995 Tim Palmer, a meteorologist with a physics background and an abiding interest in chaos, discovered a very subtle potential loophole in the derivation of Bell's inequality. Basically, it is the assumption that correlation functions are (theoretically) computable. Palmer's conclusion is that quantum indeterminacy may perhaps be replaced by certain kinds of 'hidden variable' chaotic dynamic, provided that the chaos is sufficiently nasty. Nasty enough to wreck the derivation of Bell's inequalities, nice enough to remain deterministic.

The first step is to appreciate just how nasty deterministic chaos can be.

I've told you, several times, that chaos is not as unpredictable as many people think. A chaotic attractor has its own kind of stability. If you perturb a point a small distance away from its attractor, it will return rapidly to the attractor again. So you *can* predict that the point will stay on its attractor.

I lied.

Well, a tiny bit. What I said is true when there is only one attractor. And it is *often* true when there are more. However, there is at least one important case when it is not true, which is when there are at least two attractors and one of them has a 'riddled basin'.

The basin of an attractor is the set of all points in phase space that are attracted to it. The simplest image for this is that the attractor sits at the bottom of some kind of bowl, and the phase point rolls downhill until it hits it. In this image, the basin of an attractor entirely surrounds it. If you have several attractors, their basins are like a series of valleys, separated by nice, smooth ridges. The 'basin boundaries' are smooth curves along the tops of the ridges.

However, nonlinear dynamics can be much more complicated than that. In 1992 Jay Alexander, I. Kan, James Yorke and Z. You discovered that the basin may look more like a colander than a bowl – riddled with holes, hence the name 'riddled basin'. Unlike a colander, those holes are not round; instead they are long, thin swirly streaks that reach right in to the attractor (Figure 151). Points in those streaks do *not* move towards the attractor: instead, they are repelled away. The boundaries of riddled basins are fractals, not smooth curves.

It turns out that riddled basins are not at all exotic: they show up reliably when the dynamical system obeys a few perfectly reasonable conditions. Their existence is closely tied to the fact – mentioned in Chapter 15 – that chaotic attractors usually contain unstable periodic points. Indeed riddling can take a more extreme form, in which there are two competing attractors, both have riddled basins, and each basin fills up the holes in the other. Such *intertwined basins* are also perfectly commonplace in the world of nonlinear dynamics. Indeed they can occur with more than two competing attractors.

A system with two attractors whose basins are intertwined is *seriously* unpredictable. You can predict that eventually any chosen initial point will end up on either one attractor or the other, but you can't predict which. As close as you like to initial points that end up on attractor 1 there exist initial points that go to attractor 2, and vice versa. Now the butterfly effect doesn't just move points around on the same attractor – it can switch them from one attractor to the other.

It's a bit like predicting that a rolled die will either come up 1, 2, 3, 4, 5, or 6, without saying which. Indeed a deterministic die behaves very much as

Figure 151   Riddled basin for the steady state attractor of a pendulum whose bob has been replaced by a rotating double arm. Points coloured black belong to the basin. The lower figure shows a close-up of one small region.

if it has *six* attractors, the steady states corresponding to its six faces, all of whose basins are intertwined. For technical reasons that can't quite be true, but it is true that deterministic systems with intertwined basins are wonderful substitutes for dice; in fact they're super-dice, behaving even more 'randomly' – apparently – than ordinary dice. Super-dice are so chaotic that they are uncomputable. Even if you know the equations for the system perfectly, then given an initial state, you cannot calculate which attractor it will end up on. The tiniest error of approximation – and there will always be such an error – will change the answer completely.

You can, however, calculate probabilities, much as I did for the spinning coin with its striped phase space. Now the black and white stripes are replaced by the basins of the two attractors. The probability that an initial state lying in some small region ends up on a particular attractor is a computable number associated with that chosen region. For practical purposes, you can compute the probabilities by 'Monte Carlo simulation': that is, try a million randomly chosen points inside the region and see what proportion of them ends up on the desired attractor. For theoretical understanding you must appeal to things known as 'invariant measures', which are very similar to probabilities. Although your Monte Carlo calculation will get many individual fates wrong, because super-dice are uncomputable, the proportion of hits depends on what fraction of the chosen region meets the relevant basin, and in that case the errors tend to cancel each other out. (This is typical of statistics: if you estimate voting intentions, say, by giving a questionnaire to a sample of the population, then the average intentions over the sample are a much better indicator of the overall true intentions than any of the individual answers. This remains true even if some respondents deliberately lie.)

Systems with intertwined basins are in principle fully deterministic, so they can be represented by mathematical equations without any explicit random terms. They are practically uncomputable: given an initial state, you cannot calculate with any confidence where it will go. But they are statistically computable: given an ensemble (physicists like this word – it means 'set', or, in geometric terms, 'region') of initial states, you can calculate the probability of ending up on any particular attractor.

Palmer's idea is to use this kind of system to provide hidden variables that determine how a quantum state changes when you observe it. Think of an initial point in the hidden variables' phase space as the quantum state before you start to observe it, and the attractors for the hidden variables as representing the possible eigenstates. Because of the statistical computability of such systems, you get well-defined quantum probabilities, consistent with experiments. But the mathematics used to derive the Bell inequality relies on writing down several expressions that use the *actual values* of the hidden variables, and comparing them. Since the dynamics of the hidden variables is uncomputable, those expressions make no sense – so you can't compare them. This is the loophole through which the Bell inequality escapes.

Palmer's actual theory is more specific and more subtle. He considers not only the mathematical computability of the internal dynamic, but the possibility that nature somehow can 'compute' it by extra mathematical means. As, perhaps, it would have to do if it were to implement a theoretically uncomputable theory!

Bell's inequality still escapes.

In fact, the EPR paradox might escape as well. Let me float an idea past you – not Tim Palmer's this time: I wouldn't want him to be blamed for it. When there is an internal 'hidden variable' dynamic, electrons whose states are synchronized when they are close together may remain synchronized as they speed apart. All you need do is to put their hidden variables into identical states (or maybe symmetrically related states, such as 'spin up'/'spin down'). Since they are obeying the same *deterministic* mathematical rules for the evolution of their internal states, those states will remain identical. It is not necessary for 'information' to pass between them instantaneously, faster than light, for the synchrony to be maintained. Instead of 'spooky action at a distance' we have a kind of dynamic memory. There are problems to be overcome with this idea, of course, and one of them is the butterfly effect. Any slight initial lack of synchrony will rapidly grow. One possible way round this is to allow synchronizing signals to pass between the electrons at a speed below that of light. Then each will influence the other after a time delay. It is known that such time-delayed signals can keep identical chaotic systems in synchrony. Another possibility is to quantize the internal variables so that their values are discrete. Then they really can be perfectly synchronized, like two identical digital computers with a common clock rate running the identical computation. (The internal dynamic can still be chaotic: if initial states are not actually identical, they can still diverge exponentially.) At the very least, this suggestion shows that there are possible theoretical scenarios that are very different from anything we could have imagined without the concept of chaos.

You may feel that Palmer's idea is all very well, but what use is an uncomputable theory? Sorry, wrong question. The Copenhagen interpretation of quantum mechanics is even more uncomputable – instead of an internal dynamic it employs some capricious act of fate – and the Copenhagen interpretation is generally considered useful. Palmer's approach is in any case statistically computable, and in quantum mechanical experiments, that's as good as you'll ever get anyway. Indeed, Palmer's approach is the kind of thing that, at least in spirit, you *must* employ if you want to render the quantum dice deterministic by the introduction of hidden variables. Otherwise Bell's inequality will bite. The precise details are probably wrong – there are many possibilities for internal dynamics with intertwined basins, and Palmer chooses a specific one merely for the sake of argument – but the spirit of the approach is mathematically sound. And Einstein would definitely have approved of the philosophy.

None of which impels nature to work that way, of course. But what Palmer's work establishes is that the Holy Grail of a deterministic, but chaotic, explanation of quantum indeterminacy is *not* made unachievable by the Bell inequality.

And that's worth knowing.

World of If

I often wonder how different today's science would have been if chaos had been discovered *before* quantum mechanics. (Actually that's not likely, because the marvellous computers that make chaos so obvious to everybody rely upon quantum effects to make their circuitry function – but let's pretend. You can *find* chaos without computers, that's what mathematicians did in the sixties. It just required computers to convince everybody else.) Now, instead of Einstein protesting that God doesn't play dice, he would probably have suggested that God *does* play dice. Nice, classical, deterministic dice. But – of course – chaotic dice. The mechanism of chaos provides a wonderful opportunity for God to run His universe with deterministic laws, yet simultaneously to make fundamental particles seem probabilistic.

Whether this approach would have established itself is, naturally, debatable. Palmer's work shows that it could at least have got started, and if we can get round the Panda Principle we may yet find out whether it can deliver the full goods. I can't help thinking that physicists – with Einstein and Schrödinger shouting enthusiastic encouragement – would have tried long and hard to construct a deterministic but chaotic theory of the microscopic world, and that they would have abandoned this idea for a merely probabilistic theory with the utmost reluctance.

17

Farewell, Deep Thought

'You're really not going to like it,' observed Deep Thought.
'Tell us!'
'All right,' said Deep Thought. 'The Answer to the Great Question . . .'
'Yes . . . !'
'Of Life, the Universe and Everything . . .' said Deep Thought.
'Yes . . . !'
'Is . . .' said Deep Thought, and paused.
'Yes . . . !'
'Is . . .'
'Yes . . . !!! . . . ?'
'Forty-two,' said Deep Thought, with infinite majesty and calm.

Douglas Adams, The Hitch Hiker's Guide to the Galaxy

Suppose Laplace's 'Vast Intellect' were indeed to follow his instructions, to 'condense into a single formula the movement of the greatest bodies of the universe and that of the lightest atom' and then 'submit its data to analysis'. Would it get any more sensible an answer than Douglas Adams's characters Loonquawl and Phouchg did in *The Hitch Hiker's Guide to the Galaxy*?

Vast and Considerable Intellect

Probably not.

Let me leave aside certain material considerations, which some would argue have no philosophical relevance – though I wonder if they aren't the essence of the matter. Namely, I'll ignore the awkward question of what Vast Intellect would write Its equations *on* – given that It must deal with at least six variables – position and velocity – for every particle in the Universe, and thus need more paper and ink than could be constructed if the entire universe were composed of those substances. As an anonymous 17th-century poet says:

> If all the world were paper,
> And all the seas were inke,
> And all the trees were bread and cheese,
> What should we do for drinke?

I also won't ask what manner of brain Vast Intellect would require to store, let alone think about, Its Master Equation for Life, the Universe, and Everything. A brain bigger than the universe, clearly implying that Vast Intellect must stand outside the universe and peer in. Not a bad idea, on grounds similar to Heisenberg's Uncertainty Principle – if Vast Intellect were part of the universe, then every time It pondered the value of $dx_{7345232115}/dt$, It would change the very thing It was pondering (Figure 152).

If we accept that Vast Intellect is truly omniscient, then Laplace has a pretty good point. If the universe really does obey deterministic mathematical laws, then Vast Intellect can use them to predict what the universe will do.

But that's a pretty nebulous piece of philosophy, an excellent example of how to get nonsense by going to extremes. If we want to draw conclusions with implications on the human scale, rather than the superhuman, then we have to set more realistic requirements. The picture changes dramatically.

I have in mind a slightly lesser being than Laplace's ideal – Considerable Intellect, let's say. It has enormous brain-power, more than the human race put together. (Come to think of it, when you put the human race together, its brainpower seems to be negative. But you know what I mean.) 'Big. Really Big. You just won't believe how vastly hugely mindbogglingly big,' to quote Adams again. Furthermore, to load the dice even more decisively in Considerable Intellect's favour, I'll pit it (small 'i' in deference to Vast Intellect) against a greatly reduced problem. A miniature universe, well within comprehensible bounds, where not only Considerable Intellect but indeed any competent human mathematician can not only write the equations down in principle, but can do so in practice. Namely, Hill's reduced model of the three-body problem: Neptune, Pluto, and a grain of dust.

As Poincaré's *Celestial Mechanics* despairingly observes, this problem leads to chaos, in the form of homoclinic tangles. When dynamics is chaotic, it can only be predicted accurately if the initial conditions are known to infinite precision. But it takes an infinite memory to store a number to infinite precision. In short, Considerable Intellect can't even get started.

And that's the message for us teachable apes. When the dynamics of a system goes chaotic, there's a trade-off between the precision to which we know its current state, and the period of time over which we can say what – in detail – it will do. And the precision of observations has to be almost impossibly good to make even medium-term predictions.

On the other hand, we *can* still make very accurate predictions – not of

Figure 152   Vast Intellect's dilemma.

the exact long-term behaviour, but of its general qualitative nature. We can impose quantitative limits on it; and we can determine its statistical features.

If you can't win, move the goalposts.

Designer Chaos

Chaos has many lessons to teach us. Its prime message is a general one: 'Don't jump to conclusions.' Irregular phenomena do *not* require complicated equations, or equations with explicit random terms.

That message cuts both ways.

First, the 'loss' side of the balance sheet. Even if you're fortunate enough and clever enough to have devised good equations, you still may have trouble understanding the system that they model. Even if the equations are very simple, the behaviour of the system may not be. Whether or not something is complex depends upon what questions you ask and what point of view you adopt.

On the 'profit' side we find the same remark. A phenomenon that *looks* complicated may not really be. It might be governed by a simple – but chaotic – model. Now we're getting into Designer Chaos: using know-how about typical types of dynamics to build plausible models.

Sometimes it works. The heartbeat, measles epidemics, and perhaps the tumbling of Hyperion are examples where it does. After a flirtation with chaos we emerge with a better understanding of the physical problem, and one that we really can *use*.

Sometimes it doesn't. I see no evidence that chaotic dynamics is likely to improve the quality of weather-forecasts. Its main contribution to date is to suggest that we're asking a silly question. Forecasts over a few days, maybe a week – that's fine. A month? Not a hope.

That's a personal belief. Some genius could blow it away tomorrow. Maybe other methods can succeed where solving the equations for the weather is doomed to fail. Time will tell. I know what I've got my money on.

Unrepeatable Experiments

Chaos forces us to revise the conventional idea of an experimental test. Conventionally, you start with a theory, make predictions, and perform an experiment to falsify them. If it doesn't falsify them, you say you've verified the prediction, and you assume – a pragmatic view rather than a logically sound one – that the theory is right.

Fine. Last night I did an experiment to see whether water flows uphill, and it did. Physics is dead.

You don't believe me, do you? Let me tell you about the experiment . . .

What's that? *Do it again?* Sorry, I can't do that . . .

You're not buying this, are you? Quite right too. In order to carry conviction, an experiment must be *repeatable*. If two different scientists do the same experiment in two different laboratories, they ought to get the same results. Of course, any effects that might change the results must be taken into account and eliminated. It's a lot hotter in Bombay than it is in Novosibirsk: if temperatures matter, the Indian scientist has to do the experiment in a refrigerator and the Russian has to turn up the heating.

But a chaotic trajectory, from a given initial condition, is a non-repeatable experiment. Indeed it's a non-repeatable *prediction*, as the tale of the two supercomputers makes clear. You might argue that on a given *make* of computer, the 'experiment' *is* repeatable. But different laboratories should surely be permitted to use different equipment.

So chaos tells us that even when our theory is deterministic, not all of its predictions lead to repeatable experiments. Only those that are robust under small changes of initial conditions are good candidates for tests. The topology of the attractor, say, or its fractal dimension.

That means that we can test whether, say, a chaotic model of turbulence accurately describes the way the fluid as a whole behaves; but we can't test whether a given fluid particle really is obeying the dynamical equations of Navier and Stokes. Not directly, not the way Galileo tested his theory of motion under gravity. Some details of the theory are beyond practical tests.

All of this demands – and has received – a response from experimentalists. We've seen examples throughout the last few chapters. Experimental methods must be redesigned to study chaotic systems. In fact one of the great contributions of chaos is that experimentalists now present their data in much more geometric and meaningful ways – attractors rather than power spectra, Poincaré sections rather than time-series.

Sleepwalk to Chaos

There are other morals to be drawn, not specific to chaotic dynamics.

Arthur Koestler, in his book *The Sleepwalkers*, portrays scientific discovery as a series of inspired blunders. When important new ideas are found, hardly anyone appreciates them; the people who make them misunderstand what they mean; and progress comes by a combination of accident and serendipity.

Of course, that's a very crude paraphrase. And science wouldn't get very far if all it could do was sleepwalk. The developmental side of science, one of its greatest strengths, exploits unexpected discoveries – accidental or not – in a conscious fashion, and turns them into something with more than curiosity value.

But the tale of chaos is not without its sleepwalkers. Many of the key discoveries reported in our tale have the same unreal air. The people doing the research were misunderstood, couldn't get support, persisted despite – rather than because of – the scientific establishment. Against that, to the establishment's credit, we must set a willingness to change tack completely when the new and unorthodox ideas began to prove themselves. One can wish for a little more exercise of imagination, but scientific conservatism has its place. Pioneers must expect to hack their jungles alone, otherwise science would spend all its time sponsoring half-baked crackpots.

One striking common thread that runs through all of the early work in chaos is that the people doing it were, at heart, mathematicians. Not all of them by profession, mind you. Lorenz was a meteorologist, Hénon an astronomer, Feigenbaum a physicist, May a biologist. But they all let their mathematical instincts guide them, when too much concentration on the 'real world' would have destroyed any confidence that their work could ever be anything better than an oversimplification. If you look for the physics in Lorenz's equations, it's virtually non-existent. Better approximations to the true dynamics don't do anything like Lorenz's – as his colleagues pointed out to him at the time. Decades later one of them, Willem Malkus, said wryly: 'Of course, we completely missed the point. Ed wasn't thinking in terms of our physics at all. He was thinking in terms of some sort of generalized or abstracted model which exhibited behaviour that he intuitively felt was characteristic of some aspects of the external world.'

In other words, Lorenz was thinking like a mathematician, not a meteorologist.

Campaign for Real Mathematics

The discovery of chaos required many things and many people. It needed pure mathmaticians to develop the topological approach to qualitative dynamics, and to ask sufficiently general questions. It needed physicists to link the answers to the real world. It needed experimentalists to check that the theories made sense. It needed electronic engineers to design and build computers with good graphics and powerful number-crunching capabilities.

Which contribution was the most important?

Silly question. Which do you consider most important: your heart, your lungs, or your brain?

Take one away, and you're dead. It's the *combination* that counts.

But, speaking as a mathematician, I do want to say one thing. People outside mathematics often criticize the subject for lack of contact with reality. The story of chaos is just one of many currently unfolding, which show that this criticism is misplaced. It's like criticizing a lung because it can't pump blood.

If you take a 'goal-oriented' viewpoint you'd expect a breakthrough in the understanding of turbulence, say, to come from an intensive programme of research by fluid dynamicists. In fact these were not the crucial ingredients for the strange attractor breakthrough – for such it is, however many questions it may leave unanswered. The crucial theoretical ideas came from topology, a subject not hitherto noted for its relevance to fluid flow. The crucial experimental tool was the laser, which at the time was widely underestimated, 'a solution looking for a problem'. And the experimentalists who used that tool were physicists who had earned their colours working on phase transitions, not fluids.

Science is a complicated, interlocking structure. Ideas can come from anywhere. A good idea is like an infectious disease: it spreads. No one can predict what it will lead to, no one can confine it within prescribed bounds. Ideas do not come with little labels attached:

> **WARNING – Topology.**
> **Avoid contact with the real world.**

Unfortunately, many people tacitly assume that they do.

To criticize mathematics for its abstraction is to miss the point entirely. *Abstraction is what makes mathematics work.* If you concentrate too closely on too limited an application of a mathematical idea, you rob the mathematician of his most important tools: analogy, generality, and simplicity. Mathematics is the ultimate in technology transfer. It was true in Euler's day: the analogy between electrostatics and fluid dynamics was obvious to a mathematician, absurd to anyone else. It remains true today: we've just seen how a method devised to study chaos in turbulent flows works equally well on measles epidemics.

However, technology transfer needs more than just the technology. Someone has to transfer it. So, while mathematicians should be encouraged to continue doing whatever it is that mathematicians do – whether or not the outside world can understand a word of it – it will remain just an art-form unless enough people are willing to make the effort to apply it to problems outside mathematics. The story of chaos is full of such people. They come from all subjects – physics, biology, engineering, chemistry, physiology, astronomy, as well as mathematics. They are the true 'applied mathematicians', and they do what that phrase ought to mean.

They take mathematics . . .

. . . and apply it.

Plus ça change . . .

Chaos is still a hot topic, but whenever a hot topic hits the scientific headlines, it always turns out that somewhere in the distant past there were people who knew about it. In some sense.

With hindsight, you can often see things that weren't anything like as clear at the time. The trick is not so much to know something, but to *know you know it*. That is, to appreciate that it's important, and to have a context in which to put it.

Earlier ages saw parts of this picture – but never put them together. They didn't have the motivation to ask the right questions, the techniques to find the answers. They saw isolated details, but never the Big Picture.

But it's clear that Poincaré, in particular, saw more than his contemporaries appreciated. To establish this, I'm going to give a rather long quotation from one of Poincaré's essays. You'll find much of the above discussion within it, even though it's almost a century old. Its title: *Chance*.

> A very slight cause, which escapes us, determines a considerable effect which we cannot help seeing, and then we say this effect is due to chance. If we could know exactly the laws of nature and the situation of the universe at the initial instant, we should be able to predict exactly the situation of this same universe at a subsequent instant. But even when the natural laws should have no further secret for us, we could know the initial situation only *approximately*. If that permits us to foresee the subsequent situation *with the same degree of approximation*, this is all we require, we say that the phenomenon has been predicted, that it is ruled by laws. But this is not always the case; it may happen that slight differences in the initial conditions produce very great differences in the final phenomena; a slight error in the former would make an enormous error in the latter. Prediction becomes impossible and we have the phenomenon of chance.
>
> Why have the meteorologists such difficulty in predicting the weather? Why do the rains, the storms themselves seem to us to come by chance, so that many persons find it quite natural to pray for rain or shine, when they would think it ridiculous to pray for an eclipse? We see that great perturbations generally happen in regions where the

atmosphere is in unstable equilibrium. The meteorologists are aware that this equilibrium is unstable, that a cyclone is arising somewhere; but where they cannot tell; one-tenth of a degree more or less at any point, and the cyclone bursts here and not there, and spreads its ravages over countries which it would have spared. This we could have foreseen if we had known that tenth of a degree, but the observations were neither sufficiently close nor sufficiently precise, and for this reason all seems due to the agency of chance.

The game of roulette does not take us as far as might seem from the preceding example. Assume a needle to be turned on a pivot over a dial divided into a hundred sectors alternately red and black. If it stops on a red sector, I win; if not, I lose. The needle will make, suppose, ten or twenty turns, but it will stop sooner or not so soon, according as I shall have pushed it more or less strongly. It suffices that the impulse vary only by a thousandth or a two thousandth to make the needle stop over a black sector or the following red one. These are differences the muscular sense cannot distinguish and which elude even the most delicate instruments. So it is impossible for me to foresee what the needle I have started will do, and this is why my heart throbs and I hope everything from luck.

And Poincaré offers some thoughts on the implications for experiment, which again echo what I've just said:

When we wish to check a hypothesis, what should we do? We cannot verify all its consequences, since they would be infinite in number; we content ourselves with verifying certain ones and if we succeed we declare the hypothesis confirmed.

The phase space of the universe, like that of the coin in Chapter 16, is also striped by its fates. Billions of dimensions of phase space, with billion-dimensional stripes, to be sure; but that just makes things worse. This would be true even if the universe were a regular non-chaotic system. When chaos strikes, the stripes grow infinitely thin, and mix together like spaghetti and sauce, compounding the effective indeterminacy.

All deterministic bets are off. The best we can do is probabilities.

In this sense, dice are a bad metaphor for genuine chance, but a much better one for deterministic chaos.

On the other hand, what *is* genuine chance? Poincaré pointed out that roulette, too, is deterministic. Maybe there's no such thing as a genuinely random event. All is predetermined; but we're too stupid to see the pattern. Within any given closed system, immutable law prevails. Chance events occur when an outside influence, not accounted for in those laws, disturbs their orderly functioning.

No truly closed system, free of outside influences, exists; and in this sense, random disturbances may always occur. However, they're random in a slightly unsatisfactory way. Given enough information, you feel you could have seen them coming.

The chance events due to deterministic chaos, on the other hand, occur even within a closed system determined by immutable laws. Our most cherished examples of chance – dice, roulette, coin-tossing – seem closer to chaos than to the whims of outside events. So, in this revised sense, dice are a good metaphor for chance after all. It's just that we've refined our concept of randomness. Indeed, the deterministic but possibly chaotic stripes of phase space may be the true source of probability.

Quantum uncertainty may be like this. An infinitely intelligent being with perfect senses – God, Vast Intellect, or Deep Thought – might actually be able to predict exactly when a given atom of radium will decay, a given electron shift in its orbit. But, with our limited intellects and imperfect senses, we may never be able to find the trick.

Indeed, because we're *part* of the universe, our efforts to predict it may interfere with what it was going to do. This kind of problem gets very hairy and I don't want to pursue what may well be an infinite regress: I don't know how a computer would function if its constituent atoms were affected by the results of its own computations.

Conservation of Complexity

Chaos teaches us that simple rules may give rise to complex behaviour. This discovery has a very positive aspect: it means that systems that we have hitherto considered too complicated to understand may in fact be governed by simple rules. It also has a negative one: being able to write down the rules for a system may not of itself provide much understanding. Either way, there is no real choice: chaos exists, both aspects are *true*, and our job as rational human beings is to make the best of things given this new knowledge. If you tore up every book and article on chaos, it would still be there. The only difference would be that you wouldn't *know* it was there.

Personally, I don't consider that an advance.

Since the first edition of this book was published a kind of converse to chaos theory has come to prominence, which has equally profound implications for our understanding of rule-based systems. I'm going to say only a little about it here, because to do it justice would require another book (one that Jack Cohen and I have already written under the title *The Collapse of Chaos*). Mainly

I'll try to give you some of its flavour and explain how it relates to chaos.

This new companion to chaos is called complexity theory, and it is particularly associated with the Santa Fe Institute, founded in 1984 by the Nobel-winning Caltech particle physicist Murray Gell-Mann. Complexity theory centres upon the tendency of complicated systems to exhibit simple behaviour – simple at *some* level of description, but not that of the system's components. The philosophical core of complexity theory is the concept of *emergence*, in which a system may transcend its components, so that 'the whole is greater than the sum of its parts'. For instance, complexity theorists see a stock-market crash as an emergent response of a complex monetary system to the actions of vast numbers of individual investors. No single investor can cause the market to crash, no single investor actually *wants* the market to crash. Yet, when the interactions between investors happen to head along a particular nonlinear dynamic path, their collective responses reinforce each other, and a crash is the inevitable outcome.

It is no coincidence that complexity and chaos are related. Both are part and parcel of the theory of nonlinear dynamics, one of *the* great success stories of late 20th-century science. As soon as you permit the flexibility of non-linearity in your models of nature, you will encounter these two phenomena. And if you don't permit that flexibility – well, you're a pachydermologist operating in a world that you think consists solely of large grey creatures with big floppy ears, but which actually contains beasts that you haven't even dreamed of. Ignorance may be bliss, but it's bliss purchased at a price that's not worth paying.

Both chaos and complexity challenge a deep-seated assumption about cause and effect that I shall call 'conservation of complexity'. In this view simple rules always imply simple behaviour. If you accept this then it follows that all complicated behaviour must arise from complicated rules. This assumption, commonly tacit, has directed major movements in science. For example, it is why we think of the complexity of living things as a puzzle: where does the complexity 'come from'?

Until very recently hardly anybody would have dared to suggest that it need not come from anywhere.

One problem with the conservation of complexity is that if simplicity is inherited directly from rules to behaviour then it is difficult to reconcile a complex universe with the simplicity of its rules. The usual answer is that the universe's complexity arises from the interaction of large numbers of simple components: it is complex in the way that a dictionary or a telephone directory is complex. But recently the idea that complexity is conserved has received a series of mathematical challenges. One, as we have seen, is the discovery of chaos, in which complexity arises through the *nonlinear* interaction of *small* numbers of simple components. Another is complexity theory, which emphasizes the converse: that highly complex interactions taking place in systems composed of many individual elements often conspire to create large-scale but simple patterns – emergent phenomena.

Most of the great mysteries of existence seem to be emergent phenomena. Mind, consciousness, biological form, social structure . . . It is tempting to leap to the conclusion that chaos and complexity must hold all the answers to these mysteries. In *The Collapse of Chaos* Jack Cohen and I argue that, at least as currently conceived, they do not and cannot. The role of chaos and complexity has been crucial and positive: they have caused us to start asking sensible questions and to stop making naïve assumptions about the sources of complexity or pattern. They have provided answers to many specific scientific questions and opened up new ways of thinking about them. They have even led to commercial payoff. But as far as the big questions of life, the universe, and everything go, they represent only a tiny first step along a difficult path.

What is a complex system? There isn't an agreed mathematical definition, but the general idea is that it should not be possible to describe the behaviour of the system *concisely*, even though it has definite elements of organization. Complex systems are neither ordered nor random, but combine elements of both kinds of behaviour in a very elusive but striking manner.

There are many kinds of complex system. The complexity may be purely spatial – the system exhibits complicated patterns, but the patterns don't change over time. An example is the DNA molecule, with its double helix, and its intricate sequence of 'codons' (about a billion in humans) that prescribe the chemical computations needed to produce life. Or the complexity may be purely temporal – the spatial structure is simple at any instant, but it changes in a complicated way as time passes. An example is the market price of some fixed commodity, say gold, which at any instant is just a single number, but which changes erratically over quite short intervals of time. Or the system may be complex in both space and time, like the human brain, with its billion nerve cells all hooked together in an organized but complicated way, and its ever-changing patterns of electrical signals. A complex system can also be *adaptive*, responding to outside influences, or even the results of its own behaviour, and 'learning' from them – changing itself in response. Examples include ecosystems, and evolving species. Other examples of complex systems include a rainforest, a living cell, an entire cat, or a national economy.

A complex system, in this sense, is not just 'complicated'. The long-chain molecules in toffee are just as *complicated* as those in DNA: they contain as many atoms, and it would take just as much space to list the structure in detail. But they are not as *complex*: they lack the organized aspects of DNA. The molecules in toffee are basically just random assemblies of atoms of carbon, oxygen, and hydrogen, subject to a few general rules about how many of each occur in a given region. The difference between complexity and complication is like that between the text of *Hamlet* and a table of random numbers.

One area where complexity is shedding some light is the theory of evolution. Biologists have long been puzzled by the ability of living systems to become ever more organized – in apparent violation of the second law of thermodynamics, which states that any closed system in thermodynamic equilibrium will become more and more disordered. Part of the answer to the puzzle is that living systems are neither closed nor in equilibrium; they take in energy from outside sources, and they are perpetually on the move. But oceans take in energy and move too, so these features alone do not explain the strangely *purposeful* behaviour of living creatures. Other evolutionary puzzles include sudden rapid changes in species, and mass extinctions. Charles Darwin was a gradualist: he said that 'nature does not make jumps'. But the fossil record is littered with jumps. Some, like the demise of the dinosaurs, are now thought to have been caused by outside catastrophes – in this case the infamous K/T meteorite that crashed to Earth 65 million years ago near the coast of Mexico – but not all. Complexity theorists have carried out computer experiments in 'artificial life' which show that, although these kinds of behaviour may seem surprising to us, they are actually very common. It is our intuition, not the universe, that behaves strangely. For instance, you can set up an artificial world of self-copying strings of 0's and 1's inside a computer, let them compete for memory space, and allow occasional random mutations. Over a period of time you will find more and more complicated self-reproducing strings, parasitic strings that hijack other strings' reproductive techniques, social creatures that have to get together to reproduce . . . just as Darwin observed in *real* life. You often see sudden bursts of diversity, with new 'species' forming, and sudden mass extinctions, all happening spontaneously as a consequence of the computer's simple rules of behaviour. We don't yet know *why* this kind of thing happens – but computer experiment after computer experiment shows that it does, that it's common, and that the only mystery is our own lack of understanding.

There is *no* 'law of conservation of complexity' that tells us that a simple system can never become more complicated on its own.

Complexity builds on the concept of chaos. In fact a basic buzzword – or more accurately buzzphrase – in complexity theory is the 'edge of chaos'. Some systems behave in very simple ways, the clockwork universe of regularly ticking cogs. Systems that become chaotic behave in far more complicated ways, the extreme being the totally random motion of gas molecules. Poised in between are more interesting kinds of behaviour – complicated but with hints of pattern. These complex but organized systems seem to be exactly on the transition between order and chaos – and that's the 'edge of chaos'. The suggestion is that selection or learning drives them towards this boundary. Systems that are too simple do not survive in a competitive environment, because more sophisticated systems can outwit them by exploiting their regularities. (If a courier firm always collects the payroll from the bank at 10 am on Fridays, following exactly the same route, then thieves can stage a robbery easily.) Systems that are too random also do not survive, because they never achieve anything coherent. (If the couriers follow totally random paths, it takes them so long to reach the bank that some other firm has got the job by the time they finally arrive.) So it pays, in survival terms, to be as complicated as possible, *without* becoming totally structureless. Evolutionary systems are forced to poise themselves on the edge of chaos.

Complexity theory is explaining what types of system tend to complicate and organize themselves, why they do it, and whereabouts such behaviour lives in the dynamical spectrum from total order to total chaos. Its future aim is to develop a coherent and comprehensive range of techniques for understanding the complex systems that we find in nature, and to codify their behaviour in a simple set of basic principles.

The Goat in the Machine

That's what the complexity theorists say, but others disagree. The heart of the discussion is not so much what the universe really is, as how human beings can understand it. It is about explanations, not essences. The traditional view of explanation in science is relatively simple: an explanation of some phenomenon is a deduction of that phenomenon from natural laws. We summarize this view in Figure 153, a schematic representation of the corresponding mental process: we look down a 'mental funnel' from a natural phenomenon and 'see' the underlying rules. I prefer that word because 'laws' has overtones of ultimate truth, and no area of science can legitimately claim that status. Notice that the arrow of explanation runs upwards, from rules to phenomena, whereas the arrow of discovery runs in the opposite direction. I will standardize diagrams to show just the arrow of explanation. This paradigm of 'explanation' has emerged over a period of centuries, beginning its spectacular rise to prominence in the work of Isaac Newton and his contemporaries. As scientists looked down more and more mental funnels they started to find more and more common rules. For example, the rules known as 'quantum mechanics' are found down the funnel leading from chemistry, where they explain chemical bonding, and they are also found down the funnel from cosmology, where they explain the origin of the universe in the Big Bang. It is very impressive to find the same rules explaining such

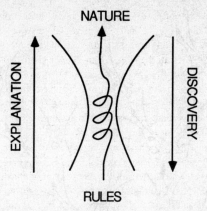

Figure 153   The process of scientific explanation illustrated by the image of a mental funnel.

different things, and this leads to a general feeling that rules that are common to many funnels must be more 'fundamental' than those that are not.

Most of the funnels in modern science eventually terminate in two systems of rules: quantum mechanics and general relativity. Unfortunately these two systems are mutually contradictory. Quantum mechanics is indeterminate and considers matter to be ultimately indivisible, whereas general relativity is a determinate theory of continuous space and time. The contradiction is philosophical rather than operational, in that only one of these viewpoints is appropriate to most questions, so we do not have to face the contradiction head on. Nonetheless, the mismatch means that neither theory can be truly fundamental. One way out of this impasse would be to find a 'deeper' set of rules that explained both quantum mechanics and general relativity. This long-sought synthesis has been dubbed the Theory of Everything, on the grounds that it lies at the bottom of *all* mental funnels (Figure 154).

Belief in a Theory of Everything is essentially fundamentalist. The philo-sophical stand taken by this conception of explanation is respectable and worthy: it is called *reductionism*. It leads to a model of science as a hierarchy of rules, each valid on an appropriate level of explanation. The rules on a given level are all consequences at least to some useful degree of approximation of those on lower levels. Deeper levels are more fundamental, and if a Theory of Everything exists, it lies at the deepest level of all. In an extreme but common view, it is not just a useful approximation to the truth: it *is* the truth.

To what extent does the reductionist vision of science correspond to how the universe actually operates?

The complexities of the universe arise in an enormous variety of ways. Take a deep breath: that may seem simple enough, but the oxygen you are

THEORY OF
EVERYTHING

Figure 154   Is there a single Theory of Everything at the bottom of every funnel?

using was separated from water by the Sun's energy, probably through chemical reaction with chlorophyll in a leaf. It takes ten pages of complicated chemical and energetic equations to explain the biochemistry of photosynthesis, and we've already seen that science is literally unable to predict the weather patterns that brought the resulting oxygen molecules to your lungs, even though those equations are much simpler. And once the oxygen has arrived there, things become no simpler. Your lungs are intricate, idiosyncratic fractals. The diffusion of oxygen into your blood depends upon the chemical rules for the behaviour of haemoglobin. Both haemoglobin and the photosynthetic pigment chlorophyll are enormously complicated chemical machines, whose changes of shape as they function can barely be modelled on our most complex computers.

Watch a goat eyeing a rose-bush. Think of all the cones in her eye, all the connections to her brain, all her muscles as she walks; all the image-processing algorithms carried out by her visual cortex, all the control signals sent from brain to muscles. Watch her amble over and take a mouthful of leaves. The way her chewing muscles grind the leaves is barely understood by specialists, and as for what happens to the leaf pulp when it meets the bacteria in her special stomach . . .

Yet there are large-scale simplicities too. There are simple ecological models that explain how goats eating leaves have turned the Sahara from a fertile plain, providing ancient Rome with much of its food, into a sandy waste. They warn that the Greek goat/olive-tree economy is heading in the same direction – but try to explain that to a Greek olive-grower.

Everywhere you look there are things like molecules, leaves, and goats, much too complicated for mere humans even to begin to understand, and processes that can be followed only in very simplified versions by experts. Instead of seeing simplicities down the reductionist funnels, we have become trapped in the *reductionist nightmare* (Figure 155), in which the funnels keep branching forever without hitting bottom.

To some extent, these problems are unavoidable; science is 'like that'. But are we making rods for our own backs by always adopting the methodology of the reductionist nightmare? Complexity theory represents a reaction against such methods, and as such it is a welcome addition to the scientist's intellectual arsenal.

Simplicity Theory

In contrast to the reductionist nightmare, our brains have evolved a quite different approach to the world – a 'quick-and-dirty' approach, but very effective despite (or because of) that. You know exactly what we mean when we recommend that you don't let a goat loose in your rose-garden, and why. 'Goats eat leaves' is a simple thought. And it doesn't take a deep grasp of higher mathematics to keep the goat out of your garden – just a strong fence. The *detailed* analysis of that fence is far beyond the reach of any equations for the structure of materials, but most of us can build a goat-proof fence. Similarly, the goat does not need to be versed in the equations of nonlinear elasticity to discover that the fence is weaker than its builders might have thought.

What we really need is simplicity theory, not complexity theory.

There is a rhetoric of reductionist science that claims that, *even if the goat doesn't know it*, immensely complicated things must go on inside the goat to make it behave as it does. When you put up a goat-proof fence, you are accessing human muscle-and-nerve programmes refined by millions of years

NATURE

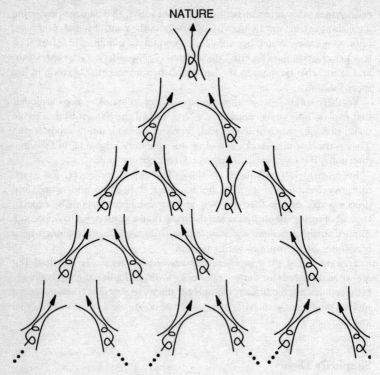

Figure 155   The reductionist nightmare.

of evolution to shape equally sophisticated wood, moving soil and erecting a topologically adequate barrier. It just *seems*, to you and the goat, that what's going on is simple: but *it isn't really*. But whether or not the deep structure of the universe matches this rhetoric precisely, we still conclude that apparent simplicities are at large in our complicated world.

I don't think that this kind of reductionist rhetoric is the only – or the best – way to understand the interactive dynamics of goats and rose-bushes. While goat behaviour may indeed 'really' be a logical consequence of the molecular structure of goats and bushes, nobody in their right mind would actually contemplate thinking about it on that level. I also don't think that the universe is really much simpler than it looks, and that people (and perhaps goats) can access the much simpler rules that underlie the apparent complexity. This is the view of science as (no more than) common sense, and the aim of the scientific method then becomes to see through the complexity and extract the underlying simplicity. The (again tacit) assumption is that once we know the laws then everything else follows. And, as I've said, the most fundamentalist version of this viewpoint is the search for a Theory of Everything.

These two ways of pursuing science, the reductionist rhetoric and common sense, are both reductionist: the difference is that the first 'reduces' the system to ever-more-complex subsystems (for instance, by modelling the molecular structure of haemoglobin or sequencing the human genome), whereas the second obtains a genuine reduction of complexity by replacing complex behaviour with simple rules. But neither is totally satisfactory as an explanation of natural phenomena, either on a philosophical level or a practical one.

I've characterized the search for a single, ultimate system of natural laws as 'fundamentalist' for two reasons. One is the claim that such a system would indeed uncover the true fundamentals of the universe. The other is that the fundamentalism of particle physicists is a mirror of fundamentalist religion. Both see a single source for all of the complexity of the universe, but for one group it is 'Theory of Everything' at the bottom, whereas for the other it is 'God' at the top. Particle physicists reduce all of nature to a Theory of Everything, and assert that this explains the entire universe; fundamentalist religions reduce all of nature to God's will, and also assert that this explains the entire universe. Physicists look inside the system to find their fundamentals; religious fundamentalists look outside it.

Both types of fundamentalism can produce a consistent story that leads from their fundamental principle to rose-bush-eating goats, and both are satisfied by it. The religious story would perhaps focus on the importance of the goat in God's scheme of creation; the scientific story would be a climb up the reductionist hierarchy. Both can tell stories that explain why goats eat rose-bushes. But we are more interested in a different question. Neither type of fundamentalist can give any coherent argument to explain why the goats' preference for rose-bushes *depends* upon their fundamental principle, in the sense that it could not have arisen without it.

In fact, particle physicists use the word 'fundamental' in two ways. In the phrase 'fundamental particles' the word indicates that these are what you get when you smash matter apart into tinier and tinier components at higher and higher energies. In contrast, the usage in phrases such as 'fundamental laws of nature' indicates an assumption that deep down *this is how nature really works*. A good example of the confusion of these two meanings is Steven Weinberg's *Dreams of a Final Theory*, a spirited defence of the dream of a Theory of Everything. In it he argues that because particle physics is fundamental (meaning 1) it is therefore an essential prerequisite for understanding everything else (meaning 2). However, this is a logical *non-sequitur*. Indeed the opposite is true. Our understanding of most natural phenomena must necessarily be independent of the fine details of fundamental particle interactions, and most other things that live sufficiently many layers inside the reductionist 'tree of everything'. Let me explain why.

Fungibility

Current science possesses *no* truly fundamental theories – not in the sense that they describe what nature actually does. They are all approximations, valid within some reasonably well-defined domain. Quantum mechanics works well at the submicroscopic level. General relativity is great for describing entire universes – but not for systems as apparently straightforward as binary stars, where it seems impossible even to set up the equations sensibly. Science is a patchwork of models, each of which has been enormously refined within its own domain. The models habitually disagree when those patches overlap. Some disagreements are relatively harmless: atomic theory and continuum fluid mechanics disagree on the fine structure of water, holding it respectively to be discrete and infinitely divisible, but on macroscopic scales continuity and discreteness effectively approximate each other. Others are fatal: for example, as I write, the best current theory of astrophysics and the best current theory of cosmology compel us to accept stars older than the universe that contains them. Today's science is a pluralist patchwork of locally valid models, not a global monolith. Indeed it succeeds *because* it is a pluralist patchwork of locally valid models.

Our concept of explanation is also a patchwork. A philosophical model that fits it well is what Richard Dawkins calls 'hierarchical reductionism', which sees scientific theories as a hierarchical structure, with some on different levels from others, corresponding to different levels of description of phenomena. (The hierarchy is not rigid and the levels need not be like layers of bricks in a wall.) For example, the complexities of ecosystems are explained by referring them back to those of organisms; organisms are explained by the growth of spatially organized proteins and other macromolecules; the complex organization of organisms is referred back to the linear complexity of their DNA code; the complexity of DNA is referred back to combinations of simpler atoms – and so on, right back to the Theory of Everything.

As Dawkins rightly remarks, it is not necessary to trace every phenomenon right back down this chain of reductions in order to understand it. Chemistry can be considered as 'given' for the purposes of understanding DNA; DNA can be taken as 'given' for the purpose of understanding protein manufacture in organisms, and so on. But what do we mean when we take something as 'given'?

Lawyers have a concept known as 'fungibility'. Things are fungible if substituting one for another has no legal implications. For example, cans of baked beans with the same manufacturer and the same nominal weight are fungible: you have no legal complaint if the shop substitutes a different can when the assistant notices that the one you've just bought is dented. The fact that the new can contains 1,346 beans, whereas the old one contained 1,347, is legally irrelevant.

That's what 'take as given' means, too. Explanations that climb the reductionist hierarchy are cascades of fungibilities. Such explanations are comprehensible, and thus convincing, only because each stage in the story relies only upon *particular simple features* of the previous stage. The complicated details a level or two down do not need to be carried upwards indefinitely. Such features are intellectual resting-points in the chain of logic. Examples include the observation that atoms can be assembled into many complex structures, making molecules possible, and the complicated but elegant geometry of the DNA double helix that permits the 'encoding' of complex 'instructions' for making organisms. The story can then continue with the computational abilities of DNA coding, onward and upward to goats, without getting enmeshed in the quantum wave functions of amino acids.

What we tend to forget, when told a story with this structure, is that it could have had many different beginnings. Anything that lets us start from the molecular level would have done just as well. A totally different subatomic theory would be an equally valid starting-point for the story, provided it led to the same general feature of a replicable molecule. Subatomic particle theory is fungible when viewed from the level of goats. It has to be, or else we would never be able to keep a goat without first doing a Ph.D. in subatomic physics.

In consequence, only the plot line of the story really appears in the explanation. The details, on any given level, might be changed considerably without this having any effect on the story line, the ending, or the degree of conviction that it engenders. I don't mean that the actual universe can do things in lots of different ways: I mean that such differences have no significant effect *on explanations*. Science is about explanations – if not it could be replaced by a catalogue of facts, but as Sir Peter Medawar once said, 'theories destroy facts', meaning that a single theory can explain many facts and thereby render them superfluous. So the truth is that most of science would be totally unaffected by the discovery of the long-sought Theory of Everything.

That doesn't mean it's not interesting, and it doesn't mean it's not important. It just means that it's not fundamental in sense 2: a prerequisite for understanding everything else. The particle physicists can safely be left to get on with the search, or not: this search is *not* essential for the health of the rest of science. I suspect that when the US Congress cancelled funding for the Superconducting Super Collider, a multibillion-dollar accelerator whose avowed purpose was to move physics closer to a Theory of Everything, they dimly recognized that there is more than one meaning to the word 'fundamental'.

Ant Country

If we don't resort to fundamentalism, we have to find a different route for the genesis of large-scale simplicities. I maintain that this route is nearly always emergence.

A simple example of emergence occurs in Langton's ant, a rule-based mathematical system invented by Chris Langton of the Santa Fe Institute. It's a simple example of how complexity theory generates new concepts and reveals new types of behaviour in simple rule-based systems. Begin with a grid of squares, which can be in one of two states: black or white. For simplicity suppose that initially they are all white. The ant starts out on the central square of the grid, heading in some selected direction – say east. It moves one square in that direction, and looks at the color of the square it lands on. If it lands on a black square then it paints it white and turns 90° to the left. If it lands on a white square it paints it black and turns 90° to the right. It keeps on following those same simple rules forever.

Despite (no, because of) their simplicity, those rules produce surprisingly complex behaviour. For the first five hundred or so steps, the ant keeps returning to the central square, leaving behind it a series of rather symmetric patterns. For the next ten thousand steps or so, the picture becomes very chaotic. Suddenly – almost as if the ant has finally made up its mind what to do – it repeatedly follows a sequence of precisely 104 steps that moves it two cells south-west, and continues this indefinitely, forming a diagonal band known as a *highway* (Figure 156). This simple large-scale feature *emerges* from the low-level rules, in the following sense. The only rigorous way to deduce the occurrence of a highway from the rules is to work through every single one of the ten thousand or so steps that lead to the start of the 104-step cycle. Then you can fairly easily explain why the cycle repeats forever, and in doing so creates a highway. So here we have a feature whose existence can currently be demonstrated rigorously *only* by following the reductionist rhetoric down its ramifying nightmare of funnels.

OK, but it takes only ten thousand steps – that's not so huge. That's true; however, some relatives of Langton's ant also build a highway – but only after tens of millions of steps. Who knows how lengthy the pre-highway stage might be in yet other ant-like systems? And there is more. Computer experiments strongly suggest that Langton's ant *always* ends up building a highway, even if you scatter finitely many black squares around the grid before it starts. Nobody has ever been able to prove this, and it certainly can't

Figure 156   Three distinct stages in the dynamics of Langton's ant. A highway is clearly visible at the bottom of the right-hand picture. It will grow for ever as the ant continues to obey its rules.

be done by reductionist rhetoric: there are infinitely many different ways to scatter black squares, so your proof would have to be infinitely long. So here we have a high-level simplicity that seems to be universal, but which cannot currently be deduced from the Theory of Everything for the system. Even though we *know* the Theory of Everything in this case. So here the Theory of Everything lacks explanatory power: it predicts everything but explains nothing.

We can take Langton's ant as a symbol for the gap between the top-down reductionism of the reductionist nightmare and the bottom-up reductionism of the Theory of Everything. Top-down analysis proceeds from nature and looks down mental funnels to see what lies inside. Bottom-up analysis proceeds from the Theory of Everything and ascends levels of description by deducing logical consequences of those laws in a hierarchical manner. I maintain that the top and the bottom *do not meet*, and this is why emergent phenomena appear to transcend the systems that give rise to them. Cohen and I call this 'no man's land' between top and bottom *Ant Country* (Figure 157).

How does the reductionist chain of logic traverse Ant Country? It doesn't – it just hopes you don't notice the gap. For example, think of Newton's law of gravity, where a result proved for a uniform mathematical sphere is applied without hesitation to a non-uniform, non-spherical planet. The mathematical rules explain the sphere's gravitational field; this explanation is transferred – by analogy, not logic – to the planet. I don't dispute that this process often *works*, and I don't think it's bad science. But it definitely breaks the logical chain of reductionism.

NATURE

ANT COUNTRY

THEORY OF EVERYTHING

Figure 157   Ant Country.

Worse, it encourages the comforting but philosophically dangerous illusion that the simplicity of the rules leads *directly* to the simplicity of elliptical orbits for real planets. Instead, the explanatory story must pass through Ant Country – where the emergent phenomena live, where complexity is created from nothing, and where systems organize themselves into more complex systems without anything equally complex telling them how to do it. Few scientists are aware that Ant Country exists, and even fewer have any intention of exploring it.

Simplexity and Complicity

Exploring Ant Country won't be easy, because to do so you need a sensible, effective theory of emergence. I don't think that tackling complexity head on – as is being attempted, for example, in the human genome project – is the answer. It merely delays the inevitable crunch when we become overwhelmed by so much information that we can't organize it. We need a new kind of theory, a more holistic one in which high-level patterns can be understood without referring them back to low-level rules. I have a name for it: 'morphomatics'. I don't have an 'it' to attach the name to.

Today's mathematics does show that such a theory might in principle be possible. Modern dynamical systems theory is both reductionist and holistic. It is reductionist when it uses differential equations to reduce the flow of fluids to the interactions of huge numbers of tiny cubes. But it also has different weapons in its armoury: large-scale qualitative principles like continuity, connectivity, and symmetry. As well as the low-level arithmetic of equations it has the high-level geometry of attractors. So in this area of mathematics, the bottom-up approach and the top-down one do meet in the middle, and weld together into something more powerful than either. Dynamical systems theory still has its own regions of Ant Country, however: its power is limited.

A comparable approach to science in general does not yet exist. Complexity theory is one step in that direction, but it suffers from too limited a view of the nature of the problem. In *The Collapse of Chaos* Cohen and I classify emergent phenomena into two qualitatively different types, which – for not entirely frivolous reasons – we call *simplexity* and *complicity*. Simplexity is emergence of the kind exemplified by the highways of Langton's ant: a large-scale pattern that occurs within a rule-based system but whose detailed deduction from the rules is enormously lengthy and uninformative (or perhaps unknown). An example from real science is the physicists' belief that the structure of crystal lattices is a consequence of the laws of quantum mechanics. There is all sorts of indirect evidence that this is true, but no known proof. Right now, complexity theory is mainly about simplexities.

Complicity is far more elusive, but also far more important. It occurs when two (or more) rule-based systems interact. It is not unusual in such circumstances to find that new high-level regularities emerge from the inter-actions. Often those interactions are unknown (since they are not part of the rules of either subsystem). A real-world example of complicity is the evolution of bloodsucking. This occurred when the rules for early mammalian anatomy (blood) interacted with the rules for an ancestor of the mosquito (which had developed an organ for sucking liquids, probably water). By a creative coincidence, the water-sucking proboscis happened to be able to penetrate human skin. This collision of two developmental spaces caused them to co-evolve in a new way, not inherent in either developmental space on its own. The result of this complicit co-evolution is an insect that is *adapted* to sucking human blood. One consequence is malaria, a more predictable simplexity erected on top of the initial complicity.

Traditional science sees regularities in nature as *direct* reflections of regular laws. That view is no longer tenable. Neither is the view that the universe rests upon a single fundamental rule system, and all we have to do is find it. Instead, there are – and must be – rules at every level of description. To some extent we ourselves choose the types of description in which the rules arise, because our brains cannot cope with raw complexity. Every human being programs its brain, and its sense organs, to extract meaningful features from its environment as it develops – especially during early childhood. Simple rules exist because simplicity emerges from complex interactions on lower levels of description. The universe is a plurality of overlapping rules. And in the gaps between the rules lies Ant Country, in which simplicity and complexity not only fail to be conserved, but transmute into one another.

This is the deep message of chaos theory and its companions. Their mathematical techniques will explain more and more puzzling phenomena, and lead to more and more nuts-and-bolts dollars-and-cents applications to keep businesses and governments happy. But the real reason for pursuing the new theories is an intellectual one. As we become accustomed to these new thought patterns, they will produce fundamental and irreversible changes in how we think about our world.

Epilogue

Dicing with the Deity

Chance is the pseudonym of God when he did not want to sign.

Anatole France

If God played dice . . .

. . . He'd win

Further Reading

Even Jehovah,
After Moses had got the Commandments
Committed to stone
Probably thought:
I always forget the things
I really intended to say.

Christopher Morley

Books, Popular

John Barrow, *Theories of Everything* (Oxford: Oxford University Press, 1991) [A very readable critical account of Theories of Everything.]

John L. Casti, *Paradigms Lost* (London: Scribners, 1989) [Informed and informative survey of changes in the scientific world-view.]

John L. Casti, *Searching for Certainty: What Scientists Can Learn about the Future* (New York: Morrow, 1990) [A well-written overview of modern reductionist science, with many indications of incompleteness and contextuality.]

Arthur C. Clarke, *The Colours of Infinity* (London: Strange Attractions, 1992) [Everything you want to know about the Mandelbrot set: gentle but thorough description of the mathematics.]

Jack Cohen and Ian Stewart, *The Collapse of Chaos* (Harmondsworth: Penguin Books, 1995) [Book three of the *Does God* trilogy: beyond chaos and complexity. 'The most startling and thought-provoking book I've read all year. I was pleased to learn that most of the things I thought I knew were wrong' – Terry Pratchett.]

Peter Coveney and Roger Highfield, *The Arrow of Time* (London: Flamingo, 1991) [Well-written but inconclusive account of time-reversal and chaos.]

Paul Davies (ed.), *The New Physics* (Cambridge: Cambridge University Press, 1989) [Background on the quantum and much else: includes beautiful essay on chaos by Joseph Ford bringing out its information-theoretic aspects.]

Freeman Dyson, *Disturbing the Universe* (New York: Basic Books, 1979) [Deep thoughts about the universe by one of the wisest leading physicists.]

Freeman Dyson, *Infinite in All Directions* (New York: Basic Books, 1988) [More of the same. Explains why the 'heat death of the universe' is a misleading image.]

J. Richard Eiser, *Attitudes, Chaos, and the Connectionist Mind* (Oxford: Blackwell, 1994) [What nonlinear dynamics may tell us about consciousness.]

Ivar Ekeland, *Mathematics and the Unexpected* (Chicago: University of Chicago Press, 1988) [Elegant and literary introduction to chance and chaos in dynamics.]

Ivar Ekeland, *The Broken Dice* (Chicago: University of Chicago Press, 1993) [Sequel, woven around Norse legends. As good as its predecessor, if not better.]

Richard P. Feynman, *QED: The Strange Theory of Light and Matter* (Harmondsworth: Penguin Books, 1990) [Crystal clear explanation of what quantum mechanics has to say about the universe. Dates from the period when he didn't worry much about what quantum theory *meant*.]

Michael J. Field and Martin Golubitsky, *Symmetry in Chaos* (Oxford: Oxford University Press, 1992) [Amazing picture-book of what chaos looks like in the presence of symmetry, with possible applications to pattern formation. State-of-the-art research mathematics which shows how chaos and stable form might coexist.]

Alan Garfinkel, *Forms of Explanation* (New Haven: Yale University Press, 1981) [What do we mean by 'explanation'? Carefully and sensibly argued.]

Ronald N. Giere, *Explaining Science* (Chicago: University of Chicago Press, 1988) [Philosophy of scientific theories.]

James Gleick, *Chaos: Making a New Science* (New York: Viking, 1987) [Brilliant on personalities, but not much solid science.]

James Gleick, *Genius: Richard Feynman and Modern Physics* (London: Little, Brown and Co., 1992) [Excellent biography that misses out a little on the playful side of one of the great physicists of this century, but illuminates his thinking on the deep issues of physics.]

John Gribbin, *In Search of Schrödinger's Cat* (London: Black Swan, 1992) [Good popular book on the meaning of quantum mechanics.]

Nina Hall (ed.), *The New Scientist Guide to Chaos* (Harmondsworth: Penguin Books, 1991) [Collection of articles by experts for the general reader. One of the best introductions to the scientific content of chaos.]

Helge S. Kragh, *Dirac: a Scientific Biography* (Cambridge: Cambridge University Press, 1990) [Excellent biography, examines Dirac's penchant for beautiful falsehood in preference to ugly truth.]

Thomas Kuhn, *The Structure of Scientific Revolutions* (Chicago: University of Chicago Press, 1962) [The book that presented science as a collection of shifting paradigms and let social scientists stop feeling guilty that they weren't achieving the same success as the physical scientists.]

Roger Lewin, *Complexity: Life at the Edge of Chaos* (New York: Macmillan, 1992) [A people-based description of the Santa Fe Institute.]

Benoît Mandelbrot, *The Fractal Geometry of Nature*, 2nd edn. (San Francisco: W. H. Freeman, 1982) [Fractals from the horse's mouth. Penetrating, elegant, infuriating, obscure.]

Michael McGuire, *An Eye for Fractals* (Redwood City, CA: Addison-Wesley Publishing Co., 1991) [Collection of photos of fractal objects in nature.]

David Peak and Michael Frame, *Chaos under Control* (New York: W. H. Freeman, 1994) [Subtitled 'The Art and Science of Complexity' with a lot about cellular automata. Highly illustrated, good humoured, refreshing.]

Heinz-Otto Peitgen and Peter H. Richter, *The Beauty of Fractals* (New York: Springer-Verlag, 1986) [The world's first mathematical coffee-table book.]

Ivars Peterson, *The Mathematical Tourist* (New York: W. H. Freeman, 1988) [Selection of topics for the layman, includes chaos and fractals.]

Ivars Peterson, *Islands of Truth* (New York: W. H. Freeman, 1990) [Sequel: includes complexity and applications of fractals to aggregation.]

Clifford A. Pickover, *Computers, Pattern, Chaos, and Beauty* (New York: St Martin's Press, 1990) [Visual adventures on the research fringes of mathematics.]

Clifford A. Pickover, *Computers and the Imagination* (New York: St Martin's Press, 1991) [More of the above.]

Ilya Prigogine and Isabelle Stengers, *Order out of Chaos* (London: Flamingo, 1985) [Non-equilibrium thermodynamics and the emergence of structure.]

Przemyslaw Prusinkiewicz and Aristid Lindenmayer, *The Algorithmic Beauty of Plants* (New York: Springer-Verlag, 1990) [Beautiful picture book showing how the mathematical structures of fractals can reproduce the branching patterns of plants. Secret order amid nature's complexity.]

Ed Regis, *Great Mambo Chicken and the Transhuman Condition: Science Slightly over the Edge* (New York: Addison-Wesley Publishing Co., 1990) [Wild, wacky, unputdownable collection of unorthodox science, including artificial life, a mainstay of complexity theory. The story of the maverick gurus.]

Rudy Rucker, *Mind Tools* (Harmondsworth: Penguin Books, 1989) [Good, exciting but occasionally frenzied pop-maths.]

David Ruelle, *Chance and Chaos* (Princeton: Princeton University Press, 1991) [Wise words from one of the mathematical founders of chaos theory, very good on time-reversibility, and astonishingly readable.]

Manfred Schroeder, *Fractals, Chaos, Power Laws* (New York: W. H. Freeman, 1991) [Thorough survey somewhere between popularization and technical.]

Julien C. Sprott, *Strange Attractors* (New York: M&T Books, 1993) [Highly illustrated instructions for investigating chaos by computer. Disk included.]

Philip Stehle, *Order, Chaos, Order* (New York: Oxford University Press, 1994) [Informed and readable history of quantum physics, using 'chaos' in the original dictionary sense.]

Ian Stewart, *Les Fractals* (Paris: Belin, 1982) [Comic book, in French.]

Ian Stewart, *From Here to Infinity* (Oxford: Oxford University Press, 1996) [New edition of *The Problems of Mathematics*. Overview of current state of mathematics for non-specialists: includes fractals, dynamical systems, chaos.]

Ian Stewart and Martin Golubitsky, *Fearful Symmetry: is God a Geometer?* (Oxford: Blackwell, 1992); (Harmondsworth: Penguin Books, 1993) [Book one of the *Does God* trilogy. A whole new way of looking at pattern, complexity, and the generation of order in nature. Symmetric chaos.]

Mitchell Waldrop, *Complexity: the Emerging Science at the Edge of Order and Chaos* (New York: Simon & Schuster, 1992) [How emergence is becoming respectable: a detailed look at the Santa Fe Institute and the theories that it is developing.]

Steven Weinberg, *Dreams of a Final Theory: the Search for the Fundamental Laws of Nature* (London: Hutchinson Radius, 1993) [A leading advocate explains what he means by a Theory of Everything. Thoughtful, fascinating; tacitly assumes that 'fundamental' in the sense of 'ultimate bits and pieces' is the same as 'fundamental' in the sense of 'foundation for everything else'.]

Books, Advanced

Ralph H. Abraham and Christopher D. Shaw, *Dynamics: the Geometry of Behavior* (4 vols.) (Santa Cruz: Aerial Press, 1988) [From the Visual Mathematics Library – comic-bookish treatment of serious dynamics.]

D. K. Arrowsmith and C. M. Place, *An Introduction to Dynamical Systems* (Cambridge: Cambridge University Press, 1990) [Excellent text for mathematics graduate students.]

Michael F. Barnsley, *Fractals Everywhere*, 2nd edn. (Boston: Academic Press, 1993) [Advanced mathematics text with a huge number of pictures: explains the principles of fractal image-compression.]

Michael F. Barnsley and Lyman P. Hurd, *Fractal Image Compression* (Wellesley, MA: A. K. Peters, 1993) [Explains the practice too, with impressive examples of 2500:1 compression of colour images.]

M. V. Berry, I. C. Percival, and N. O. Weiss (eds.) *Dynamical Chaos* (London: Royal Society, 1987) [Proceedings of one of the earliest major conferences on chaos, lots of applications.]

A. B. Cambel, *Applied Chaos Theory* (San Diego: Academic Press, 1993) [What the mathematical techniques mean and how to interpret them in applications. Lots of photos of the creators of the subject.]

Martin Casdagli and Stephen Eubank (eds.), *Nonlinear Modelling and Forecasting*, (Redwood City, CA: Addison-Wesley Publishing Co., 1992) [How to predict chaos in the short term. Mainly for professionals.]

Gregory J. Chaitin, *Information, Randomness, and Incompleteness* (Singapore: World Scientific Publishing Co. Pte. Ltd., 1987) [Penetrating papers on the meaning of 'random' and the information-theoretic cost of computations. Mostly for professionals, but some articles accessible to the lay reader.]

Predrag Cvitanovic, *Universality in Chaos*, 2nd edn. (Bristol: Adam Hilger, 1989) [Compilation of original articles from the journals.]

Robert L. Devaney, *An Introduction to Chaotic Dynamical Systems*, 2nd edn. (Redwood City, CA: Addison-Wesley Publishing Co., 1989) [Best undergraduate text on discrete dynamics that I know; focuses on Julia sets and Mandelbrot set.]

Robert L. Devaney and Linda Keen (eds.), *Chaos and Fractals* (Providence, RI: American Mathematical Society, 1989) [The mathematics behind the Mandelbrot set, with colour pictures.]

Jens Feder, *Fractals* (New York: Plenum, 1988) [Applications of fractals in the physical sciences.]

Leon Glass and Michael C. Mackey, *From Clocks to Chaos* (Princeton, NJ: Princeton University Press, 1988) [Chaos in physiology, dynamical diseases. Accessible to the non-specialist.]

John Guckenheimer and Philip Holmes, *Nonlinear Oscillations, Dynamical Systems, and Bifurcations of Vector Fields* (New York: Springer-Verlag, 1986) [Still one of the best introductions for the professional.]

Biai-Lin Hao, *Chaos II* (Singapore: World Scientific Publishing Co. Pte. Ltd., 1990) [Comprehensive reprint collection of the original papers that created the subject.]

Harold M. Hastings and George Sugihara, *Fractals: a User's Guide for the Natural Sciences*

(Oxford: Oxford University Press, 1993) [Fractal modelling and case studies, mainly in the biological sciences.]

Arun V. Holden (ed.) *Chaos* (Manchester: Manchester University Press, 1986) [Chaos and applications, especially to physiology.]

E. Atlee Jackson, *Perspectives of Nonlinear Dynamics*, 1 and 2 (Cambridge: Cambridge University Press, 1989, 1990) [Highly illustrated free-wheeling but detailed exposition for physics and mathematics students; can be read by non-specialists willing to think hard.]

S. A. Levin (ed.), *Studies in Mathematical Biology*, 1 and 2 (Washington, DC: Mathematical Association of America, 1978) [Includes good exposition of chaos in population dynamics.]

J. L. McCauley, *Chaos, Dynamics and Fractals* (Cambridge: Cambridge University Press, 1993) [Computational aspects of chaos with emphasis on the finite precision of computers.]

Tom Mullin (ed.), *The Nature of Chaos* (Oxford: Clarendon Press, 1993) [Superb collection of articles on chaos in experimental science: aimed at experts but readable by anyone interested.]

Edward Ott, *Chaos in Dynamical Systems* (Cambridge: Cambridge University Press, 1993) [All the basic ideas, for graduate students in mathematics and physics.]

Edward Ott, Tim Sauer, and James A. Yorke, *Coping with Chaos* (New York: John Wiley Inc., 1994) [Detection and control of chaos.]

Heinz-Otto Peitgen, Hartmut Jürgens, and Dietmar Saupe, *Chaos and Fractals: New Frontiers of Science* (New York: Springer-Verlag, 1992) [Massive introduction to fractals, 984 pages, 686 illustrations – 40 in colour. Accessible with high-school mathematics.]

David Ruelle (ed.) *Turbulence, Strange Attractors, and Chaos* (Singapore: World Scientific Publishing Co. Pte. Ltd., 1995) [Reprinted collection of original papers, mainly by the editor – which is only fair because he is one of *the* founders of the subject.]

Heinz Georg Schuster, *Deterministic Chaos: an Introduction* (Weinheim: Physik-Verlag, 1984) [Excellent exposition from the physicist's point of view.]

J. M. T. Thompson and P. Gray, *Chaos and Dynamical Complexity* (London: Royal Society, 1990) [Sequel to Berry et al., engineering and chemistry this time.]

J. M. T. Thompson and H. B. Stewart, *Nonlinear Dynamics and Chaos* (New York: John Wiley Inc., 1986) [The engineer's point of view.]

Donald L. Turcotte, *Fractals and Chaos in Geology and Geophysics* (Cambridge: Cambridge University Press, 1992) [Fractal forgeries of landscapes are not just visual puns: they relate the mechanisms of geological change.]

Yoshisuke Ueda, *The Road to Chaos* (Santa Cruz: Aerial Press, 1992) [If anyone tries to tell you that the early pioneers of chaos had an easy ride, read this. By the discoverer of the Ueda attractor.]

Tamás Vicsek, *Fractal Growth Phenomena* (Singapore: World Scientific Publishing Co. Pte. Ltd., 1989) [Theory of fractal cluster aggregation.]

Bruce J. West, *Fractal Physiology and Chaos in Medicine* (Singapore: World Scientific Publishing Co. Pte. Ltd., 1990) [Chaotic dynamics of the heart and brain. Readable by non-specialists.]

Magazine and Journal Articles

David Z. Albert, 'Bohm's alternative to quantum mechanics', *Scientific American* (May 1994), pp. 32–9

M. Bayliss, M. Muldoon, M. Nicol, L. Reynolds, and I. Stewart, 'The FRACMAT Test for wire coilability: a new concept in wire testing', *Wire Industry*, 62 (1995), pp. 669–74

Wallace S. Broecker, 'Chaotic climate', *Scientific American* (Nov. 1995), pp. 44–9

J. Robert Buchler, Thierry Serre, and Zoltán Kolláth, 'A chaotic pulsating star: the case of R Scuti', *Physical Review Letters*, 73 (1995), pp. 842–5

Stephen Budiansky, 'Chaos in Eden', *New Scientist* (14 Oct. 1995), pp. 33–5

Marcus Chown, 'Fly me cheaply to the Moon', *New Scientist* (7 Oct. 1995), p. 19

David J. Christini and James J. Collins, 'Controlling nonchaotic neuronal noise using chaos control techniques', *Physical Review Letters*, 75 (2 Oct. 1995), pp. 2782–5

Jack Cohen and Ian Stewart, 'Chaos, contingency, and convergence', *Nonlinear Science Today*, 1: 2 (1991), pp. 9–13

James P. Crutchfield, J. Doyne Farmer, Norman H. Packard, and Robert S. Shaw, 'Chaos', *Scientific American* (Dec. 1986), pp. 38–49

Paul Davies, 'Chaos frees the universe', *New Scientist* (6 Oct. 1990), pp. 48–51

William L. Ditto, 'Mastering chaos', *Scientific American* (Aug. 1993), pp. 62–8

Stillman Drake, 'The role of music in Galileo's experiment', *Scientific American* (June 1975), pp. 98–104

Berthold-Georg Englert, Marlan O. Scully, and Herbert Walther, 'The duality in matter and light', *Scientific American* (Dec. 1994), pp. 56–61

Michael J. Field and Martin Golubitsky, 'Symmetries on the edge of chaos', *New Scientist* (9 Jan. 1993), pp. 32–5

Alan Garfinkel, Mark L. Spano, William L. Ditto, and James N. Weiss, 'Controlling cardiac chaos' *Science*, 257 (28 Aug. 1992), pp. 1230–35

Sunetra Gupta and Roy Anderson, 'Sex, AIDS and mathematics', *New Scientist* (12 Sept. 1992), pp. 34–8

Douglas R. Hofstadter, 'Pitfalls of the uncertainty principle and paradoxes of quantum mechanics', *Scientific American* (July 1981), pp. 10–15

Roderick V. Jensen, 'Quantum chaos', *Nature*, 355 (23 Jan. 1992), pp. 311–18

Eric Kostelich, 'Symphony in chaos', *New Scientist* (8 Apr. 1995), pp. 36–9

Jim Lesurf, 'A spy's guide to chaos', *New Scientist* (1 Feb. 1992), pp. 29–33

Roger Lewin, 'A simple matter of complexity', *New Scientist* (5 Feb. 1994), pp. 37–40

Debora MacKenzie, 'The cod that disappeared', *New Scientist* (16 Sept. 1995), pp. 24–9

M. Muldoon, M. Nicol, and L. Reynolds, and I. Stewart, 'Chaos Theory in quality control of spring wire': Part I, *Wire Industry*, 62 (1995), pp. 309–11; Part II, *ibid.*, pp. 491–2; Part III, *ibid.*, pp. 492–5

Julio M. Ottino, 'The mixing of fluids', *Scientific American* (Jan. 1989), pp. 40–49

T. N. Palmer, 'A nonlinear dynamical perspective on climate change', *Weather*, 48 (Oct. 1993), pp. 314–26

T. N. Palmer, 'A local deterministic model of quantum spin measurement', *Proceedings of the Royal Society of London*, A: 451 (1995), pp. 585–608

Troy Shinbrot, Celso Grebogi, Edward Ott, and James A. Yorke, 'Using small perturbations to control chaos', *Nature*, 363 (3 June 1993), pp. 411–17

Douglas Smith, 'How to generate chaos at home', *Scientific American* (Jan. 1992), pp. 121–3

Ian Stewart, 'Chaos: does God play dice?' *Encyclopaedia Britannica Yearbook of Science and the Future 1990* (Chicago: Encyclopaedia Britannica), 1989, pp. 54–73

Ian Stewart, 'Dicing with death in the Solar System', *Analog*, 109: 9 (1989), pp. 57–73

Ian Stewart, 'Does chaos rule the cosmos?' *Discover*, 13: 11 (Nov. 1992), pp. 56–63

Ian Stewart, 'Chaos' in L. Howe and A. Wain (eds.), *Predicting the Future* (Cambridge: Cambridge University Press, 1993), pp. 24–51

Ian Stewart, 'A new order' in *Complexity*, Supplement to *New Scientist*, 1859 (6 Feb. 1993), pp. 2–3

Ian Stewart, 'Recent developments: Mathematics', World Science Report 1993 (Paris: UNESCO Publishing, 1993) pp. 176–91

Ian Stewart, 'Chaos Theory as a forecasting tool?' *Brand Strategy*, 65 (27 May 1994), pp. 13–14

Ian Stewart, 'Two's integrable, three's chaos', *New Scientist*, 1947 (15 Oct. 1994), p. 16

Ian Stewart, 'Complexity', *Encyclopaedia Britannica Yearbook of Science and Technology 1995* (Chicago: Encyclopaedia Britannica, 1996)

Ingo Titze, 'What's in a voice?', *New Scientist* (23 Sept. 1995), pp. 38–42

Videotapes

The Beauty and Complexity of the Mandelbrot Set, Science Television, American Mathematical Society, PO BOX 6248, Providence, RI, USA [Presented by John Hubbard, aimed at mathematics students.]

Chaos, Fractals and Dynamics: Computer Experiments in Mathematics, Science Television, American Mathematical Society, PO BOX 6248, Providence, RI, USA [Presented by Robert Devaney, aimed at mathematics students.]

Chaotica 1, James Crutchfield, Physics Dept., University of California, Berkeley, CA, 94720, USA [Computer animations, variety of topics.]

The Colours of Infinity, British Universities Film & Video Council, 55 Greek Street, London W1V 5LR [Arthur C. Clarke heads a star-studded cast explaining the Mandelbrot set.]

Fractals, an Animated Discussion, W. H. Freeman, 20 Beaumont Street, Oxford OX1 2NQ [Interviews with Lorenz and Mandelbrot on their contributions. Excellent discussion of basins of attraction.]

A Strange Attractor in a Chemical System, Science Television, Aerial Press, PO BOX 1360, Santa Cruz, CA 95061, USA [Short investigation of a particular application.]

Virtual Ph.D. Course: Chaos and Complexity, EuroPACE 2000, Celestijnenlaan 200A, B-3001 Heverlee, Belgium [Set of tapes covering the whole area, lots of interviews with leading figures. Printed course guides also available.]

Illustration Acknowledgements

Grateful acknowledgement is made to the following for permission to reproduce copyright material:

Addison-Wesley Publishing Co., Reading, MA.: Ralph Abraham and Jerrold E. Marsden, *Foundations of Mechanics* (1978) (Benjamin Cummings imprint) – Figure 108

Addison-Wesley Publishing Co., Redwood City, CA: Martin Casdagli and Stephen Eubank (eds.), *Nonlinear Modelling and Forecasting* (1992) – Figure 130

Ariel Press, Santa Cruz: Ralph H. Abraham and Christopher D. Shaw, *Dynamics: the Geometry of Behavior* (1988) – Figures 79, 107, 118

American Association for the Advancement of Science: *Science*, 257 (1992), p. 1233 (Alan Garfinkel et al.) – Figure 140

American Mathematical Society: *Memoirs of the American Mathematical Society*, 81 (1968), pp. 1–60 (Jürgen K. Moser) – Figure 59

American Meteorological Society: *Journal of the Atmospheric Sciences*, 20 (1963), pp. 130–41 (Edward N. Lorenz) – Figures 54, 56

American Philosophical Society: *Transactions of the American Philosophical Society*, 64 (1974) (Derek de Solla Price) – Figure 8

AT & T Bell Laboratories: *Record* (March 1986), pp. 4–10 (David M. Gay, Narenda K. Karmarkar, and K. G. Ramakrishnan) – Figure 35

Agnessa Babloyantz – Figures 143, 144, 145, 146

Belknap Press, Cambridge, MA: Stephen M. Stigler, *The History of Statistics* (1986) – Figures 16, 19

Bibliothèque Royale Albert 1er, Brussels: Portrait of Adolphe Quetelet, Odevaere E 3574 C – Figure 17

J. Robert Buchler – Figures 127, 128

Chapman & Hall Ltd., London: D. K. Arrowsmith and C. M. Place, *Ordinary Differential Equations* (1992) – Figures 37, 38, 39, 40

Cray Research Inc., Minneapolis – Figure 51

John Crutchfield – Figures 20, 80

Stillman Drake – Figure 11

W. H. Freeman, San Francisco: Morris Kline (ed.), *Mathematics in the Modern World* [1969] – Figures 2, 3, 12; Benoît Mandelbrot, *The Fractal Geometry of Nature*, 2nd edn. (1982) – Figures 89, 90, 95

The Guardian and Manchester Evening News PLC – Figure 52

Greg King and Harry Swinney – Figure 73

Longman Group Ltd., London: G. I. Barenblatt, G. Iooss, and D. D. Joseph (eds.), *Nonlinear Dynamics and Turbulence* (1983) (Pitman imprint), pp. 156–71 (J. P. Gollub) – Figure 74

Macmillan Magazines Ltd., London: *Nature*, 322, p. 791 – Figure 92; *ibid.*, 321, p. 668 – Figure 93; *ibid.*, 361, pp. 608–12 (Jacques Laskar and P. Robutel) – Figure 116

Manchester University Press: Arun V. Holden (ed.), *Chaos* (1986), pp. 158–78 (W. M. Schaffer and M. Kot) – Figures 121, 122; pp. 237–56 (Leon Glass, Alvin Shrier and Jacques Bélair) – Figure 125

Mathematical Association of America: S. A. Levin (ed.), *Studies in Mathematical Biology* (1978), pp. 317–66 (Robert M. May) – Figure 119; pp. 411–38 (G. Oster) – Figures 117, 120

Robert M. May – Figures 123, 124

Alistair Mees – Figure 130

National Aeronautics and Space Administration, Washington, DC – Figures 4, 5, 69, 81, 104

New York Institute of Technology: Peter Oppenheimer – Figures 96, 97

North-Holland Publishing Co., Amsterdam: *Physica D, Nonlinear Phenomena*, 6 (1983), pp. 385–92 (A. Arneodo, P. Coullet, C. Tresser, A. Libchaber, J. Maurer, D. d'Humières) – Figure 87

Arthur J. Olson – Figure 91

Penguin Books Ltd, Harmondsworth: D. L. Hurd and J. J. Kipling, *The Origins and Growth of Physical Science* (1964) – Figure 9

Royal Library, Windsor Castle: Leonardo da Vinci drawing, RL 12660V – Figure 68

Royal Society: *Proceedings of the Royal Society of London*, A:413 (1987), reprinted as M. V. Berry, I. O. Percival, N. Weiss (eds.), *Dynamical Chaos*, pp. 9–26 (L. Glass, A. L. Goldberger, M. Courtemance, and A. Schreier) – Figure 126; pp. 109–30 (Jack Wisdom) – Figures 106, 110, 111, 112

Colin Sparrow – Figure 55

Springer-Verlag Inc., New York: Heinz-Otto Peitgen and Peter H. Richter, *The Beauty of Fractals* (1992) – Figures 23, 99; Heinz-Otto Peitgen, Hartmut Jürgens, and Dietmar Saupe, *Chaos and Fractals: New Frontiers of Science* (1992) – Figures 100, 101, 102, 103

University of Maryland Chaos Group – Figure 151

John Wiley Inc., New York: Carl B. Boyer, *A History of Mathematics* (1968) – Figures 10, 15; J. M. T. Thompson and H. B. Stewart, *Nonlinear Dynamics and Chaos* (1986) – Figures 27, 46, 61, 65, 67, 70, 75, 78

World Scientific Publishing Co. Pte. Ltd., Singapore: *Fractal Physiology and Chaos* (1990) (Bruce J. West) – Figure 141

Index

READ MORE IN PENGUIN

In every corner of the world, on every subject under the sun, Penguin represents quality and variety – the very best in publishing today.

For complete information about books available from Penguin – including Puffins, Penguin Classics and Arkana – and how to order them, write to us at the appropriate address below. Please note that for copyright reasons the selection of books varies from country to country.

In the United Kingdom: Please write to *Dept. EP, Penguin Books Ltd, Bath Road, Harmondsworth, West Drayton, Middlesex UB7 ODA*

In the United States: Please write to *Consumer Sales, Penguin Putnam Inc., P.O. Box 12289 Dept. B, Newark, New Jersey 07101-5289.* VISA and MasterCard holders call 1-800-788-6262 to order Penguin titles

In Canada: Please write to *Penguin Books Canada Ltd, 10 Alcorn Avenue, Suite 300, Toronto, Ontario M4V 3B2*

In Australia: Please write to *Penguin Books Australia Ltd, P.O. Box 257, Ringwood, Victoria 3134*

In New Zealand: Please write to *Penguin Books (NZ) Ltd, Private Bag 102902, North Shore Mail Centre, Auckland 10*

In India: Please write to *Penguin Books India Pvt Ltd, 11 Community Centre, Panchsheel Park, New Delhi 110017*

In the Netherlands: Please write to *Penguin Books Netherlands bv, Postbus 3507, NL-1001 AH Amsterdam*

In Germany: Please write to *Penguin Books Deutschland GmbH, Metzlerstrasse 26, 60594 Frankfurt am Main*

In Spain: Please write to *Penguin Books S. A., Bravo Murillo 19, 1° B, 28015 Madrid*

In Italy: Please write to *Penguin Italia s.r.l., Via Benedetto Croce 2, 20094 Corsico, Milano*

In France: Please write to *Penguin France, Le Carré Wilson, 62 rue Benjamin Baillaud, 31500 Toulouse*

In Japan: Please write to *Penguin Books Japan Ltd, Kaneko Building, 2-3-25 Koraku, Bunkyo-Ku, Tokyo 112*

In South Africa: Please write to *Penguin Books South Africa (Pty) Ltd, Private Bag X14, Parkview, 2122 Johannesburg*

BY THE SAME AUTHOR

Game, Set and Math
Enigmas and Conundrums

How can two people play a GAME of poker over the telephone without the possibility of cheating? How do you cope mathematically with contorted worms, drunken tennis-players and snakes that eat their own tails? How can you distinguish plausible fallacies from unbelievable facts?

If you want to know all the answers, SET off on this journey to the frontiers of MATHs!

This lively books introduces the different kinds of infinity, explains how to build your own virus, explores the brighter ideas of Pascal and Fermat, and shows why parallel lines meet – in Rome! Twelve mind-bending puzzles for the different days of Christmas make the perfect dessert in a banquet of mathematical delights.

and with Martin Golubitsky:

Fearful Symmetry
Is God a Geometer?

Symmetric causes can have asymmetric effects. This paradox is called symmetry-breaking, and the extraordinary discoveries of recent years have shown that it plays a major role in the formation of patterns present in everyday life. Starting with an analysis of the different forms of symmetry, the authors show how a few key principles can help us understand everything from cats' movements, corn circles and snow crystals to subatomic particles and the overall structure of the cosmos.

'Symmetry-breaking is an important unifying idea in modern abstract mathematics, so it's a tribute to the authors' expository gifts that much of the book reads like a captivating travelogue' John Allen Paulos, *The New York Times Book Review*